PEARSON   ALWAYS LEARNING

Colleen Belk • Virginia Borden Maier

Biology: Science for Life with Physiology

Second Custom Edition for Portland State University

Taken from:
Biology: Science for Life with Physiology, Fourth Edition
by Colleen Belk and Virginia Borden Maier

Cover Art: Courtesy of Photodisc/Getty Images.

Taken from:

Biology: Science for Life with Physiology, Fourth Edition
by Colleen Belk and Virginia Borden Maier
Copyright © 2013, 2010, 2007 by Pearson Education, Inc.
Boston, Massachusetts 02116

This special edition published in cooperation with Pearson Learning Solutions.

Pearson Learning Solutions, 501 Boylston Street, Suite 900, Boston, MA 02116
A Pearson Education Company
www.pearsoned.com

Printed in the United States of America

2 3 4 5 6 7 8 9 10 V092 16 15 14 13 12

000200010271686845

SI

PEARSON   ISBN 10: 1-256-79863-0
ISBN 13: 978-1-256-79863-7

About the Authors

Colleen Belk and **Virginia Borden Maier** collaborated on teaching biology to nonmajors for over a decade together at the University of Minnesota–Duluth. This collaboration has continued through Virginia's move to St. John Fisher College in Rochester, New York, and has been enhanced by their differing but complementary areas of expertise. In addition to the non-majors course, Colleen Belk teaches general biology for majors, genetics, cell biology, and molecular biology courses. Virginia Borden Maier teaches general biology for majors, evolutionary biology, zoology, plant biology, ecology, and conservation biology courses.

After several somewhat painful attempts at teaching the breadth of biology to non-majors in a single semester, the two authors came to the conclusion that they needed to find a better way. They realized that their students were more engaged when they understood how biology directly affected their lives. Colleen and Virginia began to structure their lectures around stories they knew would interest students. When they began letting the story drive the science, they immediately noticed a difference in student interest, energy, and willingness to work harder at learning biology. Not only has this approach increased student understanding, but it has also increased the authors' enjoyment in teaching the course—presenting students with fascinating stories infused with biological concepts is simply a lot more fun. This approach served to invigorate their teaching. Knowing that their students are learning the biology that they will need now and in the future gives the authors a deep and abiding satisfaction.

Preface

To the Student

As you attended your classes in high school or otherwise worked to prepare yourself for college, you were probably unaware that an information explosion was taking place in the field of biology. This explosion, brought on by advances in biotechnology and communicated by faster, more powerful computers, has allowed scientists to gather data more quickly and disseminate it to colleagues in the global scientific community with the click of a mouse. Every discipline of biology has benefited from these advances, and today's scientists collectively know more than any individual could ever hope to understand.

Paradoxically, as it becomes more and more challenging to synthesize huge amounts of information from disparate disciplines within the broad field of biology, it becomes more vital that we do so. The very same technologies that led to the information boom, coupled with expanding human populations, present us with complex ethical questions. These questions include whether it is acceptable to clone humans, when human life begins and ends, who owns living organisms, what our responsibilities toward endangered species are, and many more. No amount of knowledge alone will provide satisfactory answers to these questions. Addressing these kinds of questions requires the development of a scientific literacy that surpasses the rote memorization of facts. To make decisions that are individually, socially, and ecologically responsible, you must not only understand some fundamental principles of biology but also be able to use this knowledge as a tool to help you analyze ethical and moral issues involving biology.

To help you understand biology and apply your knowledge to an ever-expanding suite of issues, we have structured each chapter of *Biology: Science for Life with Physiology* around a compelling story in which biology plays an integral role. Through the story you not only will learn the relevant biological principles but also will see how science can be used to help answer complex questions. As you learn to apply the strategies modeled by the text, you will begin developing your critical thinking skills.

By the time you finish this book, you should have a clear understanding of many important biological principles. You will also be able to critically evaluate which information is most reliable instead of simply accepting all the information you hear or read about. Even though you may not be planning to be a practicing biologist, well-developed critical thinking skills will enable you to make decisions that affect your own life, such as whether to take nutritional supplements, and decisions that affect the lives of others, such as whether to believe the DNA evidence presented to you as a juror in a criminal case.

It is our sincere hope that understanding how biology applies to important personal, social, and ecological issues will convince you to stay informed about such issues. On the job, in your community, at the doctor's office, in the voting booth, and at home watching the news or surfing the web, your knowledge of the basic biology underlying so many of the challenges that we as individuals and as a society face will enable you to make well-informed decisions for your home, your nation, and your world.

To the Instructor

By now you are probably all too aware that teaching non-majors students is very different from teaching biology majors. You know that most of these students will never take another formal biology course; therefore, your course may be the last chance for these students to see the relevance of science in their everyday lives and the last chance to appreciate how biology is woven throughout the fabric of their lives. You recognize the importance of engaging these students because you know that these students will one day be voting on issues of scientific importance, holding positions of power in the community, serving on juries, and making health care decisions for themselves and their families. You know that your students' lives will be enhanced if they have a thorough grounding in basic biological principles and scientific literacy. To help your efforts toward this end, this text is structured around several themes.

Themes Throughout *Biology: Science for Life with Physiology*

The Story Drives the Science. We have found that students are much more likely to be engaged in the learning process when the textbook and lectures capitalize on their natural curiosity. This text accomplishes this by using a story to drive the science in every chapter. Students get caught up in the story and become interested in learning the biology so they can see how the story is resolved. This approach allows us to cover the key areas of biology, including basic chemistry, the unity and diversity of life, cell structure and function, classical and molecular genetics, evolution, and ecology, in a manner that makes students want to learn. Not only do students want to learn, but also this approach allows students both to connect the science to their everyday lives and to integrate the principles and concepts for later application to other situations. The story approach will give you flexibility in teaching and will support you in developing students' critical thinking skills.

The Process of Science. This book also uses another novel approach in the way that the process of science is modeled. The first chapter is dedicated to the scientific method and hypothesis testing, and each subsequent chapter weaves the scientific method and hypothesis testing throughout the story. The development of students' critical thinking skills is thus reinforced for the duration of the course. Students will see that the application of the scientific method is often the best way to answer questions raised in the story. This practice not only allows students to develop their critical thinking skills but also, as they begin to think like scientists, helps them understand why and how scientists do what they do.

Integration of Evolution. Another aspect of *Biology: Science for Life with Physiology* that sets it apart from many other texts is the manner in which evolutionary principles are integrated throughout the text. The role of evolutionary processes is highlighted in every chapter, even when the chapter is not specifically focused on an evolutionary question. For example, when discussing infectious diseases, the evolution of antibiotic-resistant strains of bacteria is addressed. The physiology unit includes an essay on evolution in each chapter. These essays illustrate the importance of natural selection in the development of various organs and organ systems across a wide range of organisms. With evolution serving as an overarching theme, students are better able to see that all of life is connected through this process.

Pedagogical Elements

Open the book and flip through a few pages and you will see some of the most inviting, lively, and informative illustrations you have ever seen in a biology text. The illustrations are inviting because they have a warm, hand-drawn quality that is clean and uncluttered. Most important, the illustrations are informative not only because they were carefully crafted to enhance concepts in the text but also because they employ techniques like the "pointer" that help draw the students' attention to the important part of the figure (see p. 14). Likewise, tables are more than just tools for organizing information; they are illustrated to provide attractive, easy references for the student. We hope that the welcoming nature of the art and tables in this text will encourage nonmajors to explore ideas and concepts instead of being overwhelmed before they even get started.

In addition to lively illustrations of conventional biology concepts, this text also uses analogies to help students understand difficult topics. For example, the process of

translation is likened to baking a cake (see p. 199). These analogies and illustrations are peppered throughout the text.

Students can reinforce and assess what they are learning in the classroom by reading the chapter, studying the figures, and answering the end-of-chapter questions. We have written these questions in every format likely to be used by an instructor during an exam so that students have practice in answering many different types of questions. We have also included "Connecting the Science" questions that would be appropriate for essay exams, class discussions, or use as topics for term papers.

In an effort to accommodate the wide range of teaching and learning styles of those using our book, each chapter includes a section entitled **A Closer Look**. Preceding each A Closer Look section is a general overview of the topic sufficient to provide nonmajors students with a basic understanding. A Closer Look then provides details that add depth to this understanding. This feature provides instructors with teaching flexibility; for example, some instructors are satisfied with a general overview of cellular respiration, including the reactants and products, and similar basics of metabolism. Others want to teach electron transport and oxidative phosphorylation. **Our format** facilitates both approaches; one can use the overview only or the overview and the Closer Look section. The Closer Look section is differentiated from the rest of the text, making it easy to include or omit from students' assigned reading.

Each chapter contains several **Stop and Stretch** questions designed to provide rest stops where students can assess whether they have understood the material they just completed reading. These questions require students to synthesize material they have read and apply their knowledge in a new situation. Many of these questions are designed to familiarize students with data analysis and help them develop skill at interpreting and understanding scientific data. Answers to these questions are provided in the appendix.

Another feature that aids in student learning is **Visualize This,** which consists of questions in figure captions that test a student's understanding of the concepts contained in the art. These questions are designed to encourage students to spend extra time with complex figures and develop a more sophisticated understanding of the concepts they present. Visualize This questions are included with three or more figures in each chapter and answers to these questions can be found in the appendix.

Each chapter ends with a feature called **Savvy Reader**. We developed this feature in response to the desire of so many of the professors teaching this course to help students learn to critically evaluate science presented in the media. We all want our students to become better, more informed consumers of science. Each Savvy Reader offers an excerpt or a summary of an article or advertisement taken from any of a variety of media sources. After reading the excerpt, the student then answers a series of questions written with the goal of helping them critically evaluate the scientific information presented in the article. Students learn to evaluate claims made in the popular press about issues ranging from the benefits of particular health care products to whether our friendships are based on similarities in genetic makeup.

Improvements in the Fourth Edition

The positive feedback garnered by previous editions assured us that presenting science alongside a story works for students and instructors alike. In the fourth edition, we have rearranged some of the content and added a new chapter and story line. Chapter 3 addresses the question of dietary supplementation while covering the essential content of nutrients and membrane transport. Chapter 4 revisits the question of body fat and health and now includes the content of cellular respiration. Chapter 5 reviews the latest information regarding the science of global warming and continues to cover the basics of photosynthesis. Story lines have been updated or replaced in other chapters compared to previous editions. Chapter 12 on Species and Races now examines the use of DNA to determine genealogy.

With the previous editions, we focused on improving flexibility for instructors via **A Closer Look** chapter subsections, helping students assess their understanding via **Stop and Stretch** and **Visualize This** questions , and providing students with opportunities to be better consumers of popular media with the **Savvy Reader**. In this edition, we have provided instructors and students with guidance in the form of **Learning Outcomes**. These outcomes describe the knowledge or skills that students should have upon completing the reading and associated activities in the text and ancillary materials. The chapter summary is organized around the Learning Outcomes, and all questions and exercises are tagged to a Learning Outcome. Exercises in the Mastering Biology platform are also tagged to these Learning Outcomes, providing students ample opportunity to master these important concepts.

Finally, users of earlier editions will notice a new style for phylogenetic trees in the text—horizontal rather than vertical, to match the standard presentation in the field of evolutionary biology.

Supplements

A group of talented and dedicated biology educators teamed up with us to build a set of resources that equip nonmajors with the tools to achieve scientific literacy that will allow them to make informed decisions about the biological issues that affect them daily. The student resources offer several ways of reviewing and reinforcing the concepts and facts covered in this textbook. We provide instructors with a test bank, Instructor Guide, and the Instructor Resource DVD which includes an updated and expanded suite of lecture presentation materials, and a valuable source of ideas for educators to enrich their instruction efforts. Available in print and media formats, the *Biology: Science for Life with Physiology* resources are easy to navigate and support a variety of learning and teaching styles.

We believe you will find that the design and format of this text and its supplements will help you meet the challenge of helping students both succeed in your course and develop science skills—for life.

As with previous editions, the overall goal of the text remains to provide a thorough overview of the essentials of biological science while trying to avoid overloading students with information. We worked closely with instructors using prior editions as well as other reviewers to pinpoint essential content to include in this edition while staying true to the book's philosophy of learning science in a story format. The development of the fourth edition has truly been a collaborative process among ourselves, the students and instructors who used prior editions, and our many thoughtful reviewers. We look forward to learning about your experience with *Biology: Science for Life with Physiology*, 4th edition.

Supplement Authors

Instructor Guide
Jill Feinstein   *Richland Community College*

Mastering Biology Quiz and Test Items
Catherine Podeszwa   *University of Minnesota–Duluth*

PowerPoint Lectures
Jill Feinstein   *Richland Community College*

"Because science, told as a story, can intrigue and inform the non-scientific minds among us, it has the potential to bridge the two cultures into which civilization is split—the sciences and the humanities. For educators, stories are an exciting way to draw young minds into the scientific culture."

—E.O. WILSON

A Captivating Narrative

Every chapter introduces biology topics through a story, explaining important biological processes with concrete examples and applications.

CHAPTER 3

Is It Possible to Supplement Your Way to Better Health?

Nutrients and Membrane Transport

Do supplemented waters improve hydration?

Real-life Examples

Each chapter story draws from real-life experiences, engaging the reader and making the biology more approachable. For example, Chapter 3 frames the discussion of nutrients and membrane transport around a story line on the current craze in nutritional supplements.

What Reviewers Have to Say

" A great text—right level for non-majors students with clear examples and explanation. I really like the link with the Learning Objectives, which I have not yet seen in other texts. "

—Cara Shillington, Instructor, *Eastern Michigan University*

" This is a well-written text that makes learning about biology a truly fun experience. The authors do a remarkable job of keeping the topics clear and understandable while at the same time making the concepts interesting by incorporating some very interesting stories that are both relevant and fascinating. It's not easy to write a text for students who are often resistant to concepts like evolution and natural selection; this text makes our job as instructor that much simpler. Well done. "

—Andrew Goliszek, Instructor, *North Carolina A&T University*

Updated
Table of Contents

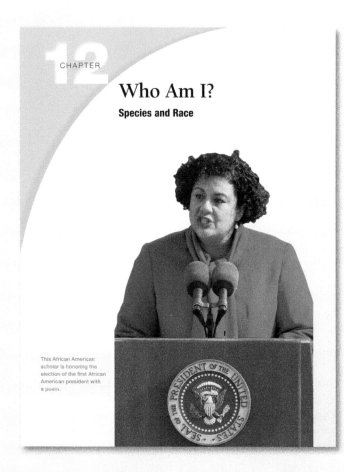

CHAPTER 12

Who Am I?
Species and Race

This African American scholar is honoring the election of the first African American president with a poem.

January 20, 2009, represented a watershed moment in American history, the inauguration of the first African American president, Barack Obama. Regardless of how you feel about his presidency, everyone can appreciate how Obama's election represented the culmination of a major transformation in American society, less than six generations removed from the legal enslavement of Africans in this country. When the African American scholar and poet Elizabeth Alexander read a verse she penned for the inauguration entitled "Praise Song for the Day," she explicitly referenced how far the nation had come toward viewing people of her race as worthy of equal rights and opportunities:

> Say it plain: that many have died for
>   this day.
> Sing the names of the dead who
>   brought us here,
> who laid the train tracks, raised the
>   bridges,
> picked the cotton and the lettuce, built
>   brick by brick the glittering edifices
> they would then keep clean and work inside of.
> Praise song for struggle, praise song for
>   the day.
> Praise song for every hand-lettered
>   sign,
>   the figuring-it-out at kitchen tables.

African slavery was in part justified by a belief that black Africans were somehow biologically inferior to white Europeans. Even after slavery ended, the racism that helped excuse it lived on in social policies that kept blacks confined to segregated facilities and communities. Dr. Alexander's professional career has been dedicated to improving race relations and helping reduce the lingering

LEARNING OUTCOMES

LO1 Define *biological species*, and list the mechanisms by which reproductive isolation is maintained by biological species.

LO2 Describe the three steps in the process of speciation.

LO3 Explain how a "race" within a biological species can be defined using the genealogical species concept.

LO4 List the evidence that modern humans are a young species that arose in Africa.

LO5 List the genetic evidence expected when a species can be divided into unique races.

LO6 Describe how the Hardy-Weinberg theorem is used in studies of population genetics.

LO7 Summarize the evidence that indicates that human races are not deep biological divisions within the human species.

LO8 Provide examples of traits that have become common in certain human populations due to the natural selection these populations have undergone.

LO9 Define *genetic drift*, and provide examples of how it results in the evolution of a population.

LO10 Describe how human and animal behavior can cause evolution via sexual selection and assortative mating.

Less than a year after the inauguration, she learned that she was genetically more European than African.

In fact, she shares a recent ancestor with the European American comedian, Stephen Colbert.

How does this new genetic information fit into our experience of human races?

275

CHAPTER 19

Vaccinations: Protection and Prevention or Peril?

Immune System, Bacteria, Viruses, and Other Pathogens

Can vaccines cause autism as some celebrities claim?

LEARNING OUTCOMES

LO1 Compare the structure of bacteria to that of viruses.

LO2 Contrast bacterial replication with viral reproduction.

LO3 List the structures involved in the first line of defense against infection, and describe how they function in protecting the body against pathogens.

LO4 List the participants in the second line of defense against infection, and describe their functions.

LO5 List the two different cell types involved in the third line of defense against infection, and compare how they recognize and eliminate antigens.

LO6 Explain allergic reactions.

LO7 Explain the process immune cells undergo to be able to respond to millions of different antigens.

LO8 Describe the process used to determine whether a cell is foreign or native to the body and what happens when this process fails.

LO9 Explain the role of memory cells in helping to protect against infection.

LO10 Explain the role of helper T cells in allowing infection by the AIDS virus.

LO11 Describe the mechanism by which vaccines help confer immunity.

Should you get a flu shot? For many college students, this is one of the first health care decisions that will be made without much parental input. University health services personnel want you to, but you hear conflicting information from fellow students. The flu shot, like all vaccines, contains substances that resemble the disease-causing organisms. When you are vaccinated, your body is tricked into preparing for an attack of the disease in case one does occur.

It's not just college students who are vaccinated. Newborns and children under 6 years of age in the United States are vaccinated against 15 different diseases. The first vaccinations happen before a baby even leaves the hospital to go home and continue every few months until the child is a year and a half old. Another series of vaccinations occurs when the child is school aged, and a few additional vaccinations are given during adolescence. Virtually all parents would prefer not to subject their child to these vaccinations, which are often administered via uncomfortable injections, not to mention the inconvenience of taking children to the doctor for so many visits. But for most parents, the promise of protection from diseases makes the choice an easy one.

There are, however, a growing number of parents who are questioning whether to vaccinate their children. Fueled by Internet obtained information and championed by celebrities who believe vaccines have hurt their children, many parents are skipping some or all vaccinations. Many college students also choose not to get the flu shot, believing the shot might make them ill or that it's not such a big deal to get the flu. To understand the risks and benefits of vaccinations, it is first necessary to learn about the causes of diseases that vaccines attempt to prevent.

Should you get a flu shot?

Why would a male be vaccinated against cervical cancer?

Are parents justified in not allowing their babies to be vaccinated?

467

Flexible
Engaging Content

Two unique features break down complex materials and give instructors flexibility in course coverage.

An Overview: DNA Fingerprinting

All of these questions were answered using **DNA fingerprinting.** This technique allows unambiguous identification of people in the same manner that traditional fingerprinting has been used in the past. To begin this process, it is necessary to isolate the DNA to be fingerprinted. Scientists can isolate DNA from blood, semen, vaginal fluids, a hair root, skin, and even (as was the case in Ekaterinburg) degraded skeletal remains.

Because often there is not much DNA present, it must first be copied to generate larger quantities for analysis. Since each person has a unique sequence of DNA, particular regions of DNA, which vary in size, are selected for copying. When the copied DNA fragments are separated and stained, a unique pattern is produced. *See the next section for A Closer Look at DNA Fingerprinting.*

A Closer Look

Heavily revised in this edition for ease of use, An Overview explains the basics of a difficult biological concept and occurs just before the related A Closer Look section, which provides the details.

A Closer Look:
DNA Fingerprinting

When very small amounts of DNA are available, as is often the case, scientists can make many copies of the DNA by first performing a DNA-amplifying reaction.

Polymerase Chain Reaction (PCR)

The **polymerase chain reaction (PCR)** is used to amplify, or produce many copies of, a particular region of DNA (Figure 8.12). To perform PCR, scientists place four essential components into a tube: (1) the **template,** which is the double-stranded DNA to be copied; (2) the **nucleotides** adenine (A), cytosine (C), guanine (G), and thymine (T), which are the individual building-block subunits of DNA; (3) short single-stranded pieces of DNA called **primers,** which are complementary to the ends of the region of DNA to be copied; and (4) an enzyme called *Taq* polymerase.

Taq **polymerase** is a DNA polymerase that uses one strand of DNA as a template for the synthesis of a daughter strand that carries complementary nucleotides (A:T base pairs are complementary, as are G:C base pairs) (Chapter 6). This enzyme was given the first part of its name (*Taq*) because it was first isolated from *Thermus aquaticus*, a bacterium that lives in hydrothermal vents and can withstand very high temperatures. The second part of the enzyme's name (polymerase) describes its synthesizing activity—it acts as a DNA polymerase.

The main difference between human DNA polymerase and *Taq* polymerase is that the *Taq* polymerase is resistant to extremely high temperatures, temperatures at which human DNA polymerase would be inactivated. The heat-resistant qualities of *Taq* polymerase thus allow PCR reactions to be run at very high temperatures. High temperatures are

① PCR is used to amplify, or make copies of, a specific region of DNA. During a PCR reaction, primers (short stretches of single-strand DNA), free nucleotides, and template DNA are mixed with heat-tolerant *Taq* polymerase.

② The DNA is heated to separate, or denature, the two strands. As the mixture cools, the primers bond to the region of the DNA template they complement.

③ The *Taq* polymerase uses the primers to initiate synthesis.

④ A copy of the DNA template is assembled.

⑤ The mixture is heated again. The process is repeated many times, doubling the DNA amount each time.

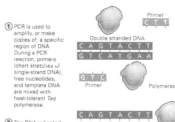

Figure 8.12 **The polymerase chain reaction (PCR).** PCR is used to amplify, or make copies of, DNA. Each round of PCR doubles the number of DNA molecules. This type of exponential growth can yield millions of copies of DNA.

focus on EVOLUTION

Plant Reproduction

Land plants have a unique life cycle. Instead of one multicellular "adult" form as is found in animals, plants have two, which alternate with each other. In fact, the life cycle of plants is commonly referred to as **alternation of generations.** The two life-forms result because the product of meiosis in plants is not a sperm or egg (gamete); it is a haploid structure (containing half the typical number of chromosomes) called a **spore,** which undergoes mitosis to produce a multicellular plant called the **gametophyte** (or gamete-producing plant). The gametophyte produces sperm or eggs (or both) by mitosis. When gametes fuse, they produce a diploid embryo that divides and eventually produces spores by meiosis. This stage in the life cycle is called the **sporophyte.** The generalized life cycle is illustrated in Figure E23.1. Because the algal ancestors of land plants were haploid for most of their lives, but sexual reproduction requires a short period of diploidy, the alternation between two forms is a vestige of evolutionary history. In fact, the trend in land plant evolution has been that the haploid gametophyte generation has become increasingly reduced.

In mosses and other nonvascular plants, the gametophyte generation is the most prominent form, reflecting their close relationship with the green algae. Sperm and eggs are produced in structures at the tips of the plant, and sperm swim in a film of water, such as a drop of rain, to reach the eggs. The embryo develops within the female reproductive organ, eventually producing a sporophyte that emerges from the tip of the gametophyte and is dependent on it for survival. Tiny spores produced in the sporophyte are dispersed on the wind and germinate into new gametophytes if they land in moist conditions.

On land, a large haploid DNA-damaging UV light, w cules, is much more of a thr have two copies of each ch not as detrimental in a diplo may be why ferns and othe arose from the nonvascula phyte. Spores are produce dispersed by wind or water tions. The small independen a spore may produce both a film of water to reach the e out of the female reproducti comes much larger and ind

The seed plants that ar next step in the evolution reduction in gametophyte persal structure. Pine tree plants, called gymnosperr tophyte, both of which an for survival. Pollen, released from small cones produced by the adult sporophyte, is actually tiny, not-yet-mature male gametophyte plants. Pollen is carried by the wind (or occasionally insects) to the female reproductive structures, which are typically produced in woody cones on the sporophyte. Here, the pollen completes maturation and releases sperm to fertilize the egg. The female gametophyte, which contains the egg, is a small structure on the surface of a cone scale. The embryo begins development inside

① Cells in the multicellular diploid plant (the sporophyte) undergo meiosis, producing a haploid spore.

Spore (1n)

② The spore germinates into a multicellular haploid plant, the gametophyte.

③ Gametophytes produce sperm and eggs by mitosis.

④ Fertilization produces a diploid sporophyte embryo, which grows into a sporophyte.

Haploid = 1n
Diploid = 2n

Figure E23.1 **Alternation of generations.**

Focus on Evolution

Within the physiology chapters, Focus on Evolution features illustrate the importance of natural selection in a wide range of organisms and allow students to see that all of life is connected through this process.

Unique Savvy Reader boxes highlight the applicability of biology in everyday life. Taken from a variety of media sources, each box contains a short excerpt, relating to chapter content, followed by critical thinking questions to help students interpret and evaluate the information.

SAVVY READER

A Toolkit for Evaluating Science in the News

The following checklist can be used as a guide to evaluating science information in the news. Although any issue that raises a red flag should cause you to be cautious about the conclusions of the story, no one issue should cause you to reject the story (and its associated research) out of hand. However, in general, the fewer red flags raised by a story, the more reliable is the report of the significance of the scientific study. Use **Table 1.2** to evaluate this extract from an article by Terry Gupta in the Nashua (New Hampshire) *Telegraph*:

[Paula] Fortier, a licensed massage therapist, shared her special brew. To boiled water and tea, she adds cayenne pepper, honey and fresh squeezed lemon. Fortier's enthusiasm may be more contagious than the flu when she describes this heated, "unbeatable" combination. "Your throat opens up," she said. "You're less irritated and you sweat out some unwanted toxins." She varies the exact measures of the ingredients according to tolerance and taste. While on the topic of hot water, one of Fortier's favorite remedies for "an achy body that feels sore all over is a hot bath in a steeping tub with 2 cups of cider vinegar for 15 minutes. This draws off toxins, reduces inflammation and leaves you feeling better."

TABLE 1.2

A guide for evaluating science in the news. For each question, check the appropriate box.

Question	Preferred answer		Raises a red flag	
1. What is the basis for the story?	Hypothesis test	☐	Untested assertion No data to support claims in the article.	☐
2. What is the affiliation of the scientist?	Independent (university or government agency)	☐	Employed by an industry or advocacy group Data and conclusions could be biased.	☐
3. What is the funding source for the study?	Government or nonpartisan foundation (without bias)	☐	Industry group or other partisan source (with bias) Data and conclusions could be biased.	☐

Revised and New Savvy Reader Boxes

Savvy Reader boxes in selected chapters have been replaced with more relevant and timely media excerpts. The following are new Savvy Readers:

- Ionized Water (Chapter 2)

- Refugees from Global Warming (Chapter 5)

- What Happens When a Species Recovers (Gray Wolf's endangered species status) (Chapter 15)

SAVVY READER

What Happens When a Species Recovers

Gray Wolf's endangered species status appears close to end

A new federal bill introduced in the U.S. House of Representatives would "de-list" the Gray Wolf from Endangered Species Act protections and allow individual states to design and carry out their own wolf management and recovery plans. The legislation is co-sponsored by Rep. Jim Matheson (D) from Utah, a state that is just beginning to see increases in wolf numbers, and Rep. Mike Ross (D-Arkansas), who co-chairs the Congressional Sportsman's Caucus, made up of advocates of hunting and fishing interests.

According to Rep. Matheson, "Scientists and wolf recovery advocates agree—the gray wolf is back. Since it is no longer endangered, it should be de-listed as a species, managed as others species are—by state wildlife agencies—and time, money, and effort can be focused where it's needed."

Rep. Ross argues that officials within the regions inhabited by particular endangered species should have a greater role in management decisions related to these species. "Excessive wolf populations are having a devastating impact on elk, moose, deer, and other species, and each state has its own unique set of challenges. Both the Obama and Bush administrations have already recommended the de-listing of wolves in many states and turning their management over to state wildlife agencies because it is the right thing to do to keep our nation's sensitive ecosystem in balance."

1. What evidence does the representative from Utah provide to support his statement that "the gray wolf is back"? Is that evidence convincing?

2. Given what you've learned in this chapter, do you think it is likely that the return of the gray wolf is having a "devastating impact" on elk and moose?

Innovative
Art Program

Designed to help students visually navigate biology, this edition's photo and art program has been revised for currency and increased clarity.

TABLE 15.1

Types of species interactions and their direct effects.			
Interaction	**Example**	**Effect on Species 1**	**Effect on Species 2**
Commensalism: Association increases the growth or population size of one species and does not affect the other.	1. Remora 2. Shark	+ As the shark feeds somewhat sloppily, the remora can collect the scraps.	0 The shark seems to suffer no negative effects from its hitchhiker.
Mutualism: Association increases the growth or population size of both species.	1. Ants 2. Acacia tree	+ The swollen thorns of the acacia provide shelter for the ants. The acacia leaves provide "protein bodies" that the ants harvest for food.	+ Ants kill herbivorous insects and destroy competing vegetation, benefiting the acacia.
Predation and Parasitism: Consumption of one organism by another.	1. Brown bear 2. Salmon	+ The brown bear catches the salmon and eats it, obtaining nourishment.	− The salmon does not survive.
Competition: Association causes a decrease or limitation in population size of both species.	1. Dandelion 2. Tomato plant	− The dandelion does not grow as well in the presence of the tomato plant. Dandelion produces fewer seeds and fewer offspring.	− The tomato plant does not grow optimally in the presence of the weed. Tomato plant produces fewer flowers and fruit.

Illustrated Tables

Unique, comprehensive tables organize information in one place and provide easy, visual references for students.

Visual Analogies

Art throughout the book offers visual analogies to teach students by comparing familiar, everyday objects and occurrences to complicated, biological concepts and processes.

Like a Tinkertoy™ connector, carbon has multiple sites for connections that allow carbon-containing molecules to take an almost infinite variety of shapes (**Figure 2.8**). See the next section for *A Closer Look* at chemical bonds.

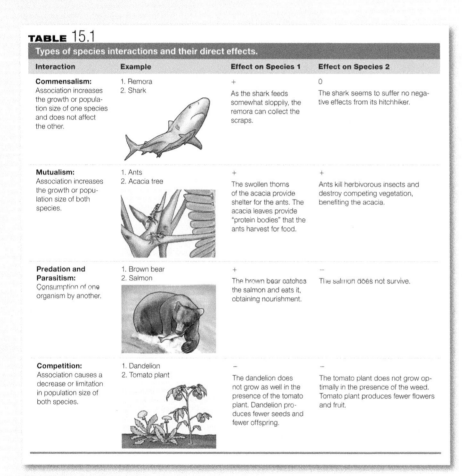

Methane (CH_4)

Carbon: The key chemical Tinkertoy® connector

Carbon dioxide (CO_2)

Glucose ($C_6H_{12}O_6$)

Figure 2.8 **Carbon, the chemical Tinkertoy™ connector.** Because carbon forms four covalent bonds at a time, carbon-containing compounds can have diverse shapes.

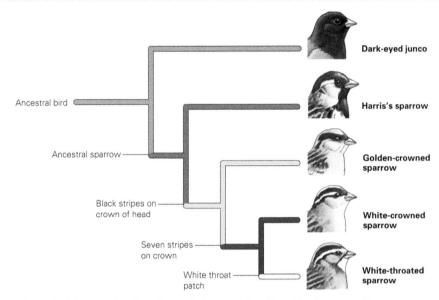

Dark-eyed junco

Ancestral bird

Harris's sparrow

Ancestral sparrow

Golden-crowned sparrow

Black stripes on crown of head

White-crowned sparrow

Seven stripes on crown

White throat patch

White-throated sparrow

Figure 13.21 Reconstruction of an evolutionary history. The relationships among these four species of sparrow can be illustrated by their shared physical traits compared to a more distant relative, the dark-eyed junco.
Visualize This: What traits did the ancestral bird likely have?

(a) Normal distribution of student height in one college class

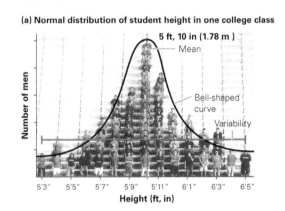

5 ft, 10 in (1.78 m)
Mean

Number of men

Bell-shaped curve

Variability

5'3" 5'5" 5'7" 5'9" 5'11" 6'1" 6'3" 6'5"
Height (ft, in)

(b) Variance describes the variability around the mean.

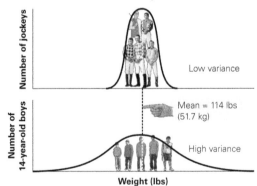

Number of jockeys

Low variance

Number of 14-year-old boys

Mean = 114 lbs (51.7 kg)

High variance

Weight (lbs)

Figure 7.16 A quantitative trait. (a) This photo of men arranged by height illustrates a normal distribution. The highest point of the bell curve is also the mean height of 5 feet, 10 inches. (b) Fourteen-year-old boys and professional jockeys have the same average weight—approximately 114 pounds. However, to be a jockey, you must be within about 4 pounds of this average. Thus, the variance among jockeys in weight is much smaller than the variance among 14-year-olds.
Visualize This: Examine Figure 7.16a closely: Does an average height of 5 feet, 10 inches in this particular population imply that most men were this height? Were most men in this population close to the mean, or was there a wide range of heights?

Visualize This

Visualize This questions, accompanying select figures, encourage students to critically evaluate illustrations to further their understanding of biology.

Integrated
Learning Outcomes

New to this edition and integrated within the text and MasteringBiology,® Learning Outcomes help professors guide students' reading and allow students to assess their understanding of the text.

In Text Pedagogical Support

LEARNING OUTCOMES

LO1 Describe the relationship between genes, chromosomes, and alleles.

LO2 Explain why cells containing identical genetic information can look different from each other.

LO3 Define *segregation* and *independent assortment* and explain how these processes contribute to genetic diversity.

LO4 Distinguish between homozygous and heterozygous genotypes and describe how recessive and dominant alleles produce particular phenotypes when expressed in these genotypes.

LO5 Demonstrate how to use a Punnett square to predict the likelihood of a particular offspring genotype and phenotype from a cross of two individuals with known genotype.

LO6 Define *quantitative trait* and describe the genetic and environmental factors that cause this pattern of inheritance.

LO7 Describe how heritability is calculated and what it tells us about the genetic component of quantitative traits.

LO8 Explain why a high heritability still does not always mean that a given trait is determined mostly by the genes an individual carries.

Each Chapter Opens with Learning Outcomes

Learning Outcomes specify the knowledge, skills, or abilities a student can expect to demonstrate after reading the chapter.

Chapter Review

Mastering BIOLOGY

Learning Outcomes

LO1 **Describe the relationship between genes, chromosomes, and alleles (Section 7.1).**
- Children resemble their parents in part because they inherit their parents' genes, segments of DNA that contain information about how to make proteins (pp. 148–149).

- Chromosomes contain genes. Different versions of a gene are called alleles (p. 150).
- Mutations in genes generate a variety of alleles. Each allele typically results in a slightly different protein product (p. 151).

Learning Outcomes Revisited at the End of the Chapter

Summaries for Learning Outcomes are provided at the end of every chapter to reinforce students' understanding of the main concepts covered.

Learning the Basics

1. **LO1** What types of drugs have helped reduce the death rate due to tuberculosis infection, and why have they become less effective more recently?

2. **LO4** Define *artificial selection*, and compare and contrast it with natural selection.

3. **LO8** Describe how *Mycobacterium tuberculosis* evolves when it is exposed to an antibiotic.

4. **LO2** Which of the following observations is not part of the theory of natural selection?

   A. Populations of organisms have more offspring than will survive; B. There is variation among individuals in a population; C. Modern organisms are unrelated; D. Traits can be passed on from parent to offspring; E. Some variants in a population have a higher probability of survival and reproduction than other variants do.

5. **LO2** The best definition of *evolutionary fitness* is _____

   A. physical health; B. the ability to attract members of the opposite sex; C. the ability to adapt to the environment, D. survival and reproduction relative to other members of the population; E. overall strength

6. **LO3** An adaptation is a trait of an organism that increases _____

   A. its fitness; B. its ability to survive and replicate; C. in frequency in a population over many generations; D. A and B are correct; E. A, B, and C are correct

shorter tusks at full adulthood than male elephants in the early 1900s. This is an example of _____

A. disruptive selection; B. stabilizing selection; C. directional selection; D. evolutionary regression; E. more than one of the above is correct

10. **LO8** Antibiotic resistance is becoming common among organisms that cause a variety of human diseases. All of the following strategies help reduce the risk of antibiotic resistance evolving in a susceptible bacterial population except _____

A. using antibiotics only when appropriate, for bacterial infections that are not clearing up naturally; B. using the drugs as directed, taking all the antibiotic over the course of days prescribed; C. using more than one antibiotic at a time for difficult-to-treat organisms; D. preventing natural selection by reducing the amount of evolution the organisms can perform; E. reducing the use of antibiotics in non-health care settings, such as agriculture

Analyzing and Applying the Basics

1. **LO3 LO6** Most domestic fruits and vegetables are a result of artificial selection from wild ancestors. Use your understanding of artificial selection to describe how domesticated strawberries must have evolved from their smaller wild relatives. What trade-offs do domesticated strawberries exhibit relative to their wild ancestors?

2. **LO3 LO5** The striped pattern on zebras' coats is considered to be an adaptation that helps reduce the

Learning Outcomes in Self-Assessment Sections

Questions in Learning the Basics and Analyzing and Applying the Basics include corresponding Learning Outcomes to encourage students to recall and formulate the most reasonable and precise answers.

Online Pedagogical Support in MasteringBiology®

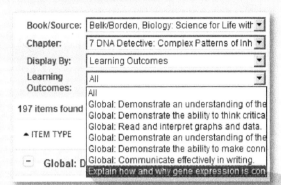

Integrated and Accessible Learning Outcomes

All MasteringBiology content, including activities and test items, are now tagged to the book specific Learning Outcomes. This unique integration of text and media enables instructors to reinforce key content in a variety of formats.

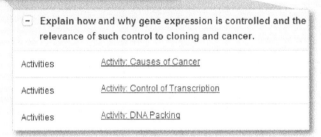

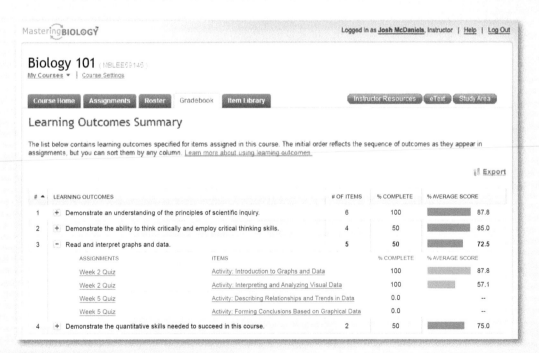

One-click Learning Outcomes Report

In MasteringBiology it's quick and easy for instructors to generate reports showing students' performance on selected Learning Outcomes. With just one click from the gradebook, instructors can see all of the Learning Outcomes associated with problems assigned in their MasteringBiology course, as well as the average score on those problems. One further click allows instructors to export this information to a Microsoft® Excel spreadsheet, which can be formatted as the instructor sees fit.

Author-created
MasteringBiology® Activities

New MasteringBiology content written by authors Colleen Belk and
Virginia Borden Maier reinforce and connect the text and media.

MasteringBiology®
www.masteringbiology.com

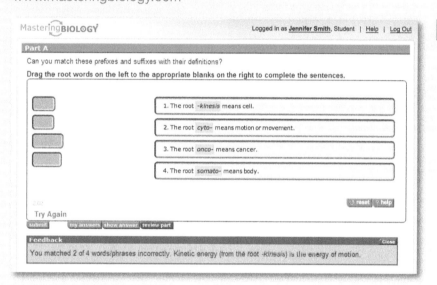

NEW! Roots to Remember

Based on the popular end of chapter
section, these activities test students'
understanding of key Latin and Greek
roots and ask students to apply the
roots in new situations.

NEW! Video-based Activities

New Video Activities support story line
content and help reinforce the narrative
and engaging nature of the text.

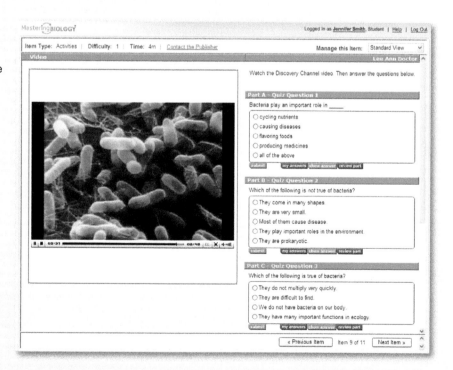

BioFlix® Animations

These animations invigorate classroom lectures and cover the most difficult biology topics, including cellular respiration, photosynthesis, and how neurons work. BioFlix include 3-D, movie-quality animations, labeled slide shows, carefully constructed student tutorials, study sheets, and quizzes.

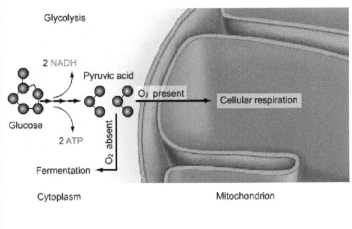

Consider the metabolism in your muscle cells. At rest or during light exercise, when oxygen is plentiful, pyruvic acid enters the Krebs cycle and continues to be metabolized through cellular respiration.

During heavy exercise, your lungs and circulatory system can't transport oxygen to your muscles rapidly enough and muscle cells run low on oxygen. The lack of oxygen forces a shift in energy metabolism toward the anaerobic pathway, fermentation. In muscle cells, fermentation results in lactate production. In some organisms, fermentation produces ethanol and carbon dioxide. However, there is no ATP production beyond the two ATP molecules generated by glycolysis.

STOP | PLAY   BACK TO INTRO

Additional Book-specific Activities

Over 25 total book-specific activities have been added to the MasteringBiology item library.

Topics include:
- Glucose Metabolism (Chapter 4)
- The Process of Speciation (Chapter 12)
- Population Growth (Chapter 14)

Dynamic
Teaching Resources

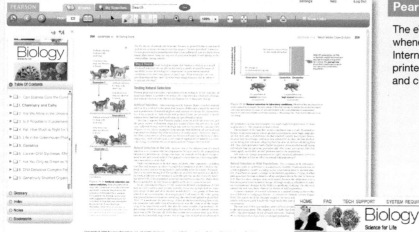

Pearson eText

The eText gives students access to the text whenever and wherever they can access the Internet. The eText pages look exactly like the printed text and include powerful interactive and customization functions.

MasteringBiology® Study Area

The Study Area of MasteringBiology provides students with access to useful materials for studying independently. Study materials include flashcards, word study tools, animations, and practice quizzes.

NEW! Pre-built Assignments

Pre-built assignments make it easier for instructors to assign MasteringBiology.

Pre-built Assignment Types:
- Engaging Homework
- Activities on Tough Topics
- Build Graphing Skills
- Current Events
- Reading Quizzes

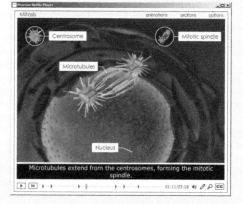

NEW! BioFlix® Player

The New Instructor's Edition DVD contains the complete BioFlix 3-D Animation Suite with a new, enhanced player. The DVD includes an "all-in-one" BioFlix lecture tool, with labels for key elements, closed captioning, drawing tool, quiz questions, and more.

Acknowledgments

Reviewers

Each chapter of this book was thoroughly reviewed several times as it moved through the development process. Reviewers were chosen on the basis of their demonstrated talent and dedication in the classroom. Many of these reviewers are already trying various approaches to actively engage students in lectures and to raise the scientific literacy and critical thinking skills among their students. Their passion for teaching and commitment to their students were evident throughout this process. These devoted individuals scrupulously checked each chapter for scientific accuracy, readability, and coverage level.

All of these reviewers provided thoughtful, insightful feedback, which improved the text significantly. Their efforts reflect their deep commitment to teaching nonmajors and improving the scientific literacy of all students. We are very thankful for their contributions.

Karen Aguirre *Coastal Carolina University*
Marcia Anglin *Miami-Dade College*
Tania Beliz *College of San Mateo*
Merri Casem *California State University, Fullerton*
Anne Casper *Eastern Michigan University*
Dani DuCharme *Waubonsee Community College*
Steve Eisenberg *Elizabethtown Community and Technical College*
Susan Fisher *Ohio State University*
Jennifer Fritz *The University of Texas at Austin*
Anthony Gaudin *Ivy Tech Community College of Indiana—Columbus/Franklin*
Mary Rose Grant *University of Missouri, St. Louis*
Richard Jacobson *Laredo Community College*
Malcolm Jenness *New Mexico Institute of Technology*
Ron Johnston *Blinn College*
Michael L'Annunziata *Pima College*
Suzanne Long *Monroe Community College*
Cindy Malone *California State University, Northridge*
Mark Manteuffel *St. Louis Community College, Flo Valley*
Roger Martin *Brigham Young University, Salt Lake Center*
Jorge Obeso *Miami Dade College, North Campus*
Nancy Platt *Pima College*
Ashley Rhodes *Kansas State University*
Bill Rogers *Ball State University*
Michael Sawey *Texas Christian University*
Debbie Scheidemantel *Pima College*
Cara Shillington *Eastern Michigan University*
Lynnda Skidmore *Wayne County Community College*
Anna Bess Sorin *University of Memphis*
Jeff Thomas *California State University, Northridge*

Derek Weber *Raritan Valley Community College*
Jennifer Wiatrowski *Pasco-Hernando Community College*

Prior Edition Reviewers

Daryl Adams, *Minnesota State University, Mankato*
Karen Aguirre, *Clarkson University*
Susan Aronica, *Canisius College*
Mary Ashley, *University of Chicago*
James S. Backer, *Daytona Beach Community College*
Ellen Baker, *Santa Monica College*
Gail F. Baker, *LaGuardia Community College*
Neil R. Baker, *The Ohio State University*
Andrew Baldwin, *Mesa Community College*
Thomas Balgooyen, *San Jose State University*
Tamatha R. Barbeau, *Francis Marion University*
Sarah Barlow, *Middle Tennessee State University*
Andrew M. Barton, *University of Maine, Farmington*
Vernon Bauer, *Francis Marion University*
Paul Beardsley, *Idaho State University*
Donna Becker, *Northern Michigan University*
David Belt, *Penn Valley Community College*
Steve Berg, *Winona State University*
Carl T. Bergstrom, *University of Washington*
Janet Bester-Meredith, *Seattle Pacific University*
Barry Beutler, *College of Eastern Utah*
Donna H. Bivans, *Pitt Community College*
Lesley Blair, *Oregon State University*
John Blamire, *City University of New York, Brooklyn College*
Barbara Blonder, *Flagler College*
Susan Bornstein-Forst, *Marian College*
Bruno Borsari, *Winona State University*
James Botsford, *New Mexico State University*
Robert S. Boyd, *Auburn University*
Bryan Brendley, *Gannon University*
Eric Brenner, *New York University*
Peggy Brickman, *University of Georgia*
Carol Britson, *University of Mississippi*
Carole Browne, *Wake Forest University*
Neil Buckley, *State University of New York, Plattsburgh*
Stephanie Burdett, *Brigham Young University*
Warren Burggren, *University of North Texas*
Nancy Butler, *Kutztown University*
Suzanne Butler, *Miami-Dade Community College*
Wilbert Butler, *Tallahassee Community College*
David Byres, *Florida Community College at Jacksonville*
Tom Campbell, *Pierce College, Los Angeles*
Deborah Cato, *Wheaton College*
Peter Chabora, *Queens College*

Bruce Chase, *University of Nebraska, Omaha*
Thomas F. Chubb, *Villanova University*
Gregory Clark, *University of Texas, Austin*
Kimberly Cline-Brown, *University of Northern Iowa*
Mary Colavito, *Santa Monica College*
William H. Coleman, *University of Hartford*
William F. Collins III, *Stony Brook University*
Walter Conley, *State University of New York, Potsdam*
Jerry L. Cook, *Sam Houston State University*
Melanie Cook, *Tyler Junior College*
Scott Cooper, *University of Wisconsin, La Crosse*
Erica Corbett, *Southeastern Oklahoma State University*
George Cornwall, *University of Colorado*
Charles Cottingham, *Frederick Community College*
James B. Courtright, *Marquette University*
Angela Cunningham, *Baylor University*
Judy Dacus, *Cedar Valley College*
Judith D'Aleo, *Plymouth State University*
Deborah Dardis, *Southeastern Louisiana University*
Juville Dario-Becker, *Central Virginia Community College*
Garry Davies, *University of Alaska, Anchorage*
Miriam del Campo, *Miami-Dade Community College*
Judith D'Aleo, *Plymouth State University*
Edward A. DeGrauw, *Portland Community College*
Heather DeHart, *Western Kentucky University*
Miriam del Campo, *Miami-Dade Community College*
Veronique Delesalle, *Gettysburg College*
Lisa Delissio, *Salem State College*
Beth De Stasio, *Lawrence University*
Elizabeth Desy, *Southwest Minnesota State University*
Donald Deters, *Bowling Green State University*
Gregg Dieringer, *Northwest Missouri State*
Diane Dixon, *Southeastern Oklahoma State University*
Christopher Dobson, *Grand Valley State University*
Cecile Dolan, *New Hampshire Community Technical College, Manchester*
Matthew Douglas, *Grand Rapids Community College*
Lee C. Drickamer, *Northern Arizona University*
Susan Dunford, *University of Cincinnati*
Stephen Ebbs, *Southern Illinois University*
Douglas Eder, *Southern Illinois University, Edwardsville*
Patrick J. Enderle, *East Carolina University*
William Epperly, *Robert Morris College*
Ana Escandon, *Los Angeles Harbor College*
Dan Eshel, *City University of New York, Brooklyn College*
Marirose Ethington, *Genessee Community College*
Deborah Fahey, *Wheaton College*
Chris Farrell, *Trevecca Nazarene University*
Richard Firenze, *Broome Community College*
Lynn Firestone, *Brigham Young University*
Brandon L. Foster, *Wake Technical Community College*
Richard A. Fralick, *Plymouth State University*
Barbara Frank, *Idaho State University*
Stewart Frankel, *University of Hartford*
Lori Frear, *Wake Technical Community College*
David Froelich, *Austin Community College*

Suzanne Frucht, *Northwest Missouri State University*
Edward Gabriel, *Lycoming College*
Anne Galbraith, *University of Wisconsin, La Crosse*
Patrick Galliart, *North Iowa Area Community College*
Wendy Garrison, *University of Mississippi*
Alexandros Georgakilas, *East Carolina University*
Robert George, *University of North Carolina, Wilmington*
Tammy Gillespie, *Eastern Arizona College*
Sharon Gilman, *Coastal Carolina University*
Mac F. Given, *Neumann College*
Bruce Goldman, *University of Conn ticut, Storrs*
Andrew Goliszek, *North Carolina Agricultural and Technical State University*
Beatriz Gonzalez, *Sante Fe Community College*
Eugene Goodman, *University of Wisconsin, Parkside*
Lara Gossage, *Hutchinson Community College*
Tamar Goulet, *University of Mississippi*
Becky Graham, *University of West Alabama*
John Green, *Nicholls State University*
Robert S. Greene, *Niagara University*
Tony J. Greenfield, *Southwest Minnesota State University*
Bruce Griffis, *Kentucky State University*
Mark Grobner, *California State University, Stanislaus*
Michael Groesbeck, *Brigham Young University, Idaho*
Stanley Guffey, *University of Tennessee*
Mark Hammer, *Wayne State University*
Blanche Haning, *North Carolina State University*
Robert Harms, *St. Louis Community College*
Craig M. Hart, *Louisiana State University*
Patricia Hauslein, *St. Cloud State University*
Stephen Hedman, *University of Minnesota, Duluth*
Bethany Henderson-Dean, *University of Findlay*
Julie Hens, *University of Maryland University College*
Peter Heywood, *Brown University*
Julia Hinton, *McNeese State University*
Phyllis C. Hirsh, *East Los Angeles College*
Elizabeth Hodgson, *York College of Pennsylvania*
Leland Holland, *Pasco-Hernando Community College*
Jane Horlings, *Saddleback Community College*
Margaret Horton, *University of North Carolina, Greensboro*
Laurie Host, *Harford Community College*
David Howard, *University of Wisconsin, La Crosse*
Michael Hudecki, *State University of New York, Buffalo*
Michael E. S. Hudspeth, *Northern Illinois University*
Laura Huenneke, *New Mexico State University*
Pamela D. Huggins, *Fairmont State University*
Sue Hum-Musser, *Western Illinois University*
Carol Hurney, *James Madison University*
James Hutcheon, *Georgia Southern University*
Anthony Ippolito, *DePaul University*
Carl Johansson, *Fresno City College*
Thomas Jordan, *Pima Community College*
Jann Joseph, *Grand Valley State University*
Mary K. Kananen, *Penn State University, Altoona*
Arnold Karpoff, *University of Louisville*
Judy Kaufman, *Monroe Community College*

Michael Keas, *Oklahoma Baptist University*
Judith Kelly, *Henry Ford Community College*
Karen Kendall-Fite, *Columbia State Community College*
Andrew Keth, *Clarion University*
David Kirby, *American University*
Stacey Kiser, *Lane Community College*
Dennis Kitz, *Southern Illinois University, Edwardsville*
Carl Kloock, *California State, Bakersfield*
Jennifer Knapp, *Nashville State Technical Community College*
Loren Knapp, *University of South Carolina*
Michael A. Kotarski, *Niagara University*
Michelle LaBonte, *Framingham State College*
Phyllis Laine, *Xavier University*
Dale Lambert, *Tarrant County College*
Tom Langen, *Clarkson University*
Lynn Larsen, *Portland Community College*
Mark Lavery, *Oregon State University*
Brenda Leady, *University of Toledo*
Mary Lehman, *Longwood University*
Doug Levey, *University of Florida*
Lee Likins, *University of Missouri, Kansas City*
Abigail Littlefield, *Landmark College*
Andrew D. Lloyd, *Delaware State University*
Jayson Lloyd, *College of Southern Idaho*
Judy Lonsdale, *Boise State University*
Kate Lormand, *Arapahoe Community College*
Paul Lurquin, *Washington State University*
Kimberly Lyle-Ippolito, *Anderson University*
Douglas Lyng, *Indiana University/Purdue University*
Michelle Mabry, *Davis and Elkins College*
Stephen E. MacAvoy, *American University*
Molly MacLean, *University of Maine*
Charles Mallery, *University of Miami*
Ken Marr Green, *River Community College*
Kathleen Marrs, *Indiana University/Purdue University*
Matthew J. Maurer, *University of Virginia's College at Wise*
Geri Mayer, *Florida Atlantic University*
T. D. Maze, *Lander University*
Steve McCommas, *Southern Illinois University, Edwardsville*
Colleen McNamara, *Albuquerque Technical Vocational Institute*
Mary McNamara, *Albuquerque Technical Vocational Institute*
John McWilliams, *Oklahoma Baptist University*
Susan T. Meiers, *Western Illinois University*
Diane Melroy, *University of North Carolina, Wilmington*
Joseph Mendelson, *Utah State University*
Paige A. Mettler-Cherry, *Lindenwood University*
Debra Meuler, *Cardinal Stritch University*
James E. Mickle, *North Carolina State University*
Craig Milgrim, *Grossmont College*
Hugh Miller, *East Tennessee State University*
Jennifer Miskowski, *University of Wisconsin, La Crosse*
Ali Mohamed, *Virginia State University*
Stephen Molnar, *Washington University*
James Mone, *Millersville University*
Daniela Monk, *Washington State University*
David Mork, *Yakima Valley Community College*

Bertram Murray, *Rutgers University*
Ken Nadler, *Michigan State University*
John J. Natalini, *Quincy University*
Alissa A. Neill, *University of Rhode Island*
Dawn Nelson, *Community College of Southern Nevada*
Joseph Newhouse, *California University of Pennsylvania*
Jeffrey Newman, *Lycoming College*
David L.G. Noakes, *University of Guelph*
Shawn Nordell, *St. Louis University*
Tonye E. Numbere, *University of Missouri, Rolla*
Lori Nicholas, *New York University*
Erin O'Brien, *Dixie College*
Igor Oksov, *Union County College*
Kevin Padian, *University of California, Berkeley*
Arnas Palaima, *University of Mississippi*
Anthony Palombella, *Longwood University*
Marilee Benore Parsons, *University of Michigan, Dearborn*
Steven L. Peck, *Brigham Young University*
Javier Penalosa, *Buffalo State College*
Murray Paton Pendarvis, *Southeastern Louisiana University*
Rhoda Perozzi, *Virginia Commonwealth University*
John Peters, *College of Charleston*
Patricia Phelps, *Austin Community College*
Polly Phillips, *Florida International University*
Indiren Pillay, *Culver-Stockton College*
Francis J. Pitocchelli, *Saint Anselm College*
Roberta L. Pohlman, *Wright State University*
Calvin Porter, *Xavier University*
Linda Potts, *University of North Carolina, Wilmington*
Robert Pozos, *San Diego State University*
Marion Preest, *The Claremont Colleges*
Gregory Pryor, *Francis Marion University*
Rongsun Pu, *Kean University*
Narayanan Rajendran, *Kentucky State University*
Anne E. Reilly, *Florida Atlantic University*
Michael H. Renfroe, *James Madison University*
Laura Rhoads, *State University of New York, Potsdam*
Gwynne S. Rife, *University of Findlay*
Todd Rimkus, *Marymount University*
Laurel Roberts, *University of Pittsburgh*
Wilma Robertson, *Boise State University*
William E. Rogers, *Texas A&M University*
Troy Rohn, *Boise State University*
Deborah Ross, *Indiana University/Purdue University*
Christel Rowe, *Hibbing Community College*
Joanne Russell, *Manchester Community College*
Michael Rutledge, *Middle Tennessee State University*
Wendy Ryan, *Kutztown University*
Christopher Sacchi, *Kutztown University*
Kim Sadler, *Middle Tennessee State University*
Brian Sailer, *Albuquerque Technical Vocational Institute*
Jasmine Saros, *University of Wisconsin, La Crosse*
Ken Saville, *Albion College*
Louis Scala, *Kutztown University*
Daniel C. Scheirer, *Northeastern University*
Beverly Schieltz, *Wright State University*

Nancy Schmidt, *Pima Community College*
Robert Schoch, *Boston University*
Julie Schroer, *Bismarck State College*
Fayla Schwartz, *Everett Community College*
Steven Scott, *Merritt College*
Gray Scrimgeour, *University of Toronto*
Roger Seeber, *West Liberty State College*
Mary Severinghaus, *Parkland College*
Allison Shearer, *Grossmont College*
Robert Shetlar, *Georgia Southern University*
Cara Shillington, *Eastern Michigan University*
Beatrice Sirakaya, *Pennsylvania State University*
Cynthia Sirna, *Gadsden State Community College*
Thomas Sluss, *Fort Lewis College*
Brian Smith Black, *Hills State University*
Douglas Smith, *Clarion University of Pennsylvania*
Mark Smith, *Chaffey College*
Gregory Smutzer, *Temple University*
Sally Sommers, *Smith Boston University*
Anna Bess Sorin, *University of Memphis*
Bryan Spohn Florida, *Community College at Jacksonville,*
    Kent Campus
Carol St. Angelo, *Hofstra University*
Amanda Starnes, *Emory University*
Susan L. Steen, *Idaho State University*
Timothy Stewart, *Longwood College*
Shawn Stover, *Davis and Elkins College*
Bradley J. Swanson, *Central Michigan University*
Joyce Tamashiro, *University of Puget Sound*
Jeffrey Taylor, *Slippery Rock University*
Martha Taylor, *Cornell University*
Glena Temple, *Viterbo*
Tania Thalkar, *Clarion University of Pennsylvania*
Alice Templet, *Nicholls State University*
Jeffrey Thomas, *University of California, Los Angeles*
Janis Thompson, *Lorain County Community College*
Nina Thumser, *California University of Pennsylvania*
Alana Tibbets, *Southern Illinois University, Edwardsville*
Martin Tracey, *Florida International University*
Jeffrey Travis, *State University of New York, Albany*
Michael Troyan, Pennsylvania State University
Robert Turgeon, *Cornell University*
Michael Tveten, *Pima Community College, Northwest Campus*
James Urban, *Kansas State University*
Brandi Van Roo, *Framingham State College*
John Va ghan, *St. Petersburg Junior College*
Martin Vaughan, *Indiana State University*
Mark Venable, *Appalachian State University*
Paul Verrell, *Washington State University*
Tanya Vickers, *University of Utah*
Janet Vigna, *Grand Valley State University*
Sean Walker, *California State University, Fullerton*
Don Waller, *University of Wisconsin, Madison*
Tracy Ware, *Salem State College*
Jennifer Warner, *University of North Carolina, Charlotte*
Lisa Weasel, *Portland State University*

Carol Weaver, *Union University*
Frances Weaver, *Widener University*
Elizabeth Welnhofer, *Canisius College*
Marcia Wendeln, *Wright State University*
Shauna Weyrauch, *Ohio State University, Newark*
Wayne Whaley, *Utah Valley State College*
Howard Whiteman, *Murray State University*
Vernon Wiersema, *Houston Community College*
Gerald Wilcox, *Potomac State College*
Peter J. Wilkin, *Purdue University North Central*
Robert R. Wise, *University of Wisconsin, Oshkosh*
Michelle Withers, *Louisiana State University*
Art Woods, *University of Texas, Austin*
Elton Woodward, *Daytona Beach Community College*
Kenneth Wunch, *Sam Houston State University*
Donna Young, *University of Winnipeg*
Michelle L. Zjhra, *Georgia Southern University*
John Zook, Ohio University
Michelle Zurawski, *Moraine Valley Community College*

Student Focus Group Participants
California State University, Fullerton:

Danielle Bruening
Leslie Buena
Andrés Carrillo
Victor Galvan
Jessica Ginger
Sarah Harpst
Robin Keber
Ryan Roberts
Melissa Romero
Erin Seale
Nathan Tran
Tracy Valentovich
Sean Vogt

Mahetzi Hernandez
Heidi McMorris
Daniel Minn
Sam Myers
David Omut
Jonathan Pistorino
James W. Pura
Samantha Ramirez
Tiffany Speed
Tristan Terry

Fullerton College:
Michael Baker

The Book Team

We are indebted to our editor Star MacKenzie. Star is a delight to work with— insightful, funny, kind, and generous with her time. Star is truly committed to producing an excellent book that meets the needs of students and instructors. We feel fortunate to have had her guidance throughout the project.

We also owe a great deal of thanks to our organized and thoughtful development editor, Susan Teahan. Susan is a creative and careful editor whose enthusiasm for the stories and interest in the underlying science added much to this revision.

This book is dedicated to our families, friends, and colleagues who have supported us over the years. Having loving families, great friends, and a supportive work environment enabled us to make this heartfelt contribution to nonmajors biology education.

COLLEEN BELK AND
VIRGINIA BORDEN MAIER

Brief Contents

Contents

Unit One
Chemistry and Cells

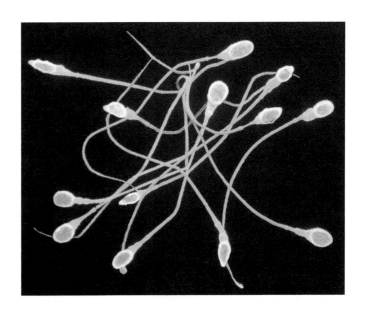

CHAPTER 1

Can Science Cure the Common Cold?

Introduction to the Scientific Method

Another cold! What can I do?

Take massive doses of Vitamin C?

We have all been there—you just recover from one bad head cold and on a morning soon after you notice that scratchy feeling in your throat that signals a new one is about to begin. It is always at the worst time, too, when you have an important exam coming up, a term paper due, and a packed social calendar. Why are you sick yet again? What can you do about it?

If you ask your friends and relatives, you will hear the usual advice on how to prevent and treat colds: Take massive doses of vitamin C. Suck on zinc lozenges. Drink plenty of echinacea tea. Meditate. Get more rest. Exercise vigorously every day. Put that hat on when you go outside! You are left with an overwhelming list of options, often contradictory and some counter to common sense. If you keep up with health news, you may be even more confused. One website reports that a popular over-the-counter cold treatment is effective, while a local TV news story details the risks of using this remedy and highlights its ineffectiveness. How do you decide what to do?

Drink echinacea tea?

Faced with this bewildering situation, most people follow the advice that makes the most sense to them, and if they find they still feel terrible, they try another remedy. Testing ideas and discarding ones that don't work is a kind of "everyday science." However, this technique has its limitations—for example, even if you feel better after trying a new cold treatment, you can't know if your recovery occurred because the treatment was effective or because the cold was ending anyway.

What professional scientists do is a more refined version of this everyday science, using strategies that help eliminate other possible explanations for a result. And while some fields of science may use unintelligible words or complicated and expensive equipment, the basic process

How would a scientist determine which advice is best?

for testing ideas is simple and universal to all areas of science. An understanding of this process can help you evaluate information about many issues that may concern and intrigue you—from health issues, to global warming, to the origin of life and the universe—with more confidence. In this chapter, we introduce you to the powerful process scientists use by asking the question we've considered here: Is there a cure for the common cold?

1.1 The Process of Science

The term *science* can refer to a body of knowledge—for example, the science of **biology** is the study of living organisms. You may believe that science requires near-perfect recall of specific sets of facts about the world. In reality, this goal is impossible and unnecessary—we do have reference books, after all. The real action in science is not memorizing what is already known but using the process of science to discover something new and unknown.

This process—making observations of the world, proposing ideas about how something works, testing those ideas, and discarding (or modifying) our ideas in response to the test results—is the essence of the **scientific method.** The scientific method allows us to solve problems and answer questions efficiently and effectively. Can we use the scientific method to solve the complicated problem of preventing and treating colds?

The Nature of Hypotheses

The statements our friends and family make about which actions will help us remain healthy (for example, the advice to wear a hat) are in some part based on the advice giver's understanding of how our bodies resist colds. Ideas about "how things work" are called **hypotheses.** Or, more formally, a hypothesis is a proposed explanation for one or more **observations.**

Hypotheses in biology come from knowledge about how the body and other biological systems work, experiences in similar situations, our understanding of other scientific research, and logical reasoning; they are also shaped by our creative mind (**Figure 1.1**). When your mom tells you to dress warmly to avoid colds, she is basing her advice on the following hypothesis: Becoming chilled makes you more susceptible to illness.

The hallmark of science is that hypotheses are subject to rigorous testing. Therefore, scientific hypotheses must be **testable**—it must be possible to evaluate a hypothesis through observations of the measurable universe. Not all hypotheses are testable. For instance, the statement that "colds are generated by disturbances in psychic energy" is not a scientific hypothesis because psychic energy does not have a material nature and thus cannot be seen or measured in a test.

In addition, hypotheses that require the intervention of a supernatural force cannot be tested scientifically. If something is **supernatural,** it is not constrained by any laws of nature, and its behavior cannot be predicted using our current understanding of the natural world. A scientific hypothesis must also be **falsifiable;** that is, an observation or set of observations could potentially prove it false. The hypothesis that exposure to cold temperatures increases your susceptibility to colds is falsifiable; we can imagine an observation that would cause us to reject this hypothesis (for instance, the observation that people exposed to cold temperatures do not catch more colds than people

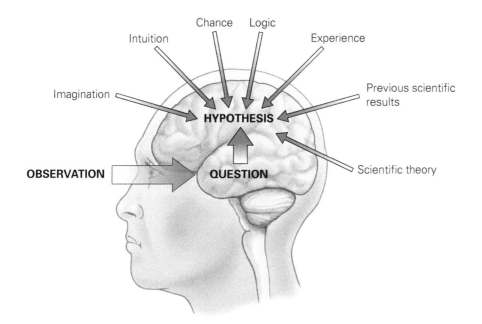

Chance Logic

Intuition

Experience

Imagination

Previous scientific
results

HYPOTHESIS

OBSERVATION

QUESTION

Scientific theory

Figure 1.1 Hypothesis generation.
All of us generate hypotheses. Many different factors, both logical and creative, influence the development of a hypothesis. Scientific hypotheses are both testable and falsifiable.

protected from chills). Of course, not all hypotheses are proved false, but it is essential in science that incorrect ideas be discarded, which occur only if it is *possible* to prove those ideas false. Lack of falsifiability is another reason supernatural hypotheses cannot be scientific. Because a supernatural force can cause any possible result of a test, hypotheses that rely on supernatural forces cannot be falsified.

Finally, statements that are value judgments, such as, "It is wrong to cheat on an exam," are not scientific because different people have different ideas about right and wrong. It is impossible to falsify these types of statements. To find answers to questions of morality, ethics, or justice, we turn to other methods of gaining understanding—such as philosophy and religion.

Scientific Theories

Most hypotheses fit into a larger picture of scientific understanding. We can see this relationship when examining how research upended a commonly held belief about diet and health—that chronic stomach and intestinal inflammation is caused by eating too much spicy food. This belief directed the standard medical practice for ulcer treatment for decades. Patients with ulcers were prescribed drugs that reduced stomach acid levels and advised to avoid eating acidic or highly spiced foods. These treatments were rarely successful, and ulcers were considered chronic, possibly lifelong, problems.

In 1982 Australian scientists Robin Warren and Barry Marshall discovered that a particular microscopic organism, specifically the bacterium *Helicobacter pylori*, was present in nearly all samples of ulcer tissue that they examined (**Figure 1.2**). From this observation, Warren and Marshall reasoned that *H. pylori* infection—invasion of the stomach wall by the bacteria—was the cause of most ulcers. Barry Marshall even tested this hypothesis on himself by consuming live *H. pylori*. He subsequently suffered from acute stomach pain.

Warren and Marshall's colleagues were at first unconvinced that ulcers could have such a simple cause. Today the hypothesis that *H. pylori* infection is responsible for most ulcers is accepted as fact. The primary reasons why this is the case? First, no reasonable alternative hypotheses about the causes of ulcers (for instance, consumption of spicy foods) has been consistently supported by hypothesis tests; and second, the hypothesis has not been rejected—that is,

(a)

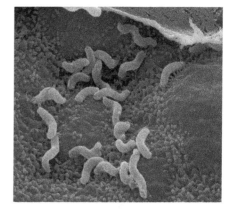

(b)

Figure 1.2 A scientific breakthrough.
(a) *Helicobacter pylori* on stomach lining (image from electron microscope). (b) Robin Warren and Barry Marshall won the 2005 Nobel Prize in Medicine for their discovery of the link between *H. pylori* and ulcers.

there have been no carefully designed experiments that show that *H. pylori* removal fails to cure most ulcers.

The third reason that the relationship between *H. pylori* and ulcers is considered fact is that it conforms to a well-accepted scientific principle, namely, the germ theory of disease. A **scientific theory** is an explanation for a set of related observations that is based on well-supported hypotheses from several different, independent lines of research. The basic premise of germ theory is that microorganisms (that is, organisms too small to be seen with the naked eye) are the cause, through infection, of some or all human diseases.

The biologist Louis Pasteur first observed that bacteria cause milk to become sour. From this observation, he reasoned that these same types of organisms could injure humans. Later, Robert Koch demonstrated a link between anthrax bacteria and a specific set of fatal symptoms in mice, providing additional evidence. Germ theory is further supported by the observation that antibiotic treatment that targets particular microorganisms can cure certain illnesses—as is the case with bacteria-caused ulcers.

In everyday speech, the word *theory* is synonymous with untested ideas based on little information. In contrast, scientists use the term when referring to well-supported ideas of how the natural world works. The supporting foundation of all scientific theories is multiple hypothesis tests.

The Logic of Hypothesis Tests

One common hypothesis about cold prevention is that taking vitamin C supplements keeps you healthy. This hypothesis is very appealing, especially given the following generally known facts:

1. Fruits and vegetables contain a lot of vitamin C.
2. People with diets rich in fruits and vegetables are generally healthier than people who skimp on these food items.
3. Vitamin C is known to be an anti-inflammatory agent, reducing throat and nose irritation.

With these facts in mind, we can state the following falsifiable hypothesis: Consuming vitamin C decreases the risk of catching a cold.

This hypothesis makes sense given the statements just listed and the experiences of the many people who insist that vitamin C keeps them healthy. The process used to construct this hypothesis is called **inductive reasoning**—combining a series of specific observations (here, statements 1–3) to discern a general principle. Inductive reasoning is an essential tool for understanding the world. However, a word of caution is in order: Just because the inductive reasoning that led to a hypothesis seems to make sense does not mean that the hypothesis is necessarily true. The example that follows demonstrates this point.

Consider the ancient hypothesis that the sun revolves around Earth. This hypothesis was induced based on the observations that the sun rose in the east every morning, traveled across the sky, and set in the west every night. For almost all of history, this hypothesis was considered to be a "fact" by nearly all of Western society. It wasn't until the early seventeenth century that this hypothesis was overturned—as the result of Galileo Galilei's observations of Venus. His observations proved false the hypothesis that the sun revolved around Earth. Galileo's work helped to confirm the more modern hypothesis, proposed by Nicolaus Copernicus, that Earth revolves around the sun.

So, even though the hypothesis about vitamin C is sensible, it needs to be tested to see if it can be proved false. Hypothesis testing is based on **deductive reasoning** or deduction. Deduction involves using a general principle to predict an expected observation. This **prediction** concerns the outcome of an action, test, or investigation. In other words, the prediction is the result we expect from a hypothesis test.

Deductive reasoning takes the form of "if/then" statements. That is, if our general principle is correct, then we expect to observe a specific outcome. A prediction based on the vitamin C hypothesis could be: *If* vitamin C decreases the risk of catching a cold, *then* people who take vitamin C supplements with their regular diets will experience fewer colds than will people who do not take supplements.

Stop & Stretch Consider the following scenario: Your coworker, Homer, is a notorious doughnut lover who has a nose for free food. You walk into the break room one morning to discover a box from the doughnut shop that is already completely empty. According to this information, what most likely happened to the doughnuts? Is this an inductive or deductive hypothesis?

Deductive reasoning, with its resulting predictions, is a powerful method for testing hypotheses. However, the structure of such a statement means that hypotheses can be clearly rejected if untrue but impossible to prove if true (**Figure 1.3**). This shortcoming is illustrated using the if/then statement concerning vitamin C and colds.

Consider the possible outcomes of a comparison between people who supplement with vitamin C and those who do not. People who take vitamin C supplements may suffer through more colds than people who do not; they may have the same number of colds as the people who do not supplement; or supplementers may in fact experience fewer colds. What does each of these results tell us about the hypothesis?

If, in a well-designed test, people who take vitamin C have more colds or the same number of colds as those who do not supplement, then the hypothesis that vitamin C provides protection against colds can be clearly rejected. But what if people who supplement with vitamin C do experience fewer colds? If this is the case, then we can only say that the hypothesis has been supported and not disproven.

Why is it impossible to say that the hypothesis that vitamin C prevents colds is true? Because there are **alternative hypotheses** that explain why people with different vitamin-taking habits vary in their cold susceptibility. In other words, demonstrating the truth of the *then* portion of a deductive statement does not prove that the *if* portion is true.

Stop & Stretch Consider this if/then statement: If your coworker Homer ate all the doughnuts, then there won't be any doughnuts left in the box that was placed in the break room 30 minutes ago. Imagine you find that the doughnuts are all gone—did you just prove that Homer ate them? Why or why not?

Consider the alternative hypothesis that frequent exercise reduces susceptibility to catching a cold. And suppose that people who take vitamin C supplements are more likely to engage in regular exercise. If both of these hypotheses

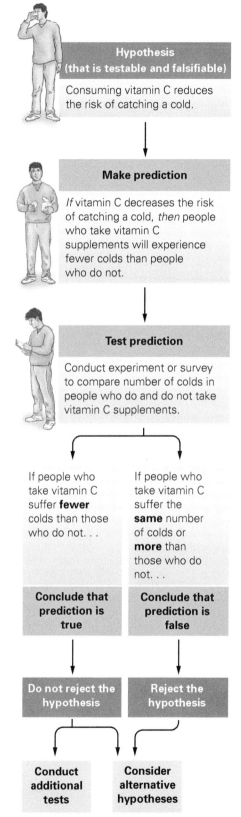

Figure 1.3 The scientific method. Tests of hypotheses follow a logical path. This flowchart illustrates the process of deduction as practiced by scientists.
Visualize This: According to this flowchart, scientists should consider alternative hypotheses even if their hypothesis is supported by their research. Explain why this is the case.

are true, then the prediction that vitamin C supplementers experience fewer colds than people who do not supplement would be true but not because the original hypothesis (vitamin C reduces the risk of colds) is true. Instead, people who take vitamin C supplements experience fewer colds because they are also more likely to exercise, and it is exercise that reduces cold susceptibility.

A hypothesis that seems to be true because it has not been rejected by an initial test may be rejected later because of a different test. This is what happened to the hypothesis that vitamin C consumption reduces susceptibility to colds. The argument for the power of vitamin C was popularized in 1970 by Nobel Prize–winning chemist Linus Pauling. Pauling based his assertion—that large doses of vitamin C reduce the incidence of colds by as much as 45%—on the results of a few studies that had been published between the 1930s and 1970s. However, repeated, careful tests of this hypothesis have since failed to support it. In many of the studies Pauling cited, it appears that alternative hypotheses explain the difference in cold incidence between vitamin C supplementers and nonsupplementers. Today, most health scientists agree that the hypothesis that vitamin C prevents colds has been convincingly falsified.

The example of the vitamin C hypothesis also highlights a challenge of communicating scientific information. You can see why the belief that vitamin C prevents colds is so widespread. If you don't know that scientific knowledge relies on rejecting incorrect ideas, a book by a Noble Prize–winning scientist may seem like the last word on the benefits of vitamin C. It took many years of careful research to show that this "last word" was, in fact, wrong.

1.2 Hypothesis Testing

The previous discussion may seem discouraging: How can scientists determine the truth of any hypothesis when there is always a chance that the hypothesis could be falsified? Even if one of the hypotheses about cold prevention is supported, does the difficulty of eliminating alternative hypotheses mean that we will never know which approach is truly best? The answer is yes—and no.

Hypotheses cannot be proven absolutely true; it is always possible that the true cause of a phenomenon may be found in a hypothesis that has not yet been tested. However, in a practical sense, a hypothesis can be proven beyond a reasonable doubt. That is, when one hypothesis has not been disproven through repeated testing and all reasonable alternative hypotheses have been eliminated, scientists accept that the well-supported hypothesis is, in a practical sense, true. "Truth" in science can therefore be defined as *what we know and understand based on all currently available information*. But scientists always leave open the possibility that what seems true now may someday be proven false.

An effective way to test many hypotheses is through rigorous scientific experiments. Experimentation has enabled scientists to prove beyond a reasonable doubt that the common cold is caused by a virus. A virus is a microscopic entity with a simple structure—it typically contains a short strand of genetic material and a few proteins encased in a relatively tough protein shell and sometimes surrounded by a membrane. A virus must infect a cell to reproduce. Of the over 200 types of viruses that are known to cause the common cold, most infect the cells in our noses and throats. The sneezing, coughing, congestion, and sore throat of a cold appear to result from the body's protective response—established by our immune system—to a viral invasion (**Figure 1.4**).

As you may know, if we survive a virus infection, we are unlikely to experience a recurrence of the disease the virus causes. For example, it is extremely rare to suffer from chicken pox twice because one exposure to the

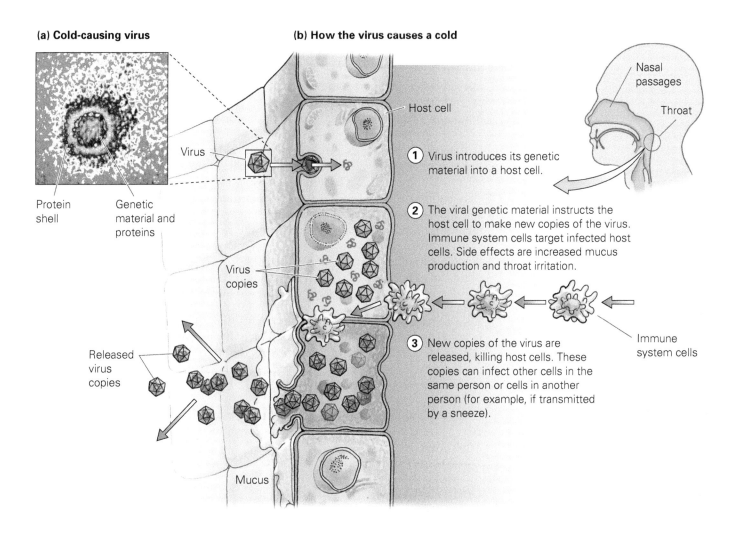

(a) Cold-causing virus

Protein shell

Genetic material and proteins

(b) How the virus causes a cold

Virus

Host cell

Nasal passages

Throat

(1) Virus introduces its genetic material into a host cell.

(2) The viral genetic material instructs the host cell to make new copies of the virus. Immune system cells target infected host cells. Side effects are increased mucus production and throat irritation.

Virus copies

(3) New copies of the virus are released, killing host cells. These copies can infect other cells in the same person or cells in another person (for example, if transmitted by a sneeze).

Immune system cells

Released virus copies

Mucus

chicken pox virus (through either infection or vaccination) usually provides lifelong immunity to future infection. The sheer number of cold viruses makes immunity to the common cold—and the development of a vaccine to prevent it—very improbable. Scientists thus focus their experimental research about common colds on methods of prevention and treatment.

The Experimental Method

Experiments are sets of actions or observations designed to test specific hypotheses. Generally, an experiment allows a scientist to control the conditions that may affect the subject of study. Manipulating the environment allows a scientist to eliminate some alternative hypotheses that may explain the result.

Experimentation in science is analogous to what a mechanic does when diagnosing a car problem. There are many reasons why a car engine might not start. If a mechanic begins by tinkering with numerous parts to apply all possible fixes before restarting the car, the mechanic will not know what exactly caused the problem (and will have an unhappy customer who is charged for unnecessary parts and labor). Instead, a mechanic begins by testing the battery for power; if the battery is charged, then she checks the starter motor; if the car still doesn't start, she looks over the fuel pump; and she continues in this manner until identifying the problem. Likewise, a scientist systematically attempts to eliminate hypotheses that do not explain a particular phenomenon.

Figure 1.4 A cold-causing virus.
(a) An electron microscopic image of a typical rhinovirus, one of the many types of viruses that cause the common cold. (b) A rhinovirus causes illness by invading nose and throat cells and using them as "factories" to make virus copies. Cold symptoms result from immune system attempts to eliminate the virus.
Visualize This: Find two points in this process where intervention by drugs or other treatment could disrupt it and lead to reduced cold symptoms.

Not all scientific hypotheses can be tested through experimentation. For instance, hypotheses about how life on Earth originated or the cause of dinosaur extinction are usually not testable in this way. These hypotheses are instead tested using careful observation of the natural world. For instance, the examination of fossils and other geological evidence allows scientists to test hypotheses regarding the extinction of the dinosaurs (Figure 1.5).

The information collected by scientists during hypothesis testing is known as **data.** The data are collected on the **variables** of the test, that is, any factor that can change in value under different conditions. In an experimental test, scientists manipulate an **independent variable** (one whose value can be freely changed) to measure the effect on a **dependent variable.** The dependent variable may or may not be influenced by changes in the independent variable, but it cannot be systematically changed by the researchers. For example, to measure the effect of vitamin C on cold prevention, scientists can vary individuals' vitamin C intake (the independent variable) and measure their susceptibility to illness on exposure to a cold virus (the dependent variable).

Stop & Stretch Some of the experiments that established the link between *H. pylori* and stomach ulcer consisted of feeding gerbils *H. pylori* and measuring the rate of ulcer development compared to untreated animals. What are the independent and dependent variables in these experiments?

Data obtained from well-designed experiments should allow researchers to convincingly reject or support a hypothesis. This is more likely to occur if the experiment is controlled.

Controlled Experiments

Control has a specific meaning in science. A **control** for an experiment is a subject similar to an experimental subject except that the control is not exposed to experimental treatment. Controlled experiments are thus designed to eliminate as many alternative hypotheses as possible.

Figure 1.5 Testing hypotheses through observation. The fossil record provides a source of data to test hypotheses about evolutionary history.

Once subjects are enlisted in an experiment, they are assigned to a control or an experimental group. If members of the control and experimental groups differ at the end of a well-designed test, then the difference is likely due to the experimental treatment.

Our question about effective cold treatments lends itself to a variety of controlled experiments on possible drug therapies. For example, an extract of *Echinacea purpurea* (a common North American prairie plant) in the form of echinacea tea has been promoted as a treatment to reduce the likelihood as well as the severity and duration of colds (**Figure 1.6**). A recent scientific experiment on the efficacy of *Echinacea* involved asking individuals suffering from colds to rate the effectiveness of a tea in relieving their symptoms. In this study, people who used echinacea tea felt that it was 33% more effective. The "33% more effective" is in comparison to the rated effectiveness of a tea that did not contain *Echinacea* extract—that is, the results from the control group.

Control groups and experimental groups must be as similar as possible to each other to eliminate alternative hypotheses that could explain the results. In the case of cold treatments, the groups should not systematically differ in age, diet, stress level, or other factors that might affect cold susceptibility. One effective way to minimize differences between groups is the **random assignment** of individuals. For example, a researcher might put all of the volunteers' names in a hat, draw out half, and designate the people drawn as the experimental group and the remaining people as the control group. As a result, each group should be a rough cross-section of the population in the study. In the echinacea tea experiment just described, members of both the experimental and control groups were female employees of a nursing home who sought relief from their colds at their employer's clinic. The volunteers were randomly assigned into either the experimental or control group as they came into the clinic.

Figure 1.6 *Echinacea purpurea*, an American coneflower. Extracts from the leaves and roots of this plant are among the most popular herbal remedies sold in the United States.

Stop & Stretch In the echinacea tea experiment, a nonrandom assignment scheme might have put the first 25 visitors to the clinic in the control group and the next 25 in the experimental group. Imagine that in this version of the experiment, the experimental group did recover in fewer days than the control group. Describe an alternative hypothesis related to the nonrandom assignment of subjects that was not eliminated by this experimental design and that could explain these results.

The second step in designing a good controlled experiment is to treat all subjects identically during the course of the experiment. In this study, all participants, whether in the control or experimental group, received the same information about the supposed benefits of echinacea tea, and during the course of the experiment, all participants were given tea to drink five to six times daily until their symptoms subsided. However, individuals in the control group received "sham tea" that did not contain *Echinacea* extract. Treating all participants the same ensures that no factor related to the interaction between subject and researcher influences the results.

The sham tea in this experiment would be equivalent to the sugar pills that are given to control subjects during drug trials. Like other intentionally ineffective medical treatments, sham tea is a **placebo.** Employing a placebo generates only one consistent difference between individuals in the two groups—in this case, the type of tea they consumed.

In the echinacea tea study, the data indicated that cold severity was lower in the experimental group compared to those who received placebo. Because their study utilized controls, the researchers can be confident that the groups

(a)

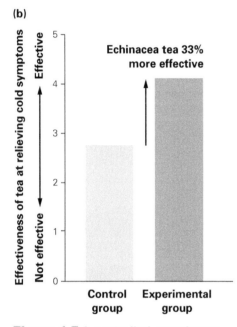

(b)

Figure 1.7 A controlled experiment. (a) In a controlled experiment testing echinacea tea as a treatment for colds, all 95 subjects were treated identically except for the type of tea they were given. (b) The results of the experiment indicated that echinacea tea was 33% more effective than the placebo.

differed because of the effect of *Echinacea*. By reducing the likelihood that alternative hypotheses could explain their results, the researchers could strongly infer that they were measuring a real, positive effect of echinacea tea on colds (**Figure 1.7**).

The study described here supports the hypothesis that echinacea tea reduces the severity of colds. However, it is extremely rare that a single experiment will cause the scientific community to accept a hypothesis beyond a reasonable doubt. Dozens of studies, each using different experimental designs and many using extracts from different parts of the plant, have investigated the effect of *Echinacea* on common colds and other illnesses. Some of these studies have shown a positive effect, but many others have shown none. In the medical community as a whole, there is significant skepticism and disagreement regarding the effectiveness of this popular herb as a cold treatment. Only through continued controlled tests of the hypothesis will we discern an accurate answer to the question, Is *Echinacea* an effective cold treatment?

Minimizing Bias in Experimental Design

Scientists and human research subjects may have strong opinions about the truth of a particular hypothesis even before it is tested. These opinions may cause participants to unfairly influence, or **bias,** the results of an experiment.

One potential source of bias is subject expectation. Individual experimental subjects may consciously or unconsciously model the behavior they feel the researcher expects from them. For example, an individual who knew she was receiving echinacea tea may have felt confident that she would recover more quickly. This might cause her to underreport her cold symptoms. This potential problem is avoided by designing a **blind experiment,** in which individual subjects are not aware of exactly what they are predicted to experience. In experiments on drug treatments, this means not telling participants whether they are receiving the drug or a placebo.

Another source of bias arises when a researcher makes consistent errors in the measurement and evaluation of results. This phenomenon is called observer bias. In the echinacea tea experiment, observer bias could take various forms. Expecting a particular outcome might lead a scientist to give slightly different instructions about which symptoms constituted a cold to subjects who received echinacea tea. Or, if the researcher expected people who drank echinacea tea to experience fewer colds, she might make small errors in the measurement of cold severity that influenced the final result.

To avoid the problem of experimenter bias, the data collectors themselves should be "blind." Ideally, the scientist, doctor, or technician applying the treatment does not know which group (experimental or control) any given subject is part of until after all data have been collected and analyzed (**Figure 1.8**). Blinding the data collector ensures that the data are **objective** or, in other words, without bias.

We call experiments **double-blind** when both the research subjects and the technicians performing the measurements are unaware of either the hypothesis or whether a subject is in the control or experimental group. Double-blind experiments nearly eliminate the effects of human bias on results. When both researcher and subject have few expectations about the outcome, the results obtained from an experiment are more credible.

Using Correlation to Test Hypotheses

Double-blind, placebo-controlled, randomized experiments represent the gold standard for medical research. However, well-controlled experiments can be difficult to perform when humans are the subjects. The requirement that both experimental and control groups be treated nearly identically means that some

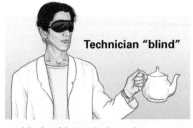

Technician "blind"

Subject "blind"

What
they
know

- **Limited knowledge** of experimental hypothesis

- **No knowledge** of which group participants belong to

- **Limited knowledge** of experimental hypothesis

- **No knowledge** of which group he or she belongs to

How
they
behave

- **No difference** in instructions to participants

- **No difference** in treatment of participants

- **No difference** in data collection

- **Unbiased** reporting of symptoms or effects of treatment

Figure 1.8 Double-blind experiments. Double-blind experiments result in data that are more objective.

people unknowingly receive no treatment. In the case of healthy volunteers with head colds, the placebo treatment of sham tea did not hurt those who received it.

However, placebo treatments are impractical or unethical in many cases. For instance, imagine testing the effectiveness of a birth control drug using a controlled experiment. This would require asking women to take a pill that may or may not prevent pregnancy while not using any other form of birth control!

Experiments on Model Systems. Scientists can use **model systems** when testing hypotheses that would raise ethical or practical problems when tested on people. For basic research that helps us understand how cells and genes function, the model systems are easily grown and manipulated organisms, such as certain species of bacteria, nematodes, and fruit flies, or even isolated cells from larger organisms that reproduce in dishes in the laboratory. In the case of research on human health and disease, model systems are typically other mammals, including human cells. Mammals are especially useful as model organisms in medical research because they are closely related to us. Like us, they have hair and give birth to live young, and thus they also share with us similarities in anatomy and physiology (**Figure 1.9**).

The vast majority of animals used in biomedical research are rodents such as rats, mice, and guinea pigs, although some areas of research require animals that are more similar to humans in size, such as dogs or pigs, or share a closer evolutionary relationship, such as chimpanzees.

The use of model systems, especially animals, allows experimental testing on potential drugs and other therapies before these methods are employed on people. Research on model organisms such as lab rats has contributed to a better understanding of nearly every serious human health threat, including cancer, heart disease, Alzheimer's disease, and AIDS (acquired immunodeficiency syndrome). However, ethical concerns about the use of animals in research persist and can complicate such studies. In addition, the results of animal studies are not always directly applicable to humans—despite a shared evolutionary history, animals still can have important differences from humans. Testing hypotheses about human health in human beings still provides the clearest answer to these questions.

(a)

(b)

(c)

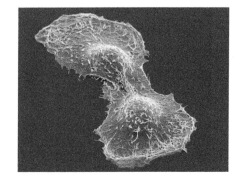

Figure 1.9 Model systems in science. (a) Research on the nematode *Caenorhabditis elegans* (also known as *C. elegans*) has led to major advances in our understanding of animal development and genetics. (b) The classic "lab rat" is easy to raise and care for and as a mammal is an important model system for testing drugs and treatments designed for humans. (c) HeLa cells are derived from a cancerous growth biopsied from Henrietta Lacks in 1951. These cells have the unique property of being able to divide indefinitely, making them valuable for research on a wide variety of subjects from HIV to human growth hormone.

Looking for Relationships Between Factors. Scientists can also test hypotheses using correlations when controlled experiments on humans are difficult or impossible to perform. A **correlation** is a relationship between two variables.

Suggestions about using meditation to reduce susceptibility to colds are based on a correlation between psychological stress and susceptibility to cold virus infections (**Figure 1.10**). This correlation was generated by researchers who collected data on subjects' psychological stress levels before giving them nasal drops that contained a cold virus. Doctors later reported on the incidence and severity of colds among participants in the study. Note that while the cold virus was applied to each participant in the study, the researchers had no influence on the stress level of the study participants—in other words, this was not a controlled experiment because people were not randomly assigned to different "treatments" of low or high stress.

Stop & Stretch   Even though the researchers did not manipulate the independent variable in this study, there are still an independent variable and a dependent variable. What are they?

Let's examine the results presented in Figure 1.10. The horizontal axis of the graph, or **x-axis**, illustrates the independent variable. This variable is stress, and the graph ranks subjects along a scale of stress level—from low stress on the left edge of the scale to high stress on the right. The vertical axis of the graph, the **y-axis**, is the dependent variable—the percentage of study participants who developed colds as reported by their doctors. Each point on the graph represents a group of individuals and tells us what percentage of people in each stress category had clinical colds.

The line connecting the 5 points on the graph illustrates a correlation—the relationship between stress level and susceptibility to cold virus infections. Because the line rises to the right, it illustrates a positive correlation. These data tell us that people who had higher stress levels were more likely to come down with colds. But does this relationship mean that high stress causes increased cold susceptibility?

To conclude that stress causes illness, we need the same assurances that are given by a controlled experiment. In other words, we must assume that the

Figure 1.10 Correlation between stress level and illness. The graph indicates that people reporting higher levels of stress became infected after exposure to a cold virus more often than did people who reported low levels of stress.
Visualize This: This graph groups people with similar, but not identical, stress index measures. Why might this have been necessary? If people with stress indices 3 and 4 have the same susceptibility to colds, does this call into question the correlation?

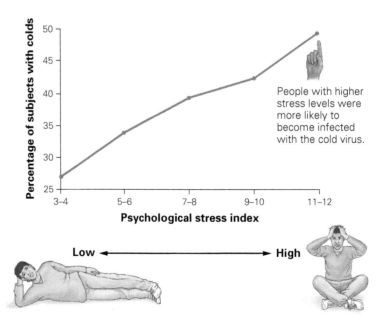

People with higher stress levels were more likely to become infected with the cold virus.

Psychological stress index

Low ⟷ High

(a) Does high stress cause high cold frequency?

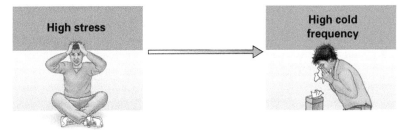

(b) Or does one of the causes of high stress also cause high cold frequency?

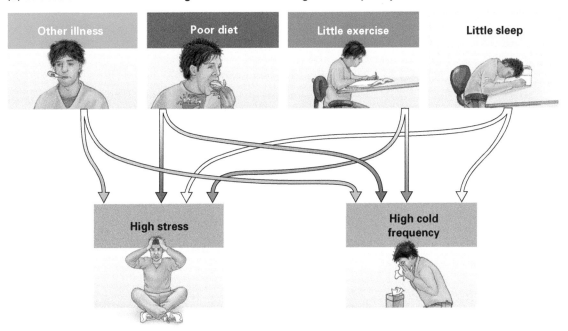

Figure 1.11 Correlation does not signify causation. A correlation typically cannot eliminate all alternative hypotheses.

individuals measured for the correlation are similar in every way except for their stress levels. Is this a good assumption? Not necessarily. Most correlations cannot control for all alternative hypotheses. People who feel more stressed may have poorer diets because they feel time-limited and rely on fast food more often. In addition, people who feel highly stressed may be in situations where they are exposed to more cold viruses. These differences among people who differ in stress level may also influence their cold susceptibility (Figure 1.11). Therefore, even with a strong correlational relationship between the two factors, we cannot strongly infer that stress causes decreased resistance to colds.

Researchers who use correlational studies can eliminate a number of alternative hypotheses by closely examining their subjects. For example, this study on stress and cold susceptibility collected data from subjects on age, weight, sex, education, and their exposure to infected individuals. None of these factors differed consistently among low-stress and high-stress groups. While this analysis increases our confidence in the hypothesis that high stress levels increase susceptibility to colds, people with high-stress lifestyles still may have important differences from those with low-stress lifestyles. It is possible that one of those differences is the real cause of disparities in cold frequency.

As you can see, it is difficult to demonstrate a cause-and-effect relationship between two factors simply by showing a correlation between them. In other words, correlation does not equal causation. For example, a commonly understood correlation exists between exposure to cold air and epidemics of the common cold. It is true that as outdoor temperatures drop, the incidence of colds increases. But numerous controlled experiments have indicated that chilling an individual does not increase his susceptibility to colds. Instead, cold outdoor temperatures mean increased close contact with other people (and their viruses), and this contact increases the number of colds we get during cold weather. Despite the correlation, cold air does not cause colds—exposure to viruses does.

Correlational studies are the main tool of epidemiology, the study of the distribution and causes of diseases. One commonly used epidemiological technique is a cross-sectional survey. In this type of survey, many individuals are both tested for the presence of a particular condition and asked about their exposure to various factors. The limitations of cross-sectional surveys include the effect of subject bias and poor recall by survey participants, in addition to all of the problems associated with interpreting correlations. **Table 1.1** provides an overview of the variety of correlational strategies employed to study the links between our environment and our health.

TABLE 1.1

Types of correlational studies.			
Name	**Description**	**Pros**	**Cons**
Ecological studies	Examine specific human populations for unusually high levels of various diseases (e.g., documenting a "cancer cluster" around an industrial plant)	Inexpensive and relatively easy to do	Unsure whether exposure to environmental factor is actually related to onset of the disease
Cross-sectional surveys	Question individuals in a population to determine amount of exposure to an environmental factor and whether disease is present	More specific than ecological study	• Expensive • Subjects may not know exposure levels • Cannot control for other factors that may be different among individuals in survey • Cannot be used for rare diseases
Case-control studies	Compare exposures to specific environmental factors between individuals who have a disease and individuals matched in age and other factors who do not have the disease	• Relatively fast and inexpensive • Best method for rare diseases	• Does not measure absolute risk of disease as a result of exposure • Difficult to select appropriate controls to eliminate alternative hypotheses • Examines just one disease possibly associated with an environmental factor
Cohort studies	Follow a group of individuals, measuring exposure to environmental factors and disease prevalence	• Can determine risk of various diseases associated with exposure to particular environmental factor	• Expensive and time-consuming • Difficult to control for alternative hypotheses • Not feasible for rare diseases
Correlational experiment	Expose individuals who experience varying levels of an independent variable to an experimental treatment	• Can control the amount of exposure to at least one environmental factor of interest	• Cannot eliminate alternative hypotheses • Only feasible for hypotheses for which an experimental treatment can be applied

1.3 Understanding Statistics

During a review of scientific literature on cold prevention and treatment, you may come across statements about the "significance" of the effects of different cold-reducing measures. For instance, one report may state that factor A reduced cold severity, but that the results of the study were "not significant." Another study may state that factor B caused a "significant reduction" in illness. We might then assume that this statement means factor B will help us feel better, whereas factor A will have little effect. This is an incorrect assumption because in scientific studies, *significance* is defined a bit differently from its usual sense. To evaluate the scientific use of this term, we need a basic understanding of statistics.

An Overview: What Statistical Tests Can Tell Us

We often use the term *statistics* to refer to a summary of accumulated information. For instance, a baseball player's success at hitting is summarized by a statistic: his batting average, the total number of hits he made divided by "at bats," the number of opportunities he had to hit. The science of **statistics** is a bit different; it is a specialized branch of mathematics used to evaluate and compare data.

An experimental test utilizes a small subgroup, or **sample,** of a population. Statistical methods can summarize data from the sample—for instance, we can describe the average, also known as the **mean,** length of colds experienced by experimental and control groups. **Statistical tests** can then be used to extend the results from a sample to the entire population.

When scientists conduct an experiment, they hypothesize that there is a true, underlying effect of their experimental treatment on the entire population. An experiment on a sample of a population can only estimate this true effect because a sample is always an imperfect "snapshot" of an entire population.

Consider a hypothesis that the average hair length of women in a college class in 1963 was shorter than the average hair length at the same college today. To test this hypothesis, we could compare a sample of snapshots from college yearbooks. If hairstyles were very similar among the women in a snapshot, you could reasonably assume that the average hair length in the college class is close to the average length in the snapshot. However, what if you see that women in a snapshot have a variety of hairstyles, from short bobs to long braids? In this case, it is difficult to determine whether the average hair length in the snapshot is at all close to the average for the class. With so much variation, the snapshot could, by chance, contain a surprisingly high number of women with very long hair, causing the average length in the sample to be much longer than the average length for the entire class (**Figure 1.12**).

A statistical test calculates the likelihood, given the number of individuals sampled and the variation within samples, that the difference between two samples reflects a real, underlying difference between the populations from which these samples were drawn. A **statistically significant** result is one that is very unlikely to be due to chance differences between the experimental and control samples, so thus likely represents a true difference between the groups.

In the experiment with the echinacea tea, statistical tests indicated that the 33% reduction in cold severity observed by the researchers was statistically significant. In other words, there is a low probability that the difference in cold susceptibility between the two samples in the experiment is due to chance. If the experiment is properly designed, a statistically significant result allows researchers to infer that the treatment had an effect. *See the next section for **A Closer Look** at statistics.*

(a) Average hair length in this snapshot is shorter...

Class of 1963

- Little variability
- High probability of reflecting average of all women in the class

(b) ...than average hair length in this snapshot...

Class of 2013

- High variability
- Low probability of reflecting average of all women in the class

...so, is hair today longer than in 1963?

Figure 1.12 The role of statistics. Statistical tests calculate the variability within groups to determine the probability that two groups differ only by chance.

A Closer Look:
Statistics

We can explore the role that statistical tests play more closely by evaluating another study on cold treatments. This study examined the efficacy of lozenges containing zinc on reducing cold severity.

Some forms of zinc can block common cold viruses from invading cells in the nasal cavity. This observation led to the hypothesis that consuming zinc at the start of a cold could decrease the severity of cold symptoms by reducing the number of cells that become infected. Researchers at the Cleveland Clinic tested this hypothesis using a sample of 100 of their employees who volunteered for a study within 24 hours of developing cold symptoms. The researchers randomly assigned subjects to control or experimental groups. Members of the experimental group received lozenges containing zinc, while members of the control group received placebo lozenges. Members of both groups received the same instructions for using the lozenges and were asked to rate their symptoms until they had recovered. The experiment was double-blind.

When the data from the zinc lozenge experiment were summarized, the statistics indicated that the mean recovery time was more than 3 days shorter in the zinc group than in the placebo group (Figure 1.13). On the surface, this result appears to support the hypothesis. However, recall the example of the snapshot of women's hair length. A statistical test is necessary because of the effect of chance.

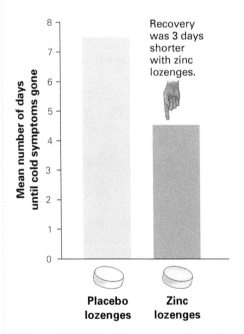

Figure 1.13 Zinc lozenges reduce the duration of colds. Fifty individuals taking zinc lozenges had colds lasting about 4½ days as opposed to approximately 7½ days for 50 individuals taking placebo.

The Problem of Sampling Error

The effect of chance on experimental results is known as **sampling error**—more specifically, sampling error is the difference between a sample and the population from which it was drawn. Similarly, in any experiment, individuals in the experimental group will differ from individuals in the control group in random ways. Even if there is *no* true effect of an experimental treatment, the data from the experimental group will never be identical to the data from the control group.

For example, we know that people differ in their ability to recover from a cold infection. If we give zinc lozenges to 1 volunteer and placebo lozenges to another, it is likely that the 2 volunteers will have colds of different lengths. But even if the zinc-taker had a shorter cold than the placebo-taker, you would probably say that the test did not tell us much about our hypothesis—the zinc-taker might just have had a less severe cold for other reasons.

Now imagine that we had 5 volunteers in each group and saw a difference, or that the difference was only 1 day instead of 3 days. How would we determine if the lozenges had an effect? Statistical tests allow researchers to look at their data and determine how likely it is—that is, the **probability**—that the result is due to sampling error. In this case, the statistical test distinguished between two possibilities for why the experimental group had shorter colds: Either the difference was due to the effectiveness of zinc as a cold treatment or it was due to a chance difference from the control group. The results indicated that there was a low probability, less than 1 in 10,000 (0.01%), that the experimental and control groups were so different simply by chance. In other words, the result is statistically significant.

A statistical measure of the amount of variability in a sample is often expressed as the **standard error.** The standard error is used to generate the **confidence interval**—the range of values that has a high probability (usually 95%) of containing the true population mean (Figure 1.14). Put simply, although the average of a sample is unlikely to be exactly the average of the population, the average plus the standard error represents the highest likely value for the population average, and the average minus the standard error represents the lowest likely average value. The confidence interval provides a way to express how much sampling error is influencing the results. A smaller confidence interval indicates that sampling error is likely to be small, and results with small confidence intervals are more likely to be statistically significant if the hypothesis is true.

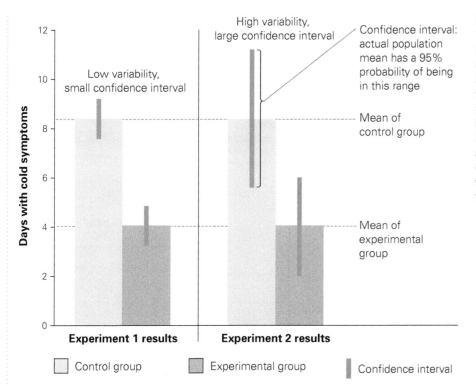

Figure 1.14 Confidence interval.
The red lines on these bar graphs represent the confidence interval for each sample mean. A more variable sample has a larger confidence interval than a less variable sample. Even though the means are the same for both experimental and control groups in these two sets of bars, only the data summarized at left illustrate a significant difference because of the greater variability in the samples illustrated on the right side.

Stop & Stretch Opinion polling before a recent election indicated that candidate A was favored by 47% of a sample of likely voters, and candidate B was favored by 51% of the sample. The standard error was 3%. Why did reporters refer to this poll as a "statistical tie"?

Factors That Influence Statistical Significance

One characteristic of experiments that influences sampling error is **sample size**—the number of individuals in the experimental and control groups. If a treatment has no effect, a small sample size could return results that appear significantly different because of an unusually large and consistent sampling error. This was the case with the vitamin C hypothesis described at the beginning of the chapter. Subsequent tests with larger sample sizes, encompassing a wider variety of individuals with different underlying susceptibilities to colds, allowed scientists to reject the hypothesis that vitamin C prevents colds.

Conversely, if the effect of a treatment is real but the sample size of the experiment is small, a single experiment may not allow researchers to determine convincingly that their hypothesis has support. Several of the experiments that tested the efficacy of *Echinacea* extract demonstrate this phenomenon. For example, one experiment performed at a Wisconsin clinic with 48 participants indicated that echinacea tea drinkers were 30% less likely to experience any cold symptoms after virus exposure as compared to individuals who received a placebo tea. However,

the small sample size of the study meant that this result was not statistically significant.

The more participants there are in a study, the more likely it is that researchers will see a true effect of an experimental treatment even if it is very small. For example, a study of over 21,000 men over 6 years in Finland demonstrated that men who took vitamin E supplements had 5% fewer colds than the men who did not take these supplements. In this case, the large sample size allowed researchers to see that vitamin E has a real, but relatively tiny, effect on cold incidence. In other words, this statistically significant result has little real-world *practical* significance; a 5% effect probably won't convince people to begin supplementing with vitamin E to prevent colds. The relationship among hypotheses, experimental tests, sample size, and statistical and practical significance is summarized in **Figure 1.15** (on the next page).

There is one final caveat to this discussion. A statistically significant result is typically defined as one that has a *probability* of 5% or less of being due to chance alone. But probability is not certainty. If all scientific research uses this same standard, as many as 1 in every 20 statistically significant results (that is, 5% of the total) is actually reporting an effect that is *not real*. In other words, some statistically significant results are "false positives," representing a surprisingly large difference between experimental and control groups that occurred only as a result of sampling error. This potential error explains why one supportive experiment is not enough to convince all scientists that a particular hypothesis is accurate.

(continued on the next page)

(A Closer Look continued)

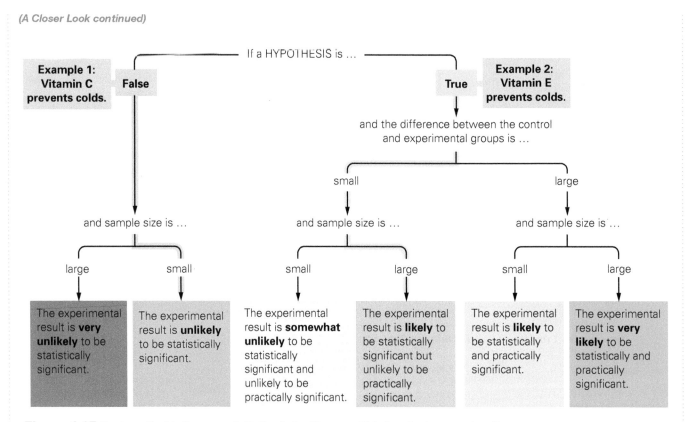

Figure 1.15 Factors that influence statistical significance. This flowchart summarizes the relationship between the true effect of a treatment and the sample size of an experiment on the likelihood of obtaining statistical significance.

Visualize This: The Nurse's Health Survey, which has followed nearly 80,000 women for over 20 years, has found that a diet low in refined carbohydrates such as white flour and sugar but relatively high in vegetable fat and protein cuts heart disease risk by 30% relative to women with a high refined carbohydrate diet. How does this result map out on the flowchart here?

Even with a statistical test indicating that the result had a probability of less than 0.01% of occurring by chance, we should begin to feel assured that taking zinc lozenges will reduce the duration of colds only after additional hypothesis tests give similar results. In fact, scientists continue to test this hypothesis, and there is still no consensus about the effectiveness of zinc as a cold treatment.

What Statistical Tests Cannot Tell Us

All statistical tests operate with the assumption that the experiment was designed and carried out correctly. In other words, a statistical test evaluates the chance of sampling error, not observer error, and a statistically significant result is not the last word on an experimentally tested hypothesis. An examination of the experiment itself is required.

In the test of the effectiveness of zinc lozenges, the experimental design—double-blind, randomized, and controlled—minimized the likelihood that alternative hypotheses could explain the results. Given such a well-designed experiment, this statistically significant result allows researchers to infer strongly that consuming zinc lozenges reduces the duration of colds.

As we saw with the experiment on vitamin E intake and cold prevention, statistical significance is also not equivalent to *significance* as we usually define the term, that is, as "meaningful or important." Unfortunately, experimental results reported in the news often use the term *significant* without clarifying this important distinction. Understanding that problem, as well as other misleading aspects of how science can be presented, will enable you to better use scientific information.

1.4 Evaluating Scientific Information

The previous sections should help you see why definitive scientific answers to our questions are slow in coming. However, a well-designed experiment can certainly allow us to approach the truth.

Primary Sources

Looking critically at reports of experiments can help us make well-informed decisions about actions to take. Most of the research on cold prevention and treatment is first published as **primary sources,** written by the researchers themselves and reviewed within the scientific community (**Figure 1.16** on the next page). The process of **peer review,** in which other scientists critique the results and conclusions of an experiment before it is published, helps increase confidence in scientific information. Peer-reviewed research articles in journals such as *Science, Nature,* the *Journal of the American Medical Association,* and hundreds of others represent the first and most reliable sources of current scientific knowledge.

However, evaluating the hundreds of scientific papers that are published weekly is a task no one of us can perform. Even if we focused only on a particular field of interest, the technical jargon used in many scientific papers may be a significant barrier to our understanding.

Instead of reading the primary literature, most of us receive our scientific information from **secondary sources** such as books, news reports, and advertisements. How can we evaluate information in this context? The following sections provide strategies for doing so, and a recurring Savvy Reader feature in this textbook will help you practice these evaluation skills.

Information from Anecdotes

Information about dietary supplements such as echinacea tea and zinc lozenges is often in the form of **anecdotal evidence**—meaning that the advice is based on one individual's personal experience. A friend's enthusiastic plug for vitamin C, because she felt it helped her, is an example of a testimonial—a common form of anecdote. Advertisements that use a celebrity to pitch a product "because it worked for them" are classic forms of testimonials.

You should be cautious about basing decisions on anecdotal evidence, which is not at all equivalent to well-designed scientific research. For example, many of us have heard anecdotes along the lines of the grandpa who was a pack-a-day smoker and lived to the age of 94. However, hundreds of studies have demonstrated the clear link between cigarette smoking and premature death. Although anecdotes may indicate that a product or treatment has merit, only well-designed tests of the hypothesis can determine its safety and efficacy.

Stop & Stretch One of the challenges of presenting science is that anecdotes make a much bigger impression on people than statistical summaries of data. Why do you think this is the case?

**Figure 1.16 Primary sources:
publishing scientific results.** Most
scientific journals require papers to
go through stringent review before publication.
Visualize This: How does having other
anonymous experts review a draft of a
scientist's paper make the final paper
more reliable?

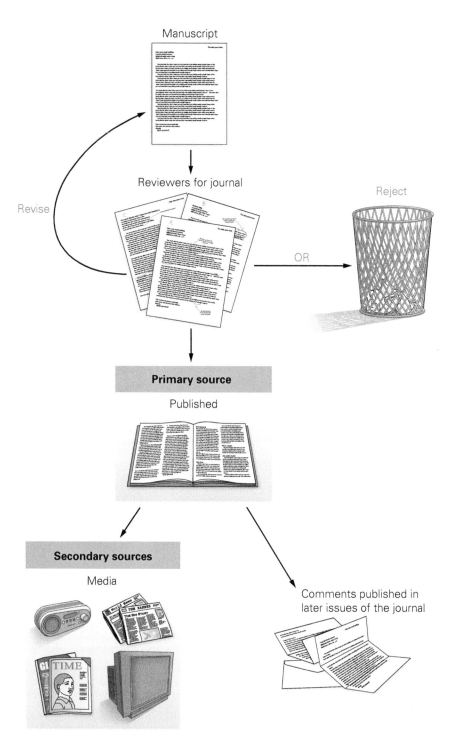

Manuscript

Reviewers for journal

Revise

OR

Reject

Primary source

Published

Secondary sources

Media

Comments published in
later issues of the journal

Science in the News

Popular news sources provide a steady stream of science information. However, stories about research results often do not contain information about the
adequacy of experimental design or controls, the number of subjects, or
the source of the scientist's funding. How can anyone evaluate the quality
of research that supports dramatic headlines such as those describing risks to
human health and the environment?

First, you must consider the source of media reports. Certainly news organizations will be more reliable reporters of fact than will entertainment tabloids,
and news organizations with science writers should be better reporters of the

substance of a study than those without. Television talk shows, which need to fill airtime, regularly have guests who promote a particular health claim. Too often these guests may be presenting information that is based on anecdotes or an incomplete summary of the primary literature.

Paid advertisements are a legitimate means of disseminating information. However, claims in advertising should be carefully evaluated. Advertisements of over-the-counter and prescription drugs must conform to rigorous government standards regarding the truth of their claims. However, lower standards apply to advertisements for herbal supplements, many health food products, and diet plans. Be sure to examine the fine print because advertisers often are required to clarify the statements made in their ads.

Another commonly used source for health information is the Internet. As you know, anyone can post information on this great resource. Typing in "common cold prevention" on a standard web search engine will return thousands of web pages—from highly respected academic and government sources to small companies trying to sell their products to individuals who have strong, sometimes completely unsupported, ideas about cures. Often it can be difficult to determine the reliability of a well-designed website, and even well-used sites such as Wikipedia may contain erroneous or misleading information. Here are some things to consider when using the web as a resource for health information:

1. Choose sites maintained by reputable medical establishments, such as the National Institutes of Health (NIH) or the Mayo Clinic.
2. It costs money to maintain a website. Consider whether the site seems to be promoting a product or agenda. Advertisements for a specific product should alert you to possible bias.
3. Check the date when the website was last updated and see whether the page has been updated since its original posting. Science and medicine are disciplines that must frequently evaluate new data. A reliable website will be updated often.
4. Determine whether unsubstantiated claims are being made. Look for references and be suspicious of any studies that are not from peer-reviewed journals.

Understanding Science from Secondary Sources

Once you are satisfied that a media source is relatively reliable, you can examine the scientific claim that it presents. Use your understanding of the process of science and of experimental design to evaluate the story and the science. Does the story about the claim present the results of a scientific study, or is it built around an untested hypothesis? Were the results obtained using the scientific method, with tests designed to reject false hypotheses? Is the story confusing correlation with causation? Does it seem that the information is applicable to nonlaboratory situations, or is it based on results from preliminary or animal studies?

Look for clues about how well the reporters did their homework. Scientists usually discuss the limitations of their research in their papers. Are these cautions noted in an article or television piece? If not, the reporter may be overemphasizing the applicability of the results.

Then, note if the scientific discovery itself is controversial. That is, does it reject a hypothesis that has long been supported? Does it concern a subject that is controversial? Might it lead to a change in social policy? In these cases, be extremely cautious. New and unexpected research results must be evaluated in light of other scientific evidence and understanding. Reports that lack comments from other experts may miss problems with a study or fail to place it in the context of other research. The Savvy Reader feature in this chapter provides a checklist of questions to ask and answer as you evaluate news reports.

Finally, the news media generally highlight only those science stories that editors find newsworthy. As we have seen, scientific understanding accumulates relatively slowly, with many tests of the same hypothesis finally leading to an accurate understanding of a natural phenomenon. News organizations are also more likely to report a study that supports a hypothesis rather than one that gives less-supportive results, even if both types of studies exist.

In addition, even the most respected media sources may not be as thorough as readers would like. For example, a recent review published in the *New England Journal of Medicine* evaluated the news media's coverage of new medications. Of 207 randomly selected news stories, only 40% that cited experts with financial ties to a drug told readers about this relationship. This potential conflict of interest calls into question the expert's objectivity. Another 40% of the news stories did not provide basic statistics about the drugs' benefits. Most of the news reports also failed to distinguish between absolute benefits (how many people were helped by the drug) and relative benefits (how many people were helped by the drug relative to other therapies for the condition). The journal's review is a vivid reminder that we need to be cautious when reading or viewing news reports on scientific topics.

Even after following all of these guidelines, you will still find well-researched news reports on several scientific studies that seem to give conflicting and confusing results. As you now know, such confusion is the nature of the scientific process—early in our search for understanding, many hypotheses are proposed and discussed; some are tested and rejected immediately, and some are supported by one experiment but later rejected by more thorough experiments. It is only by clearly understanding the process and pitfalls of scientific research that you can distinguish "what we know" from "what we don't know."

1.5 Is There a Cure for the Common Cold?

So where does our discussion leave us? Will we ever find the best way to prevent a cold or reduce its effects? In the United States alone, over 1 billion cases of the common cold are reported per year, costing many more billions of dollars in medical visits, treatment, and lost workdays. Consequently, there is an enormous effort to find effective protection from the different viruses that cause colds.

A search of medical publication databases indicates that every year nearly 100 scholarly articles regarding the biology, treatment, and consequences of common cold infection are published. This research has led to several important discoveries about the structure and biochemistry of common cold viruses, how they enter cells, and how the body reacts to these infections.

Despite all of the research and the emergence of some promising possibilities, the best prevention method known for common colds is still the old standby—keep your hands clean. Numerous studies have indicated that rates of common cold infection are 20% to 30% lower in populations who employ effective hand-washing procedures. Cold viruses can survive on surfaces for many hours; if you pick them up on your hands from a surface and transfer them to your mouth, eyes, or nose, you may inoculate yourself with a 7-day sniffle (**Figure 1.17**).

Figure 1.17 Preventing colds.
According to the Centers for Disease Control and Prevention, the best defense against infection is effective hand washing. Scrub vigorously with a liquid soap for 20 seconds, then rinse under running water for 10 seconds. Alcohol-based hand gels (without water) containing at least 60% ethanol are effective if soap and water are unavailable but will not remove visible dirt.

Of course, not everyone gets sick when exposed to a cold virus. The reason one person has more colds than another might not be due to a difference in personal hygiene. The correlation that showed a relationship between stress and cold susceptibility appears to have some merit. Research indicates that among people exposed to viruses, the likelihood of ending up with an infection increases with high levels of psychological stress—something that many college students clearly experience.

Research also indicates that vitamin C intake, diet quality, exposure to cold temperatures, and exercise frequency appear to have no effect on cold susceptibility, although along with echinacea tea and zinc lozenges, there is some evidence that vitamin C may reduce cold symptoms a bit. Surprisingly, even though medical research has led to the elimination of killer viruses such as smallpox and polio, scientists are still a long way from "curing" the common cold.

SAVVY READER

A Toolkit for Evaluating Science in the News

The following checklist can be used as a guide to evaluating science information in the news. Although any issue that raises a red flag should cause you to be cautious about the conclusions of the story, no one issue should cause you to reject the story (and its associated research) out of hand. However, in general, the fewer red flags raised by a story, the more reliable is the report of the significance of the scientific study. Use **Table 1.2** to evaluate this extract from an article by Terry Gupta in the Nashua (New Hampshire) *Telegraph*:

> [Paula] Fortier, a licensed massage therapist, shared her special brew. To boiled water and tea, she adds cayenne pepper, honey and fresh squeezed lemon. Fortier's enthusiasm may be more contagious than the flu when she describes this heated, "unbeatable" combination. "Your throat opens up," she said. "You're less irritated and you sweat out some unwanted toxins." She varies the exact measures of the ingredients according to tolerance and taste. While on the topic of hot water, one of Fortier's favorite remedies for "an achy body that feels sore all over is a hot bath in a steeping tub with 2 cups of cider vinegar for 15 minutes. This draws off toxins, reduces inflammation and leaves you feeling better."

TABLE 1.2

A guide for evaluating science in the news. For each question, check the appropriate box.

Question	Preferred answer		Raises a red flag	
1. What is the basis for the story?	Hypothesis test	☐	Untested assertion No data to support claims in the article.	☐
2. What is the affiliation of the scientist?	Independent (university or government agency)	☐	Employed by an industry or advocacy group Data and conclusions could be biased.	☐
3. What is the funding source for the study?	Government or nonpartisan foundation (without bias)	☐	Industry group or other partisan source (with bias) Data and conclusions could be biased.	☐

(continued)

TABLE 1.2 *(continued)*

A guide for evaluating science in the news. For each question, check the appropriate box.

Question	Preferred answer		Raises a red flag	
4. If the hypothesis test is a correlation: Did the researchers attempt to eliminate reasonable alternative hypotheses?	Yes	☐	No Correlation does not equal causation. One hypothesis test provides poor support if alternatives are not examined.	☐
If the hypothesis test is an experiment: Is the experimental treatment the only difference between the control group and the experimental group?	Yes	☐	No An experiment provides poor support if alternatives are not examined.	☐
5. Was the sample of individuals in the experiment a good cross section of the population?	Yes	☐	No Results may not be applicable to the entire population.	☐
6. Was the data collected from a relatively large number of people?	Yes	☐	No Study is prone to sampling error.	☐
7. Were participants blind to the group they belonged to or to the "expected outcome" of the study?	Yes	☐	No Subject expectation can influence results.	☐
8. Were data collectors or analysts blinded to the group membership of participants in the study?	Yes	☐	No Observer bias can influence results.	☐
9. Did the news reporter put the study in the context of other research on the same subject?	Yes	☐	No Cannot determine if these results are unusual or fit into a broader pattern of results.	☐
10. Did the news story contain commentary from other independent scientists?	Yes	☐	No Cannot determine if these results are unusual or if the study is considered questionable by others in the field.	☐
11. Did the reporter list the limitations of the study or studies on which he or she is reporting?	Yes	☐	No Reporter may not be reading study critically and could be overstating the applicability of the results.	☐

Chapter Review

Learning Outcomes

Mastering**BIOLOGY**

Go to the Study Area at www.masteringbiology.com for practice quizzes, myeBook, BioFlix™ 3-D animations, MP3Tutor sessions, videos, current events, and more.

LO1 **Describe the characteristics of a scientific hypothesis (Section 1.1).**

- Science is a process of testing hypotheses—statements about how the natural world works. Scientific hypotheses must be testable and falsifiable (p. 4).

LO2 **Compare and contrast the terms *scientific hypothesis* and *scientific theory* (Section 1.1).**

- A scientific theory is an explanation of a set of related observations based on well-supported hypotheses from several different, independent lines of research (p. 6).

LO3 **Distinguish between inductive and deductive reasoning (Section 1.1).**

- Hypotheses are often developed via inductive reasoning, which consists of making a number of observations and then inferring a general principle to explain them (p. 6).
- Hypotheses are tested via the process of deductive reasoning, which allows researchers to make specific predictions about expected observations (pp. 6–7).

LO4 **Explain why the truth of a hypothesis cannot be proven conclusively via deductive reasoning (Section 1.1).**

- Absolutely proving hypotheses is impossible because there may be other reasons besides the one hypothesized that could lead to the predicted result (pp. 7–8).

LO5 **Describe the features of a controlled experiment, and explain how these experiments eliminate alternative hypotheses for the results (Section 1.2).**

- Controlled experiments test hypotheses about the effect of experimental treatments by comparing a randomly assigned experimental group with a control group (pp. 10–11).
- Controls are individuals who are treated identically to the experimental group except for application of the treatment (p. 10).

LO6 **List strategies for minimizing bias when designing experiments (Section 1.2).**

- Bias in scientific results can be minimized with double-blind experiments that keep subjects and data collectors unaware of which individuals belong in the control or experimental group (p. 12).

LO7 **Define correlation, and explain the benefits and limitations of using this technique to test hypotheses (Section 1.2).**

- If performing controlled experiments on humans is considered unethical, scientists sometimes use correlations. The correlation of structure and function between humans and model organisms, such as other mammals, allows experimental tests of these hypotheses (pp. 12–13).
- Correlations cannot exclude alternative hypotheses. Thus, a correlation study can describe a relationship between two factors, but it does not strongly imply that one factor causes the other (p. 15).

LO8 **Describe the information that statistical tests provide (Section 1.3).**

- Statistics help scientists evaluate the results of their experiments by determining if results appear to reflect the true effect of an experimental treatment on a sample of a population (p. 17).
- A statistically significant result is one that is very unlikely to be due to chance differences between the experimental and control groups (p. 17).
- Even when an experimental result is highly significant, hypotheses are tested multiple times before scientists come to consensus on the true effect of a treatment (pp. 19–20).

LO9 **Compare and contrast primary and secondary sources (Section 1.4).**

- Primary sources of information are experimental results published in professional journals and peer reviewed by other scientists before publication (pp. 21–24).
- Secondary sources are typically news, books, or web sites that summarize the scientific research presented in primary sources. Secondary sources are not peer reviewed and sometimes contain poorly supported information (pp. 21–24).

LO10 **Summarize the techniques you can use to evaluate scientific information from secondary sources (Section 1.4).**

- Anecdotal evidence is an unreliable means of evaluating information, and media sources are of variable quality; distinguishing between news stories and advertisements is important when evaluating the reliability of information. The Internet is a rich source of information, but users should look for clues to a particular website's credibility (p. 21).
- Stories about science should be carefully evaluated for information on the actual study performed, the universality of the claims made by the researchers, and other studies on the same subject. Sometimes confusing stories about scientific information are a reflection of controversy within the scientific field itself (pp. 23–24).

Roots to Remember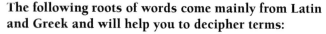

The following roots of words come mainly from Latin and Greek and will help you to decipher terms:

bio- means life. Chapter term: *biology*

deduc- means to reason out, working from facts. Chapter term: *deductive reasoning*

induc- means to rely on reason to derive principles (also, to cause to happen). Chapter term: *inductive reasoning*

hypo- means under, below, or basis. Chapter term: *hypothesis*

-ology means the study of or branch of knowledge about. Chapter term: *biology*

Learning the Basics

1. **LO5 LO6** What is the value of a placebo in an experimental design?.

2. **LO3** Which of the following is an example of inductive reasoning?

   A. All cows eat grass; **B.** My cow eats grass and my neighbor's cow eats grass; therefore all cows probably eat grass; **C.** If all cows eat grass, when I examine a random sample of all the cows in Minnesota, I will find that they all eat grass; **D.** Cows may or may not eat grass, depending on the type of farm where they live.

3. **LO1** A scientific hypothesis is _____.

   A. an opinion; **B.** a proposed explanation for an observation; **C.** a fact; **D.** easily proved true; **E.** an idea proposed by a scientist.

4. **LO2** How is a scientific theory different from a scientific hypothesis?

   A. It is based on weaker evidence; **B.** It has not been proved true; **C.** It is not falsifiable; **D.** It can explain a large number of observations; **E.** It must be proposed by a professional scientist.

5. **LO3** One hypothesis states that eating chicken noodle soup is an effective treatment for colds. Which of the following results does this hypothesis predict?

   A. People who eat chicken noodle soup have shorter colds than do people who do not eat chicken noodle soup; **B.** People who do not eat chicken noodle soup experience unusually long and severe colds; **C.** Cold viruses cannot live in chicken noodle soup; **D.** People who eat chicken noodle soup feel healthier than do people who do not eat chicken noodle soup; **E.** Consuming chicken noodle soup causes people to sneeze.

6. **LO4** If I perform a hypothesis test in which I demonstrate that the prediction I made in question 5 is true, I have _____.

   A. proved the hypothesis; **B.** supported the hypothesis; **C.** not falsified the hypothesis; **D.** B and C are correct; **E.** A, B, and C are correct.

7. **LO5** Control subjects in an experiment

   A. should be similar in most ways to the experimental subjects; **B.** should not know whether they are in the control or experimental group; **C.** should have essentially the same interactions with the researchers as the experimental subjects; **D.** help eliminate alternative hypotheses that could explain experimental results; **E.** all of the above.

8. **LO6** An experiment in which neither the participants in the experiment nor the technicians collecting the data know which individuals are in the experimental group and which ones are in the control group is known as

   A. controlled; **B.** biased; **C.** double-blind; **D.** falsifiable; **E.** unpredictable.

9. **LO7** A relationship between two factors, for instance between outside temperature and the number of people with active colds in a population, is known as a(n)

   A. significant result; **B.** correlation; **C.** hypothesis; **D.** alternative hypothesis; **E.** experimental test.

10. **LO9** A primary source of scientific results is

    A. the news media; **B.** anecdotes from others; **C.** articles in peer-reviewed journals; **D.** the Internet; **E.** all of the above.

11. **LO8** A story on your local news station reports that eating a 1-ounce square of milk chocolate each day reduces the risk of heart disease in rats, and that this result is statistically significant. This means that

    A. People who eat milk chocolate are healthier than those who do not; **B.** The difference between chocolate- and non-chocolate-eating rats in heart disease rates was greater than expected by chance; **C.** Rats like milk chocolate; **D.** Milk chocolate reduces the risk of heart disease; **E.** Two ounces of milk chocolate per day is likely to be even better for heart health than 1 ounce.

12. **LO10** What features of the story on milk chocolate and heart health described in Question 11 should cause you to consider the results less convincing?

    A. The study was sponsored by a large milk chocolate manufacturer; **B.** A total of ten rats were used in the study; **C.** The only difference between the rats was that subjects of the experimental group received chocolate along with their regular diets, while subjects of the control group received no additional food; **D.** The reporter notes that other studies indicate that milk chocolate does not have beneficial effect on heart health; **E.** All of the above.

Analyzing and Applying the Basics

1. **LO7** There is a strong correlation between obesity and the occurrence of a disease known as type 2 diabetes—that is, obese individuals have a higher instance of diabetes than nonobese individuals do. Does this mean that obesity causes diabetes? Explain.

2. **LO5 LO6** In an experiment examining vitamin C as a cold treatment, students with cold symptoms who visited the campus medical center either received vitamin C or were treated with over-the-counter drugs. Students then reported on the length and severity of their colds. Both the students and the clinic health providers knew which treatment students were receiving. This study indicated that vitamin C significantly reduced the length and severity of colds. Which factors make this result somewhat unreliable?

3. **LO7** Brain-derived neurotrophic factor (BDNF) is a substance produced in the brain that helps nerve cells (neurons) to grow and survive. BDNF also increases the connectivity of neurons and improves learning and mental function. A 2002 study that examined the effects of intense wheel-running on rats and mice found a positive correlation between BDNF levels and running distance. What could you conclude from this result?

Connecting the Science

1. Much of the research on common cold prevention and treatment is performed by scientists employed or funded by drug companies. Often these companies do not allow scientists to publish the results of their research for fear that competitors at other drug companies will use this research to develop a new drug before they do. Should our society allow scientific research to be owned and controlled by private companies?

2. Should society restrict the kinds of research performed by government-funded scientists? For example, many people believe that research performed on tissues from human fetuses should be restricted; these people believe that such research would justify abortion. If most Americans feel this way, should the government avoid funding this research? Are there any risks associated with *not* funding research with public money?

Answers to **Stop & Stretch, Visualize This, Savvy Reader,** and **Chapter Review** questions can be found in the **Answers** section at the back of the book.

Are We Alone in the Universe?

Water, Biochemistry, and Cells

Cartoon images aside, we know of no other intelligent life in our solar system.

I s there life on other planets? Scientists have found persuasive, if not conclusive, evidence of life on Mars. When the National Aeronautics and Space Administration (NASA) scientist David McKay first proposed that this cold, dry, rather harsh planet could harbor life, people were astounded.

The evidence of life found by Dr. McKay and his team did not in any way resemble the cartoon images often used to depict Martians. Instead, what these scientists found was evidence of life in a 3.6-billion-year-old, potato-sized rock. They believe that the rock broke off the surface of Mars after being hit by an asteroid. After floating through space for 16 million years, it fell to Earth, landing in Antarctica about 13,000 years ago and remaining there until discovered by scientists. This meteorite, drably named ALH84001, appeared to contain the same features that scientists use to demonstrate the existence of life in 3.6-billion-year-old Earth rocks—there were fossils, various minerals that are characteristic of life, and evidence of complex chemicals typically produced by living organisms.

While many scientists debate the assertion that this rock provides evidence of life on Mars, the announcement served to inject new energy into Mars exploration. Since then, multiple robotic rovers and mapping satellites have been sent to the planet, with the goal of someday sending astronauts there. While there are many reasons to explore Mars, the question that remains most intriguing—and is a significant focus of several of these missions—is whether life ever existed there.

The fascination about potential Martian life speaks to a fundamental question that many humans share: Are the creatures on Earth the only living organisms in the universe? Our galaxy is filled with countless stars

Some evidence suggests that Mars may harbor life.

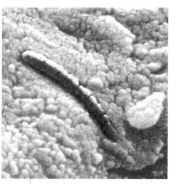

A meteorite ejected from Mars and found in Antarctica may contain lifelike forms.

Will the Mars rover find evidence of life outside Earth?

and planets, and the universe teems with galaxies. Even if we find no convincing evidence of life on Mars, there are a seemingly infinite number of places to look for other living beings. In this chapter, we discuss the characteristics and requirements of life and examine techniques that scientists use to search the universe for other living creatures.

2.1 What Does Life Require?

Because the galaxy likely contains billions of planets, scientists looking for life elsewhere seek to identify the range of conditions under which they would expect life to arise. What is it that scientists look for when identifying a planet (or moon) as a candidate for hosting life?

A Definition of Life

In science-fiction movies, alien life-forms are often obviously alive and even somewhat familiar looking. But in reality, living organisms may be truly alien; that is, they may look nothing like organisms we are familiar with on Earth. So how would we determine whether an entity found on another planet was actually alive?

Surprisingly, biologists do not have a simple definition for a "living organism." A list of the attributes found in most earthly life-forms includes growth, movement, reproduction, response to external environmental stimuli, and **metabolism** (all of the chemical processes that occur in cells, including the breakdown of substances to produce energy, the synthesis of substances necessary for life, and the excretion of wastes generated by these processes). However, this definition could apply to things that no one considers to be living. For example, fire can grow, consume energy, give off waste, move, reproduce by sending off sparks, and change in response to environmental conditions. And some organisms that are clearly living do not conform to this definition. Male mules grow, metabolize, move, and respond to stimuli, but they are unable to reproduce.

Figure 2.1 Homeostasis. Black-capped chickadees can maintain a core body temperature of 108°F (~42°C) during the day, even when the air temperature is well below zero.

If we examine more closely the characteristics of living organisms on Earth, we will see that all organisms contain a common set of biological molecules, are composed of cells, and can maintain **homeostasis,** that is, a roughly constant internal environment despite an ever-changing external environment (**Figure 2.1**). The ability to maintain homeostasis requires complex feedback mechanisms between multiple sensory and physiological systems and is possible only in living organisms. In addition, populations of living organisms can evolve, that is, change in average physical characteristics over time. If we search the universe for planets that could support life similar to that found on Earth—and thus organisms that we would clearly identify as "living"—the list of planetary requirements becomes more stringent. In particular, an Earth-like planet should have abundant liquid **water** available.

The Properties of Water

Water is a requirement for life. Although Mars does not currently appear to have any liquid water, ice is found at its poles (**Figure 2.2a**), and features of its surface indicate that it once contained salty seas and flowing water (**Figure 2.2b**).

(a) Frozen water

(b) Running water

Figure 2.2 Water on Mars. (a) This image from the Mars rover indicates that frozen water exists on Mars. (b) This photograph, taken by the European Mars Express orbiter, shows a channel on Mars that may have been formed by running water.

The presence of liquid water on Mars would fulfill an essential prerequisite for the appearance of life. But why is water such an important feature?

Water is made up of two elements: hydrogen and oxygen. **Elements** are the fundamental forms of matter and are composed of atoms that cannot be broken down by normal physical means such as boiling.

Atoms are the smallest units that have the properties of any given element. Ninety-two natural elemental atoms have been described by chemists, and several more have been created in laboratories. Hydrogen, oxygen, and calcium are examples of elements commonly found in living organisms. Each element has a one- or two-letter symbol: H for hydrogen, O for oxygen, and Ca for calcium, for example.

Atoms are composed of subatomic particles called **protons, neutrons, and electrons.** Protons have a positive electric charge; these particles and the uncharged neutrons make up the **nucleus** of an atom. All atoms of a particular element have the same number of protons, giving the element its **atomic number.** The negatively charged electrons are found outside the nucleus in an "electron cloud." Electrons are attracted to the positively charged nucleus (**Figure 2.3**). A *neutral atom* has equal numbers of protons and electrons. Electrically charged **ions** do not have an equal number of protons and electrons. In this case, the atom is not neutral and is instead charged.

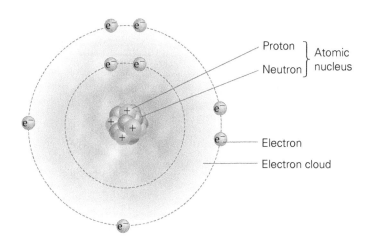

Figure 2.3 Atomic structure. An oxygen atom contains a nucleus made up of 8 protons and 8 neutrons. Orbiting electrons surround the nucleus. Although the number of particles within each atom differs, all atoms have the same basic structure.

The Structure of Water. The chemical formula for water is H_2O, indicating that it contains 2 hydrogen atoms for every 1 oxygen. Water, like other molecules, consists of 2 or more atoms joined by chemical bonds. A molecule can be composed of the same or different atoms. For example, a molecule of oxygen consists of 2 oxygen atoms joined to each other, while a molecule of carbon dioxide consists of 1 carbon and 2 oxygen atoms.

Water Is a Good Solvent. Water has the ability to dissolve a wide variety of substances. A substance that dissolves when mixed with another substance is called a solute. When a solute is dissolved in a liquid, such as water, the liquid is called a solvent. Once dissolved, components of a particular solute can pass freely throughout the water, making a chemical mixture or solution.

Water is a good solvent because it is **polar,** meaning that different regions, or poles, of the molecule have different charges. The polarity arises because oxygen is more attractive to electrons—that is, it is more **electronegative**—than most other atoms, including hydrogen. As a result of oxygen's electronegativity, electrons in a water molecule spend more time near the nucleus of the oxygen atom than near the nuclei of the hydrogen atoms. With more negatively charged electrons near it, the oxygen in water carries a partial negative charge, symbolized by the Greek letter delta, δ^-. The hydrogen atoms thus have a partial positive charge, symbolized by δ^+ (**Figure 2.4**). When atoms of a molecule carry no partial charge, they are said to be **nonpolar.** The carbon-hydrogen bonds, for example, share electrons equally and are nonpolar.

Water molecules tend to orient themselves so that the hydrogen atom (with its partial positive charge) of one molecule is near the oxygen atom (with its

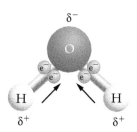

Figure 2.4 Polarity in water. Water is a polar molecule. Its atoms do not share electrons equally.
Visualize This: Toward which atom are the electrons of a water molecule pulled?

(a) Bonds between two water molecules   **(b) Bonds between many water molecules**

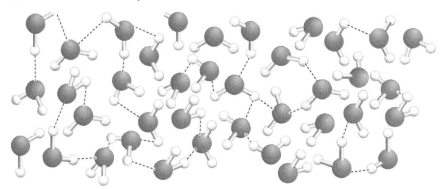

Figure 2.5 **Hydrogen bonding.** Hydrogen bonding can occur when there is a weak attraction between the hydrogen and oxygen atoms between (a) two or (b) many different water molecules.

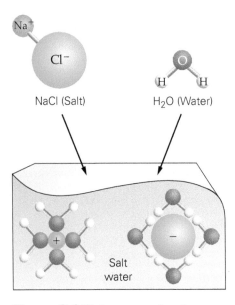

Figure 2.6 **Water as a solvent.** When salt is placed in water, the negatively charged regions of the water molecules surround the positively charged sodium ion, and the positively charged regions of the water molecules surround the negatively charged chlorine ion, breaking the bond holding sodium and chloride together and dissolving the salt.

partial negative charge) of another molecule (Figure 2.5a). The weak attraction between hydrogen atoms and oxygen atoms in adjacent molecules forms a **hydrogen bond.** Hydrogen bonding is a type of weak chemical bond that forms when a partially positive hydrogen atom is attracted to a partially negative atom. Hydrogen bonds can be intramolecular, involving different regions within the same molecule, or they can be intermolecular, between different molecules, as is the case in hydrogen bonding between different water molecules. Figure 2.5b shows the hydrogen bonding that occurs between water molecules in liquid form.

The ability of water to dissolve substances such as sodium chloride is a direct result of its polarity. Each molecule of sodium chloride is composed of 1 sodium ion (Na^+) and 1 chloride ion (Cl^-). In the case of sodium chloride, the negative pole of water molecules will be attracted to a positively charged sodium ion and separate it from a negatively charged chloride ion (Figure 2.6). Water can also dissolve other polar molecules, such as alcohol, in a similar manner. Polar molecules are called **hydrophilic** ("water-loving") because of their ability to dissolve in water. Nonpolar molecules, such as oil, do not contain charged atoms and are referred to as **hydrophobic** ("water-fearing") because they do not easily mix with water.

Water Facilitates Chemical Reactions. Because it is such a powerful solvent, water can facilitate **chemical reactions**, which are changes in the chemical composition of substances. Solutes in a mixture, called **reactants,** can come in contact with each other, permitting the modification of chemical bonds that occur during a reaction. The molecules formed as a result of a chemical reaction are known as **products.**

Water Is Cohesive. The tendency of like molecules to stick together is called **cohesion.** Cohesion is much stronger in water than in most liquids because of the sheer number of hydrogen bonds between water molecules. Cohesion is an important property of many biological systems. For instance, many plants depend on cohesion to help transport a continuous column of water from the roots to the leaves.

Water Moderates Temperature. When heat energy is added to water, its initial effect is to disrupt the hydrogen bonding among water molecules.

Therefore, this heat energy can be absorbed without changing the temperature of water. Only after the hydrogen bonds have been broken can added heat increase the temperature. In other words, the initial input of energy is absorbed.

The flowing water that was once found on Mars is now only in the form of ice. Until scientists can land on Mars and collect ice samples for analysis, its actual composition is a matter of conjecture. However, images taken by a NASA rover have led scientists to believe that some of the rocks on Mars were probably produced from deposits at the bottom of a body of saltwater. We know from surveying Earth's oceans that saltwater is hospitable to millions of different life-forms. In fact, most hypotheses about the origin of life on Earth presume that our ancestors first arose in the salty oceans.

Acids, Bases, and Salts

Salts are produced by the reaction of an **acid** (a substance that donates H^+ ions to a solution) with a **base** (a substance that accepts H^+ ions). The **pH** scale is a measure of the relative amounts of these ions in a solution (**Figure 2.7**). The more acidic a solution is, the higher the H^+ concentration is relative to the OH^- ions. Hydrogen ion concentration is inversely related to pH, so the higher the H^+ concentration, the lower the pH. Basic solutions have fewer H^+ ions relative to OH^- ions and thus a higher pH. These ions can react with other charged molecules and help to bring them into the water solution. At any given time, a small percentage of water molecules in a pure solution will be dissociated. There are equal numbers of these ions in pure water, so it is neutral, which on the pH scale is 7. The pH of most cells is very close to 7.

When a European Space Agency probe landed on Titan, one of Saturn's moons and a place where the chemical composition of the atmosphere may be similar to that found on early Earth, the photos transmitted by the probe indicated that liquid was present on the surface of this bitterly cold place. At atmospheric temperatures of approximately 2292°F (2180°C), the liquid is obviously not water; instead, it is most likely a mixture of ethane and methane, both nonpolar molecules. As a result, oceans on Titan are much poorer solvents than are oceans on Earth, and conditions in these oceans are probably not suitable for the evolution of life.

Organic Chemistry

The Martian meteorite ALH84001 had one characteristic that provided some evidence that the rock once contained living organisms: the presence of complex molecules containing the element carbon.

All life on Earth is based on the chemistry of the element carbon. The branch of chemistry that is concerned with complex carbon-containing molecules is called **organic chemistry.**

An Overview: Chemical Bonds. Chemical bonds between atoms and molecules involve attractions that help stabilize various configurations. In general, this involves the sharing or transfer of electrons. One of the most important elements in biology is carbon, which is often involved in chemical bonding because of its ability to make bonds with four other elements.

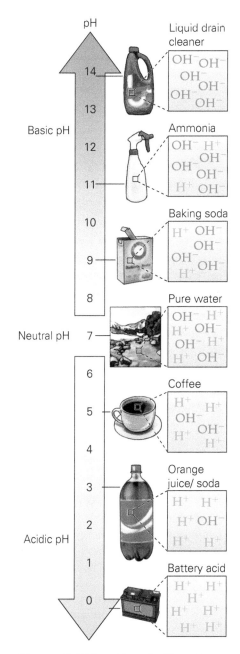

Figure 2.7 The pH scale. The pH scale is a measure of hydrogen ion concentration ranging from 0 (most acidic) to 14 (most basic). Each pH unit actually represents a 10-fold (10x) difference in the concentration of H^+ ions.
Visualize This: A substance with a pH of 5 would have how many times more H^+ ions than a substance with a pH of 7?

Like a Tinkertoy™ connector, carbon has multiple sites for connections that allow carbon-containing molecules to take an almost infinite variety of shapes (**Figure 2.8**). *See the next section for **A Closer Look** at chemical bonds.*

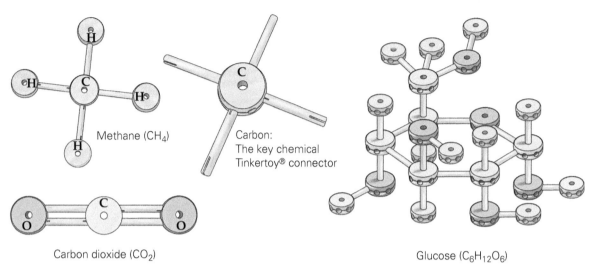

Methane (CH_4)

Carbon:
The key chemical
Tinkertoy® connector

Carbon dioxide (CO_2)

Glucose ($C_6H_{12}O_6$)

Figure 2.8 Carbon, the chemical Tinkertoy™ connector. Because carbon forms four covalent bonds at a time, carbon-containing compounds can have diverse shapes.

A Closer Look:
Chemical Bonds

The ability of elements to make chemical bonds depends on the atom's electron configuration. The electrons in the electron cloud that surrounds the atom's nucleus have different energy levels based on their distance from the nucleus. The first energy level, or **electron shell,** is closest to the nucleus, and the electrons located there have the lowest energy. The second energy level is a little farther away, and the electrons located in the second shell have a little more energy. The third energy level is even farther away, and its electrons have even more energy, and so on.

Each energy level can hold a specific maximum number of electrons. The first shell holds 2 electrons, and the second and third shells each hold a maximum of 8. Electrons fill the lowest energy shell before advancing to fill a higher energy-level shell. For example, hydrogen with its 1 electron needs only 1 more electron to fill its first shell.

Atoms with the same number of electrons in their outermost energy shell, called the **valence shell,** exhibit similar chemical behaviors. When the valence shell is full, the atom will not normally form chemical bonds with other atoms. Atoms whose valence shells are not full of electrons often combine via chemical bonds.

Atoms with 4 or 5 electrons in the outermost valence shell tend to share electrons to complete their valence shells. When atoms share electrons, a type of bond called a **covalent bond** is formed.

Carbon, with its 4 valence electrons, is said to be tetravalent (**Figure 2.9a**). In other words, it can form up to four bonds. Carbon can form 4 single bonds, 2 double bonds, 1 double bond and 2 single bonds, and so on, depending on the number of electrons needed by the atom that is its partner. **Figure 2.9b** shows carbon covalently bonded to 4 hydrogens to produce methane, an organic compound that has been found on Mars and Titan. Covalent bonds are symbolized by

(a) Electron configuration of carbon (C)

(b) Methane (CH_4)

Figure 2.9 Covalent bonding. (a) Carbon has 4 unpaired electrons in its valence shell and is thus able to make up to 4 chemical bonds. (b) Methane consists of carbon covalently bonded to 4 hydrogens.

(a) Methane   **(b) Ethylene**

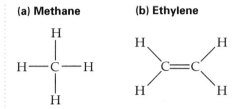

Figure 2.10 Single and double bonds. (a) Covalent bonds are symbolized by a short line indicating a shared pair of electrons. (b) Double covalent bonds involve two pairs of shared electrons, symbolized by two horizontal lines.

a short line indicating a shared pair of electrons (Figure 2.10a). When an element such as carbon enters into bonds involving two *pairs* of shared electrons, this is called a double bond. A carbon-to-carbon double bond is symbolized by two horizontal lines (Figure 2.10b).

Atoms with 1, 2, or 3 electrons in their valence shell tend to lose electrons and therefore become positively charged ions, while atoms with 6 or 7 electrons in the valence shell tend to gain electrons and become negatively charged ions. Positively and negatively charged ions associate into a type of bond called the ionic bond.

Stop & Stretch What does not make sense (chemically) about the structure below?

$$H-C=C-H$$

Ionic bonds form between charged atoms attracted to each other by similar, opposite charges. For example, the sodium atom forms an ionic bond with a chlorine atom to produce table salt (sodium chloride) when the sodium atom gives up an electron and the chlorine atom gains 1 (Figure 2.11). More than 2 atoms can be involved in an ionic bond. For instance, calcium will react with 2 chlorine atoms to produce calcium chloride ($CaCl_2$). This is because calcium has 2 electrons in its valence shell—when it loses these it has 2 more protons than electrons, giving it a double-positive charge. Each chlorine atom, with 7 electrons, picks only 1 more electron to have a stable outer shell and a single negative charge. Thus, 2 chlorine ions will be attracted to a single calcium ion.

Ionic bonds are about as strong as covalent bonds. They can be more easily disrupted, however, when mixed with certain liquids containing electrical charges. Water is one liquid that causes ions in molecules to dissociate or fall apart.

The simple organic molecules found in the Martian meteorite that appear to have formed on Mars are carbonates, molecules containing carbon and oxygen, and **hydrocarbons,** made up of chains and rings of carbon and hydrogen. Carbonates and hydrocarbons can form under certain natural conditions even without the presence of life. However, the meteorite lacked convincing evidence of **macromolecules,** large organic molecules made of many subunits, that are known to be produced only by living organisms.

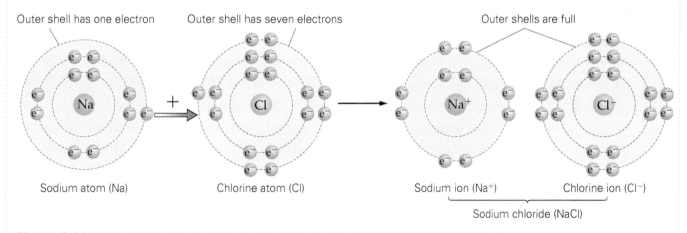

Figure 2.11 Ionic bonding. Ionic bonds form when electrons are transferred between charged atoms.

Structure and Function of Macromolecules

The macromolecules present in living organisms are carbohydrates, proteins, lipids, and nucleic acids. To date, every living Earth organism, whether bacteria, plant, or animal, has been found to contain these same macromolecules.

Carbohydrates. Sugars, or **carbohydrates,** provide the major source of energy for daily activities. Carbohydrates also play important structural roles in cells. The simplest carbohydrates are composed of carbon, hydrogen, and oxygen in the ratio (CH_2O). For example, the carbohydrate glucose is symbolized as 6(CH_2O) or $C_6H_{12}O_6$. Glucose is a simple sugar, or monosaccharide, that consists of a single ring-shaped structure. Disaccharides are two rings joined together. Table sugar, called sucrose, is a disaccharide composed of glucose and fructose, a sugar found in fruits.

Joining many individual subunits, or monomers, together produces polymers. Polymers of sugar monomers are called **polysaccharides** (Figure 2.12). Plants use tough polysaccharides in their cell walls as a sort of structural skeleton. The polysaccharide cellulose, found in plant cell walls, is the most abundant carbohydrate on Earth. The external skeletons of insects, spiders, and lobsters are composed of the polysaccharide chitin, and the cell walls that surround bacterial cells are rich in structural polysaccharides.

According to David McKay and his colleagues, the particular set of hydrocarbons found in the Martian meteorite is identical to the set formed when carbohydrates break down in certain bacteria on Earth. These trace remains of possible Martian carbohydrates are an important piece of evidence that scientists use to argue that Mars once harbored Earth-like life. Evidence of the presence of proteins on the meteorite is less convincing.

Proteins. Living organisms require **proteins** for a wide variety of processes. Proteins are important structural components of cells; they are integral to the

Figure 2.12 Carbohydrates. Monosaccharides, which tend to form ring structures in aqueous solutions, are individual sugar molecules. Disaccharides are 2 monosaccharides joined together, and polysaccharides are long chains of sugars joined together by covalent bonds. The monosaccharide glucose and the disaccharide sucrose are important sources of energy, and cellulose plays a structural role in plant cell walls.

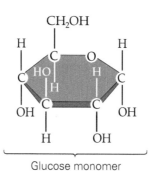

Glucose monomer

Glucose   Fructose

Sucrose—a disaccharide

Cellulose—a polysaccharide

structure of cell membranes and make up half the dry weight of most cells. Some cells, such as animal muscle cells, are largely composed of proteins. Proteins called **enzymes** accelerate and help regulate all the chemical reactions that build up and break down molecules inside cells. The catalytic power of enzymes (their ability to drastically increase reaction rates) allows metabolism to occur under normal cellular conditions. Proteins can also serve as channels through which substances are brought into cells, and they can function as hormones that send chemical messages throughout an organism's body.

Proteins are large molecules made of monomer subunits called **amino acids.** There are 20 commonly occurring amino acids. Like carbohydrates, amino acids are made of carbons, hydrogens, and oxygens; these form the amino acid's carboxyl group ($-COO^-$). In addition, amino acids have nitrogen as part of an amino (NH_2^+) group along with various side groups. Side groups are chemical groups that give amino acids different chemical properties (**Figure 2.13a**).

Polymers of amino acids can be joined together in various sequences called polypeptides. A covalent bond joins adjacent amino acids to each other. **Figure 2.13b** shows three amino acids—valine, alanine, and phenylalanine—joined by peptide bonds. Precisely folded polypeptides produce specific proteins in much the same manner that children can use differently shaped beads to produce a wide variety of structures (**Figure 2.13c**). Each amino acid side group has unique chemical properties, including being polar or nonpolar. Because each protein is composed of a particular sequence of amino acids, each protein has a unique shape and therefore specialized chemical properties.

Scientists have found no evidence of proteins in the Martian meteorite, although one group of investigators did report the presence of tiny amounts

(a) General formula for amino acid   **(b) Peptide bond formation**

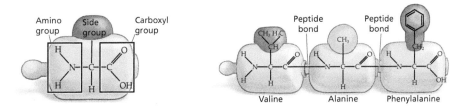

Valine   Alanine   Phenylalanine

(c) Protein

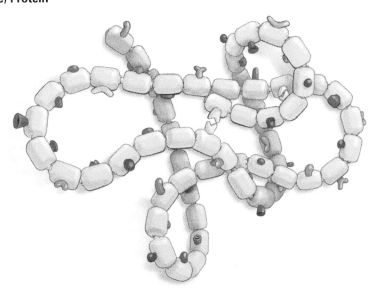

Figure 2.13 Amino acids, peptide bonds, and proteins. (a) All amino acids have the same backbone but different side groups. (b) Amino acids are joined together by peptide bonds. Long chains of these are called polypeptides. (c) Polypeptide chains fold upon themselves to produce proteins.

of three amino acids within the rock. However, it may be the case that these amino acids are contaminants; that is, they are present in the meteor because the meteor has been on protein-rich Earth for several thousand years. In addition, some amino acids are known to form under conditions where life is not present, so the presence of amino acids is not necessarily evidence of life.

Lipids. One type of organic molecule, abundant in living organisms, that has not been found in the Martian meteorite is lipids. **Lipids** are partially or entirely hydrophobic organic molecules made primarily of hydrocarbons. Important lipids include fats, steroids, and phospholipids.

Fat. The structure of a **fat** is that of a 3-carbon glycerol molecule with up to 3 long hydrocarbon chains attached to it (**Figure 2.14a**). Like the hydrocarbons present in gasoline, these can be burned to produce energy. The long hydrocarbon chains are called **fatty acid** tails of the fat. Fats are hydrophobic and function in storing energy within living organisms.

Steroids. **Steroids** are composed of 4 fused carbon-containing rings. Cholesterol (**Figure 2.14b**) is one steroid that is probably familiar; its primary function in animal cells (plant cells do not contain cholesterol) is to help maintain the fluidity of membranes. Other steroids include the sex hormones testosterone, estrogen, and progesterone, which are produced by the sex organs and have effects throughout the body.

Phospholipids. **Phospholipids** are similar to fats except that each glycerol molecule is attached to 2 fatty acid tails (not 3, as you would find in a dietary fat). The third bond in a phospholipid is to a phosphate head group. The phosphate head group consists of a phosphorous atom attached to four

Figure 2.14 Three types of lipids. (a) Fats are composed of a glycerol molecule with 3 hydrocarbon-rich fatty acid tails attached. (b) Cholesterol is a steroid common in animal cell membranes. (c) Phospholipids are composed of a glycerol backbone with 2 fatty acids attached and 1 phosphate head group. The cartoon drawing to the right shows how phospholipids are often depicted.

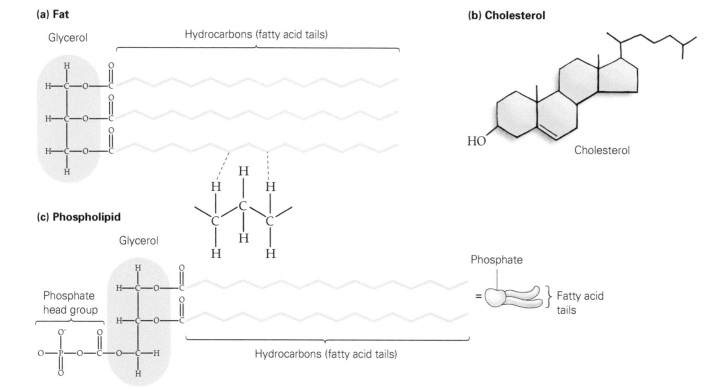

oxygen atoms and is hydrophilic. Thus a phospholipid has a hydrophilic head and two hydrophobic tails (**Figure 2.14c**). Phospholipids often have an additional head group, attached to the phosphate, that also confers unique chemical properties on the individual phospholipid. Phospholipids are important constituents of the membranes that surround cells and that designate compartments within cells.

Even if the Martian meteorite contained unambiguous traces of carbohydrates, proteins, and lipids, the source of these molecules would not clearly be a living organism. Evidence of a mechanism for passing traits to the next generation would also be required. The hereditary, or genetic, information common to all life on Earth is in the form of nucleic acids.

Nucleic Acids. Nucleic acids are composed of long strings of monomers called **nucleotides.** A nucleotide is made up of a sugar, a phosphate, and a nitrogen-containing base. There are two classes of nucleic acids in living organisms. **Ribonucleic acid (RNA)** plays a key role in helping cells synthesize proteins (and is discussed in detail in further chapters). The nucleic acid that serves as the primary storage of genetic information in nearly all living organisms is **deoxyribonucleic acid (DNA).** **Figure 2.15** (on the next page) shows the three-dimensional structure of a DNA molecule and zooms inward to the chemical structure. You can see that DNA is composed of two curving strands that wind around each other to form a double helix. The sugar in DNA is the 5-carbon sugar deoxyribose. The nitrogen-containing bases, or **nitrogenous bases,** of DNA have one of four different chemical structures, each with a different name: **adenine (A), guanine (G), thymine (T),** and **cytosine (C).** Nucleotides are joined to each other along the length of the helix by covalent bonds.

Nitrogenous bases form hydrogen bonds with each other across the width of the helix. On a DNA molecule, an adenine (A) on one strand always pairs with a thymine (T) on the opposite strand. Likewise, guanine (G) always pairs with cytosine (C). The term **complementary** is used to describe these pairings. For example, A is complementary to T, and C is complementary to G. Therefore, the order of nucleotides on one strand of the DNA helix predicts the order of nucleotides on the other strand. Thus, if one strand of the DNA mole-cule is composed of nucleotides AACGATCCG, then we know that the order of nucleotides on the other strand is TTGCTAGGC.

As a result of this **base-pairing rule** (A pairs with T; G pairs with C), the width of the DNA helix is uniform. There are no bulges or dimples in the structure of the DNA helix because A and G, called **purines,** are structures composed of two rings; C and T are single-ring structures called **pyrimidines.** A purine always pairs with a pyrimidine and vice versa, so there are always 3 rings across the width of the helix. A-to-T base pairs have 2 hydrogen bonds holding them together. G-to-C pairs have 3 hydrogen bonds holding them together.

Each strand of the helix thus consists of a series of sugars and phosphates alternating along the length of the helix, the **sugar-phosphate backbone.** The strands of the helix align so that the nucleotides face "up" on one side of the helix and "down" on the other side of the helix. For this reason, the two strands of the helix are said to be antiparallel.

The overall structure of a DNA molecule can be likened to a rope ladder that is twisted, with the sides of the ladder composed of sugars and phosphates (the sugar-phosphate backbone) and the rungs of the ladder composed of the nitrogenous-base sequences A, C, G, and T. The structure of DNA was determined

(a) DNA double helix is made of two strands.   **(b) Each strand is a chain of of antiparallel nucleotides.**

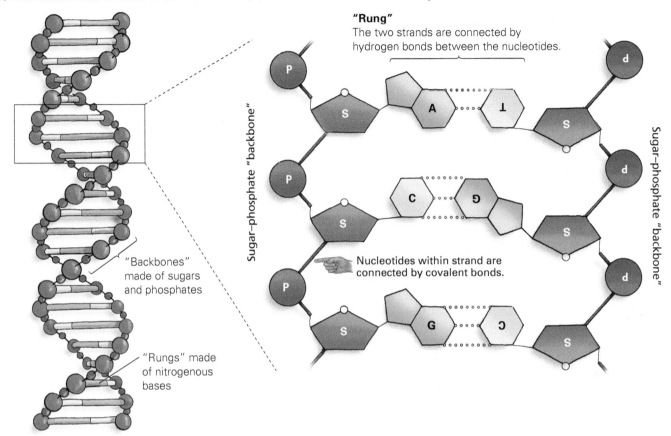

"Backbones" made of sugars and phosphates

"Rungs" made of nitrogenous bases

(c) Each nucleotide is composed of a phosphate, a sugar, and a nitrogenous base.

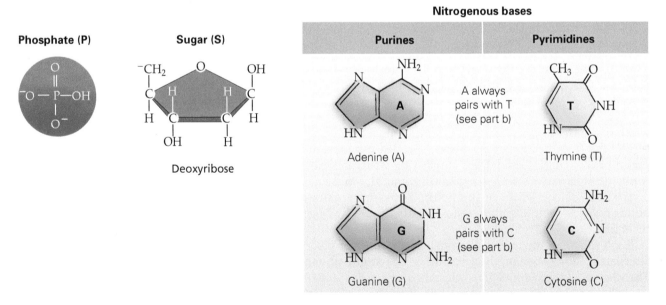

Figure 2.15 DNA structure. (a) DNA is a double-helical structure composed of nucleotides. (b) Each strand of the helix is composed of repeating units of sugars and phosphates, making the sugar-phosphate backbone, and of nitrogenous bases across the width of the helix. (c) Each nucleotide is composed of a phosphate, a sugar, and a nitrogenous base. Adenine and guanine are purines, which have a double-ring structure; cytosine and thymine are pyrimidines, which have a single-ring structure.

Visualize This: Look at part (a) of this figure. What do the red and purple spheres represent? Why are there two colors and sizes of nitrogenous bases?

by a group of scientists in the 1950s, most notably James Watson and Francis Crick (**Figure 2.16**).

How Might Macromolecules on Other Planets Differ? Many scientists argue that the fundamental constituents described here—carbohydrates, proteins, lipids, and nucleic acids—will be essentially similar wherever life is found. They will readily admit that the finer details are very likely to differ, however. For example, all proteins known on Earth contain only 20 different amino acids, despite an infinite number of possibilities. Presumably, proteins on other planets could contain completely different amino acids and many more than 20.

Not all scientists agree with the hypothesis that life on other planets will be based on carbon-containing organic compounds. Carbon is not the only chemical Tinkertoy™ connector; other elements, including silicon, can also make connections with four other atoms. Silicon is also relatively abundant in the universe and could theoretically form the backbone of an alternative organic chemistry. The basic constituents of silicon-based life could be very different from the chemical building blocks of life on Earth.

Even if all life in the universe is based on carbon chemistry, it is very unlikely that the suite of organisms found on another planet will look much like life on our planet. However, understanding the history of life on Earth also provides insight into the possible nature of life elsewhere in the universe.

Figure 2.16 The DNA model.
American James Watson (left) and Englishman Francis Crick are shown with the three-dimensional model of DNA they devised while working at the University of Cambridge in England.

2.2 Life on Earth

One of the most dramatic features of the Martian meteorite is the presence of fossils that look remarkably like the tiniest living organisms known from Earth. The largest of these fossils is less than 1/100th of the diameter of a human hair, and most are about 1/1000th of the diameter of a human hair—small enough that it would take about 1000 laid end to end to span the dot at the end of this sentence. Some are egg-shaped, while others are tubular. These fossils appear similar to the simplest and most ancient of known organisms and are the strongest piece of evidence supporting the hypothesis that Mars once was home to living organisms.

David McKay and his colleagues argue that the fossil structures in the Martian meteorite are the remains of tiny cells. A **cell** is the fundamental structural unit of life on Earth, separated from its environment by a membrane and sometimes an external wall. Bacteria are single cells, which perform all of the activities required for life. More complex organisms can be composed of trillions of cells working together and do not have any cells that could survive and reproduce independently.

Prokaryotic and Eukaryotic Cells

All cells can be placed into one of two categories, prokaryotic or eukaryotic, based on the presence or absence of certain cellular structures. Scientists believe that the first prokaryotic cells appeared on Earth over 3.5 billion years ago and that the first eukaryotes appeared about 1.7 billion years later (Chapter 13).

Bacteria are **prokaryotic** cells, which are much smaller than eukaryotic cells (**Figure 2.17a** on the next page). Prokaryotes do not have a **nucleus,** a separate membrane-bound compartment to hold the genetic material. They also do not contain any membrane-bound internal compartments. They do however, have a **cell wall** that helps them maintain their shape (**Figure 2.17b**). According to the fossil record, prokaryotic cells predate the more complex eukaryotic cells (**Figure 2.17c**). The fossils in the Martian meteorite resemble modern prokaryotic cells known as nanobacteria and show no evidence of similar walls.

Eukaryotic cells have a nucleus and other internal structures with specialized functions, called **organelles,** that are surrounded by membranes.

(a) Different sizes: prokaryotic (red) vs. eukaryotic (white) cells

(c) Eukaryotic cell features

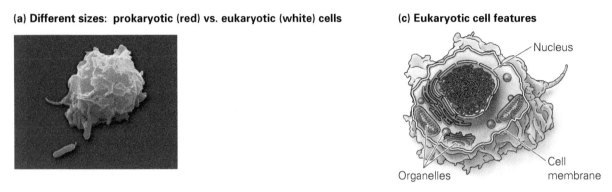

(b) Prokaryotic cell features

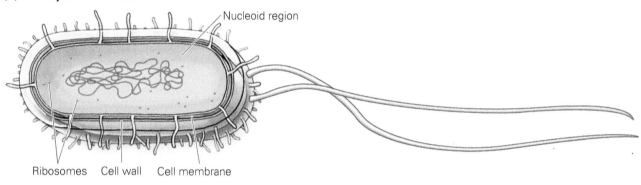

Figure 2.17 Prokaryotic and eukaryotic cells. (a) Prokaryotic cells are typically about 1/10 the diameter of a eukaryotic cell, as evidenced by the size of the two bacterial cells and a white blood cell. (b) Prokaryotic cells are structurally less complex than (c) eukaryotic cells.

Eukaryotic organisms include single-celled organisms such as amoebas and yeast as well as multicellular plants, fungi, and animals.

Many scientists dispute David McKay's interpretation that structures in ALH84001 are cellular and therefore evidence of life on Mars. In fact, similar structures can be formed in the absence of life by certain minerals under extremes of heat and pressure. If the Martian fossils are indeed cells, they should contain features similar to those found inside earthly cells.

Cell Structure

Each living cell can be considered a veritable factory, working to break down nutrients and to build required components. Let's look at the outside of the cell and then examine the structure and function of various internal cell components. **Table 2.1** describes the structures and functions of most cellular organelles in detail.

Plasma Membrane. All cells are enclosed by a structure called a **plasma membrane.** The plasma membrane defines the outer boundary of each cell, isolates the cell's contents from the environment, and serves as a semipermeable barrier that determines which nutrients are allowed into and out of the cell. Membranes that enclose structures inside the cell are usually referred to as cell membranes, while the outer boundary is the plasma membrane.

Internal and external membranes are composed in part of phospholipids. The chemical properties of these lipids make membranes flexible and self-sealing. When phospholipid molecules are placed in a watery solution, such as in a cell, they orient themselves so that their hydrophilic heads are exposed to the water and their hydrophobic tails are away from the water. They cluster into a form called a **phospholipid bilayer,** in which the tails of the phospholipids interact with themselves and exclude water, while the heads maximize their exposure to the surrounding water both inside and outside the membrane. The bilayer of phospholipids is stuffed with proteins that carry out enzymatic functions, serve as receptors for outside substances, and help transport substances throughout the cell.

A Fluid Mosaic of Lipids and Proteins. All of the lipids and most of the proteins in the plasma membrane are free to bob about, sliding from one location in

TABLE 2.1

Cell components. Illustrations and descriptions of cell components and their functions.

Component	Function
Plasma membrane 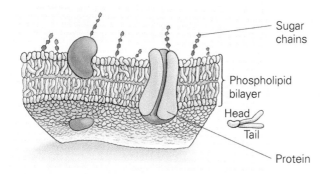	All cells are surrounded by a plasma membrane. It is composed of a bilayer of phospholipids perforated by proteins. Proteins in the bilayer help transport substances across the hydrophobic core of the membrane.
Cell wall 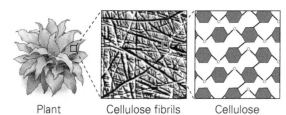	The cell wall is found outside the plasma membrane of plant and bacterial cells. The cell wall in plants is rich in the polysaccharide cellulose. Cellulose is assembled into strong fibrils and embedded in a matrix.
Nucleus 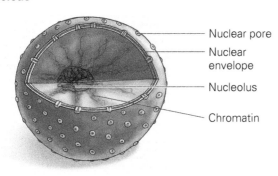	Eukaryotic cells contain a nucleus. The nucleus is a spherical structure surrounded by two membranes, together called the nuclear envelope. The nuclear envelope is studded with nuclear pores that regulate traffic into and out of the nucleus. Inside the nucleus is chromatin, composed of DNA and proteins. The nucleolus is where ribosomes are produced.
Mitochondrion 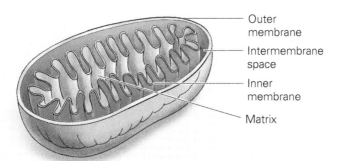	Plant and animal cells contain mitochondria, energy-producing organelles surrounded by two membranes. The inner and outer mitochondrial membranes are separated by the intermembrane space. The highly convoluted inner membrane carries many of the proteins involved in producing ATP. The matrix of the mitochondrion is the location of many of the reactions of cellular respiration.

In the Plasma membrane illustration: Sugar chains, Phospholipid bilayer, Head, Tail, Protein

In the Cell wall illustration: Plant, Cellulose fibrils, Cellulose

In the Nucleus illustration: Nuclear pore, Nuclear envelope, Nucleolus, Chromatin

In the Mitochondrion illustration: Outer membrane, Intermembrane space, Inner membrane, Matrix

(continued)

TABLE 2.1 *(continued)*

Cell components. Illustrations and descriptions of cell components and their functions.

Component	Function
Chloroplast 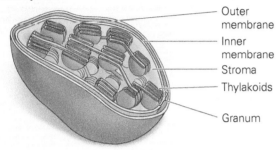 Outer membrane Inner membrane Stroma Thylakoids Granum	An important organelle present in plant cells, the chloroplast uses the sun's energy to convert carbon dioxide and water into sugars. Each chloroplast has an outer membrane, an inner membrane, a liquid interior called the stroma, and a network of membranous sacs called thylakoids that stack on one another to form structures called grana (singular: granum). Chloroplasts also contain pigment molecules that give green parts of plants their color.
Lysosome 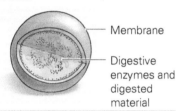 Membrane Digestive enzymes and digested material	A lysosome is a membrane-enclosed sac of digestive enzymes that degrade proteins, carbohydrates, and fats. Lysosomes roam around the cell and engulf targeted molecules and organelles for recycling.
Endoplasmic reticulum (ER) 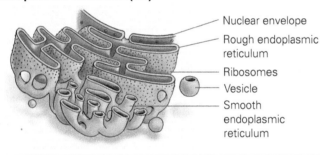 Nuclear envelope Rough endoplasmic reticulum Ribosomes Vesicle Smooth endoplasmic reticulum	The ER is a large network of membranes that begins at the nuclear envelope and extends into the cytoplasm of a eukaryotic cell. ER with ribosomes attached is called rough ER. Proteins synthesized on rough ER will be secreted from the cell or will become part of the plasma membrane. ER without ribosomes attached is called smooth ER. The function of the smooth ER depends on cell type but includes tasks such as detoxifying harmful substances and synthesizing lipids. Vesicles are pinched-off pieces of membrane that transport substances to the Golgi apparatus or plasma membrane.
Golgi apparatus 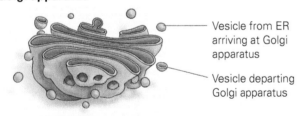 Vesicle from ER arriving at Golgi apparatus Vesicle departing Golgi apparatus	The Golgi apparatus is a stack of membranous sacs. Vesicles from the ER fuse with the Golgi apparatus and empty their protein contents. The proteins are then modified, sorted, and sent to the correct destination in new transport vesicles that bud off from the sacs.
Ribosomes  Ribosomes	Ribosomes are found in eukaryotic and prokaryotic cells. Ribosomes are built in the nucleolus and shipped out of the nucleus through nuclear pores to the cytoplasm, where they are used as workbenches for protein synthesis. They can be found floating in the cytoplasm or tethered to the ER.

TABLE 2.1

Cell components. Illustrations and descriptions of cell components and their functions.

Component	Function
Centrioles Microtubule triplet	Centrioles are barrel-shaped rings composed of micro-tubules that help move chromosomes around when a cell divides. Centrioles are involved in microtubule formation during cell division and the formation of cilia and flagella.
Cytoskeletal elements Microfilaments — Intermediate filaments — Microtubules	Cytoskeletal elements are protein fibers in the cytoplasm that give shape to a cell, hold and move organelles (including transport vesicles), and are involved in cell movement.
Central vacuole	Plant cells also have large membrane-bound, fluid-filled vacuoles that can occupy as much as 90% of a cell's total volume. The plant vacuole contains a variety of dissolved molecules, including sugars and pigments that give color to flowers and leaves. Vacuoles also function to maintain pressure inside individual cells, which helps support the upright plant.

the membrane to another. Because lipids and proteins move about laterally within the membrane, the membrane is a **fluid mosaic** of lipids and proteins. The membrane is fluid because the composition of any one location on the membrane can change. In the same manner that a patchwork quilt is a mosaic (different fabrics making up the whole quilt), so, too, is the membrane a mosaic with different regions of membrane composed of different types of phospholipids and proteins.

Cell membranes are **semipermeable** in the sense that they allow some substances to cross and prevent others from crossing. This characteristic allows cells to maintain a different internal composition from the surrounding solution.

Nucleus. All eukaryotic cells contain a nucleus surrounded by a double nuclear membrane, which houses the DNA. Inside the nucleus is the nucleolus, which is where ribosomes are assembled.

Ribosomes. Ribosomes are workbenches where proteins are assembled. They are composed of two subunits and can be found floating free in the cytosol or attached to certain membranes.

Cytosol. Between the nucleus and the plasma membrane lies the cytosol, a watery matrix containing water, salts, and many of the enzymes required for cellular reactions. The cytosol houses the organelles. The term **cytoplasm** includes the cytosol and organelles.

Organelles. Organelles are to cells as organs are to the body. Each organelle performs a specific job required by the cell, and all organelles work together to keep an individual cell healthy and to produce the raw materials that the cell needs to survive. Each organelle is enclosed in its own lipid bilayer. Some organelles are involved in metabolism. For example, organelles called **mitochondria** help the cells convert food energy into a form usable by cells, called ATP (adenosine triphosphate), while **chloroplasts** in plant cells use energy from sunlight to make sugars. **Lysosomes** help break down substances. Other organelles are involved in producing proteins. The endoplasmic reticulum (ER) is an extensive membranous organelle that can be studded with ribosomes and involved in protein synthesis (**rough endoplasmic reticulum**) or tubular in shape and involved in lipid synthesis (**smooth endoplasmic reticulum**). Proteins that are assembled on the membranes of the rough ER can be modified and sorted in a membranous structure called the **Golgi apparatus.**

There are other important subcellular structures that are not considered organelles because they are not bounded by membranes. In addition to the ribosomes described previously, **centrioles** are involved in moving genetic material around when a cell divides. Many fibers that compose the **cytoskeleton** help maintain the cell shape.

Some organelles and subcellular structures are found in certain cell types only. For instance, in addition to having chloroplasts and a cell wall, the plant cell also has a **central vacuole** to store water, sugars, and pigments. Figure 2.18 shows an animal cell and a plant cell complete with their complement of organelles.

The Tree of Life and Evolutionary Theory

Biologists disagree about the total number of different **species,** or types of living organisms, that are present on Earth today. This uncertainty stems from lack of knowledge. Although scientists likely have identified most of the larger organisms—such as land plants, mammals, birds, reptiles, and fish—millions of species of insects, fungi, bacteria, and other microscopic organisms remain unknown to science. Amazingly, credible estimates of the number of species on Earth range from 5 million to 100 million. Despite this level of uncertainty, most biologists think that the likeliest number is near 10 million.

Theory of Evolution. While the diversity of living organisms is tremendous, there exists remarkable similarity among all known species. All have the same basic biochemistry, including carbohydrates, lipids, proteins, and nucleic acids. All consist of cells surrounded by a plasma membrane. All eukaryotic organisms (including fungi, animals, and plants) contain nearly the same suite of cellular organelles. The best explanation for the shared characteristics of all species, what biologists refer to as

Stop & Stretch Would you expect prokaryotic cells to contain ribosomes? Why or why not?

Figure 2.18 Animal and plant cells. These drawings of a generalized (a) animal cell and (b) plant cell show the locations and sizes of organelles and other structures.
Visualize This: What three structures are present in plant but not animal cells?

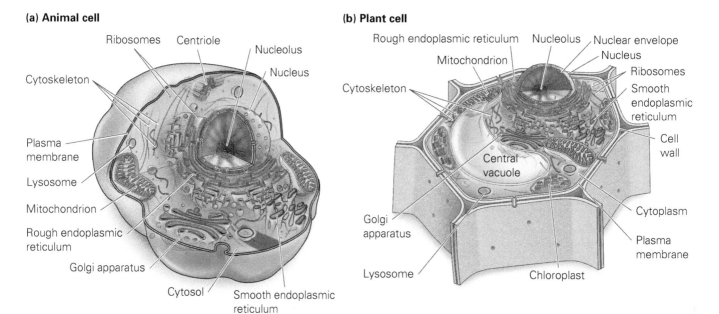

(a) Animal cell

Ribosomes
Centriole
Nucleolus
Nucleus
Cytoskeleton
Plasma membrane
Lysosome
Mitochondrion
Rough endoplasmic reticulum
Golgi apparatus
Cytosol
Smooth endoplasmic reticulum

(b) Plant cell

Rough endoplasmic reticulum
Nucleolus
Nuclear envelope
Nucleus
Mitochondrion
Ribosomes
Cytoskeleton
Smooth endoplasmic reticulum
Central vacuole
Cell wall
Golgi apparatus
Cytoplasm
Lysosome
Plasma membrane
Chloroplast

"the unity of life," is that all living organisms share a common ancestor that arose on Earth nearly 4 billion years ago. The divergence and differences among modern species arose as a result of changes in the characteristics of populations, both in response to environmental change (a process called *natural selection*) and due to chance. These ideas underlie the entire science of biology and are known as the **theory of evolution.**

A major process in the diversification of life from a single ancestor was natural selection. The basic principle of the theory of natural selection (Chapter 11) is simple: Individual organisms vary from each other, and some of these variations increase their chances of survival and reproduction. A genetic trait that increases survival and reproduction should become more prevalent over time. In contrast, less successful variants should eventually be lost from the population.

The common ancestor can be thought of as the starting place for life on Earth, and the continual divergence among species and groups of species can be thought of as life's branching. Modern organisms can therefore be arranged on a tree of life that reflects their basic unity and relationships. According to current understanding, living organisms can be grouped into three large groups: two that are prokaryotic and one containing all eukaryotes. Eukaryotes can be further grouped into several categories made up primarily of free-living, single-celled organisms (such as amoebas and algae) and the three major multicellular groups—plants, fungi, and animals (Figure 2.19). (Chapter 13 provides a deeper exploration of the diversity of life on Earth.)

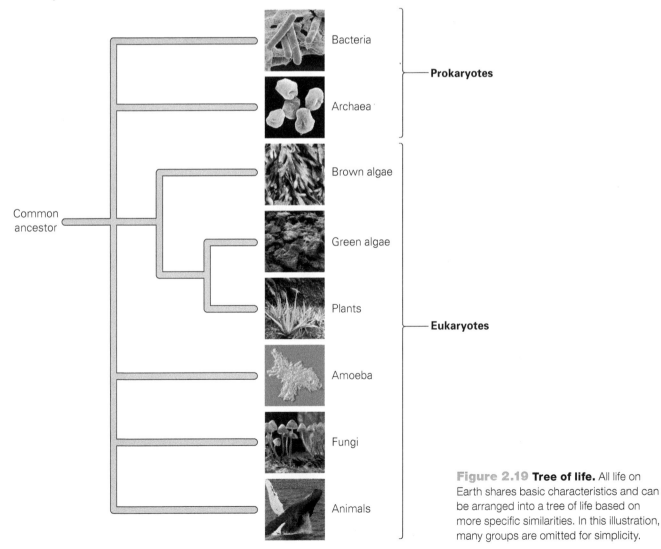

Figure 2.19 Tree of life. All life on Earth shares basic characteristics and can be arranged into a tree of life based on more specific similarities. In this illustration, many groups are omitted for simplicity.

Figure 2.20 Diversity of body form.
Not all animals are bilaterally symmetric like us, with two eyes, two ears, two arms, two legs, an obvious head and "tail," and one central axis. A sea star is radially symmetric—it can be divided into two equal halves in any direction.
Visualize This: Is an earthworm bilaterally symmetric?

Because evolutionary change results from chance events and environmental changes (including the appearance of other species), the groups of organisms present on Earth today represent only one set of an infinite number of possibilities. Similarly, life on other planets need not look identical to life on Earth. For example, instead of the common body form found in animals, called bilateral symmetry, by which bodies can be visually divided into two mirror-image halves, life on other planets could be primarily radially symmetric and thus look very different (**Figure 2.20**). In fact, it is possible that life on other planets might not be based on carbon and may not be sustained by water. No one knows what living organisms would look like on a planet where the important molecules are based on the element silicon or where the sustaining liquid is not water.

Life in the Universe. Do other living organisms exist in the universe? Given the universe's sheer size and complexity, most scientists who study this question think that the existence of life on other planets is nearly certain.

While the evidence of life in the Martian meteorite is unconvincing to many scientists, we may in our lifetimes find out that life exists, or once existed, on our planetary neighbor. What about the existence of intelligent life that could communicate with us? Some scientists argue that as a result of natural selection, the evolution of intelligence is inevitable wherever life arises. Others point to the history of life on Earth—consisting of at least 2.5 billion years, during which all life was made up of single-celled organisms—to argue that most life in the universe must be "simple and dumb." It is clear from our explorations of the solar system that none of the sun's other planets hosts intelligent life. The nearest sun-like stars that could host an Earth-like planet, Alpha Centauri A and B, are over 4 light years away—nearly 40 trillion miles. With current technologies, it would take nearly 50,000 years to reach the Alpha Centauri stars, and there is certainly no guarantee that intelligent life would be found on any planets that circle them. For all practical purposes, at this time in human history, we are still alone in the universe.

SAVVY READER

Ionized Water

Several U. S. companies sell products called water ionizers. These are plastic-encased filtering systems about the size and shape of a small suitcase, which can be installed on or under your sink at a cost of close to $1000. What follows is an example of the kinds of claims made by these manufacturers' websites:

Ionized water contains a negative electric charge that produces hydroxyl ions, and these charged particles help you to feel younger; have more energy; relieve chronic pain without medication; sleep better through the night, every night; rapidly skyrocket your energy and

stamina to do all the things you love; enjoy a hugely improved quality of life every day. If you are looking for a natural solution that most conventional doctors don't know about and you want to get rid of cancer, chronic illness, headaches, high cholesterol, diabetes, arthritis, depression, or just not feeling well enough, you need to try this amazing healing water.

Research purporting to support the anticancer properties of ionized water is offered by these manufacturers. One of the claims made is that ionized water can clean up damaging oxygen free radicals (highly reactive molecules), the concentration of which is measured in various mouse tissues after exposure to ionized water or tap water.

1. Based on what you know about water, does the manufacturer's claim that ionized water contains a negative electric charge that produces hydroxyl ions make sense?

2. Evaluate this website by listing the first two red flags that are raised. (Use the questions in the checklist provided in Chapter 1, Table 1.2.)

3. Give two examples of language used on this website that differs from the way scientists usually describe results.

4. Do the authors provide evidence that a decrease in oxygen free radicals is associated with decreased cancer?

5. What key statistical element is missing from the graph?

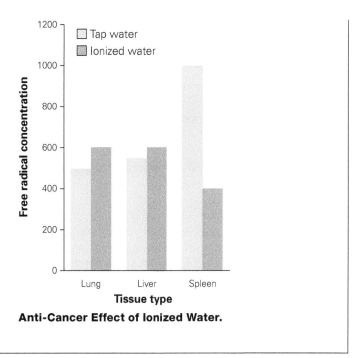

Anti-Cancer Effect of Ionized Water.

Chapter Review
Learning Outcomes

MasteringBIOLOGY

Go to the Study Area at www.masteringbiology.com for practice quizzes, myeBook, BioFlix™ 3-D animations, MP3Tutor sessions, videos, current events, and more.

LO1 **Describe the properties associated with living organisms (Section 2.1).**

- Living organisms must be able to grow, metabolize substances, reproduce, and respond to external stimuli (p. 32).
- Living organisms contain a common set of biological molecules, are composed of cells, and can maintain homeostasis and evolve (p. 32).

LO2 **List the components of water and some of the properties that make it important in living organisms (Section 2.1).**

- Water consists of 1 oxygen and 2 hydrogen atoms (p. 33).
- Water is a good solvent in part because of its polarity and ability to form hydrogen bonds. Hydrogen bonding in water facilitates chemical reactions, promotes cohesion, and allows for heat absorption (pp. 33–34).
- The polarity of water also facilitates the dissolving of salts (p. 34).
- Water has a neutral pH. The pH scale is a measure of the relative percentages of ions in a solution and ranges from 0 (acidic or rich in hydrogen ions) to 14 (basic or rich in hydroxyl ions). Water has the same number of hydrogen as hydroxyl ions (p. 35).

LO3 **Describe how atomic structure affects chemical bonding (Section 2.1).**

- Chemical bonding depends on an element's electron configuration. Electrons closer to the nucleus have less energy than those that are farther away from the nucleus (p. 36).
- Atoms that have space in their valence shell form chemical bonds (p. 36).

LO4 **Compare and contrast hydrogen, covalent, and ionic bonds (Section 2.1).**

- Hydrogen bonds are weak attractions between hydrogen atoms and oxygen atoms in adjacent molecules (p. 34).
- Covalent bonds form when atoms share electrons. These tend to be strong bonds (p. 36).
- Ionic bonds form between positively and negatively charged ions. These tend to be weaker bonds (p. 37).

LO5 **Discuss the importance of carbon in living organisms (Section 2.1).**

- Life on Earth is based on the chemistry of the element carbon, which can make bonds with up to four other elements (pp. 35–36).

LO6 **Describe the structure of carbohydrates, proteins, lipids, and nucleic acids and the roles these macromolecules play in cells (Section 2.1).**

- Carbohydrates function in energy storage and play structural roles. They can be single-unit monosaccharides or multiple-unit polysaccharides with sugar monomers joined by covalent bonds (pp. 37–38).
- Proteins play structural, enzymatic, and transport roles in cells. They are composed of amino acid monomers joined by covalent bonds (pp. 38–40).
- Lipids are partially or entirely hydrophobic and come in three different forms. Fats are composed of glycerol covalently bonded to three fatty acids. Fats store energy. Phospholipids are composed of glycerol, 2 fatty acids, and a phosphate group. They are important structural components of cell membranes. Steroids are composed of 4 fused rings. Cholesterol is a steroid found in some animal cell membranes and helps maintain fluidity. Other steroids function as hormones (pp. 40–41).
- Nucleic acids are polymers of covalently bonded nucleotides, each of which is composed of a sugar, a phosphate, and a nitrogen-containing base (pp. 41–42).

LO7 **Compare and contrast prokaryotic and eukaryotic cells (Section 2.2).**

- There are two main categories of cells: Those with nuclei and membrane-bound organelles are eukaryotes; those lacking a nucleus and membrane-bound organelles are prokaryotes (pp. 43–44).

LO8 **Describe the structure and function of cell membranes (Section 2.2).**

- The plasma membrane that surrounds cells is a semipermeable boundary composed of a phospholipid bilayer that has embedded proteins and cholesterol. Cells also have internal membranous structures (p. 47).
- Lipids and proteins can move throughout membranes. This fluidity of the membrane allows changes in the protein and lipid composition (p. 47).
- Some organisms, such as plants, fungi, and bacterial cells, have a cell wall outside the plasma membrane that helps protect these cells and maintain their shape (p. 45).

LO9 **Describe the structure and function of the subcellular organelles (Section 2.1).**

- Subcellular organelles and structures perform many different functions within the cell. Mitochondria and chloroplasts are involved in energy conversions. Lysosomes are involved in the breakdown of macromolecules. Ribosomes serve as sites for protein synthesis. Proteins can be synthesized on ribosomes attached to rough endoplasmic reticulum. Smooth endoplasmic reticulum synthesizes lipids. The Golgi apparatus

sorts proteins and sends them to their cellular destination. Centrioles help cells divide. The plant cell central vacuole stores water and other substances (pp. 45–48).

LO10 **Provide a general summary of the theory of evolution (Section 2.2).**

- There may be nearly 10 million unique life-forms on Earth. Despite all of this diversity, all life on Earth shares the same organic chemistry, genetic material, and basic cellular structures (p. 48).
- The similarities among living organisms on Earth provide support for the theory of evolution, which states that all life on Earth derives from a common ancestor. The process of evolutionary change since the origin of that ancestor led to the modern relationships among organisms, called the tree of life (pp. 49–50).

Roots to Remember

The following roots of words come mainly from Latin and Greek and will help you to decipher terms:

cyto-	and **-cyte** mean cell or a kind of cell. Chapter terms: cytosol, cytoplasm, cytoskeleton
homeo-	means like or similar. Chapter term: homeostasis
hydro-	means water. Chapter terms: hydrophilic, hydrophobic
-mer	means subunit. Chapter terms: polymer, monomer
macro-	means large. Chapter term: macromolecule
micro-	means small. Chapter term: microscopic
mono-	means one. Chapter terms: monosaccharide, monomer
-philic	means to love. Chapter term: hydrophilic
-phobic	means to fear. Chapter term: hydrophobic
-plasm	means a fluid. Chapter terms: cytoplasm, plasma membrane
poly-	means many. Chapter term: polymer

Learning the Basics

1. **LO6** List the four biological molecules commonly found in living organisms.

2. **LO9** Describe the structure and function of the subcellular organelles.

3. **LO8** Describe the structure and function of the plasma membrane.

4. **LO2** Water _____.

   A. is a good solute; **B.** dissociates into H⁺ and OH⁻ ions; **C.** serves as an enzyme; **D.** makes strong covalent bonds with other molecules; **E.** has an acidic pH.

5. LO3 Electrons _____.

A. are negatively charged; **B.** along with neutrons comprise the nucleus; **C.** are attracted to the negatively charged nucleus; **D.** located closest to the nucleus have the most energy; **E.** all of the above are true.

6. LO6 Which of the following terms is least like the others?

A. monosaccharide; **B.** phospholipid; **C.** fat; **D.** steroid; **E.** lipid.

7. LO6 Different proteins are composed of different sequences of _____.

A. sugars; **B.** glycerols; **C.** fats; **D.** amino acids.

8. LO6 Proteins may function as _____.

A. genetic material; **B.** cholesterol molecules; **C.** fat reserves; **D.** enzymes; **E.** all of the above.

9. LO6 A fat molecule consists of _____.

A. carbohydrates and proteins; **B.** complex carbohydrates only; **C.** saturated oxygen atoms; **D.** glycerol and fatty acids.

10. LO7 Eukaryotic cells differ from prokaryotic cells in that _____.

A. only eukaryotic cells contain DNA; **B.** only eukaryotic cells have a plasma membrane; **C.** only eukaryotic cells are considered to be alive; **D.** only eukaryotic cells have a nucleus; **E.** only eukaryotic cells are found on Earth.

11. LO4 Which of the following lists the chemical bonds from strongest to weakest?

A. hydrogen, covalent, ionic; **B.** covalent, ionic, hydrogen; **C.** ionic, covalent, hydrogen **D.** covalent, hydrogen, ionic; **E.** hydrogen, ionic, covalent.

12. LO10 Which of the following is not consistent with evolutionary theory?

A. All living organisms share a common ancestor; **B.** The environment affects which organism survives to reproduce; **C.** Natural selection always favors the same traits, regardless of environment; **D.** Species change over time.

Analyzing and Applying the Basics

1. LO1 A virus is made up of a protein coat surrounding a small segment of genetic material (either DNA or RNA) and a few proteins. Some viruses are also enveloped in membranes derived from the virus's host cell. Viruses cannot reproduce without taking over the genetic "machinery" of their host cell. Based on this description and biologists' definition of life, should a virus be considered a living organism?

2. LO3 Recall that any molecule containing oxygen can be polar and that carbon-hydrogen bonds are nonpolar.

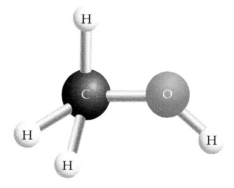

Figure 2.21 Methanol.

The structure of methanol (CH_3OH) is drawn in **Figure 2.21**. Which part of this molecule will have a partial negative charge, and which will have a partial positive charge?

3. LO5 Some scientists have argued that silicon (Si) could also be an appropriate basis for organic chemistry because it is abundant and can form bonds with many other atoms. Carbon contains 6 electrons, and silicon contains 14. Recalling that the lowest electron shell contains 2 electrons, and the next 2 shells can contain a maximum of 8, how many "spaces" does silicon have in its valence shell? How does this compare to carbon?

Connecting the Science

1. Water's characteristic as an excellent solvent means that many human-created chemicals (including some that are quite toxic) can be found in water bodies around the globe. How would our use and manufacture of toxic chemicals be different if most of these chemicals could not be dissolved and diluted in water but instead accumulated where they were produced and used?

2. Do you believe that humans should expend considerable energy and resources looking for life, even intelligent life, elsewhere in the universe? Why or why not?

Answers to **Stop & Stretch, Visualize This, Savvy Reader,** and **Chapter Review** questions can be found in the **Answers** section at the back of the book.

CHAPTER 3

Is It Possible to Supplement Your Way to Better Health?

Nutrients and Membrane Transport

Do supplemented waters improve hydration?

Nearly two-thirds of Americans take some sort of nutritional supplement, and virtually all of us have had a sports drink, vitamin water, or nutrition bar. A trip to the drug store will provide you access to hundreds of different vitamin, mineral, and plant-based herbal supplements, all claiming to enhance your health, looks, or longevity. Likewise, supermarket and convenience store aisles are filled with these conveniently packaged drinks and bars for eating on the go. Can these products help us make up for a poor diet? It seems that most Americans believe so—we spend around 6 billion dollars a year on these items. But are these products really necessary if we eat a healthy diet?

Do nutritional supplements improve health?

Are nutrition bars worth the cost?

Or is it better to just eat whole foods?

LEARNING OUTCOMES

LO1 Describe the role of nutrients in the body.

LO2 Explain the functions of water in the body.

LO3 Compare major dietary macronutrients, discussing the functions of each.

LO4 Compare dietary micronutrients, discussing the functions of each.

LO5 Describe the structure and function of the plasma membrane.

LO6 Distinguish between passive transport and active transport.

LO7 Compare the processes of endocytosis and exocytosis.

3.1 Nutrients

The food and drink that we ingest provide building-block molecules that can be broken down and used as raw materials for growth, for maintenance and repair, and as a source of energy. Another name for the substances in food that provide structural materials or energy is **nutrients.**

Macronutrients

Nutrients that are required in large amounts are called **macronutrients.** These include water, carbohydrates, proteins, and fats.

Water and Nutrition. Most animals can survive for several weeks with no nutrition other than water. However, survival without water is limited to just a few days. Besides helping the body disperse other nutrients, water helps dissolve and eliminate the waste products of digestion. Water helps to maintain blood pressure and is involved in virtually all cellular activities.

A decrease below the body's required water level, called **dehydration,** can lead to muscle cramps, fatigue, headaches, dizziness, nausea, confusion, and increased heart rate. Severe dehydration can result in hallucinations, heat stroke, and death. The body, including the brain, relies on water in the circulatory system to deliver nutrients as well as some oxygen. When water levels are low, the delivery system becomes less effective.

In addition, evaporation of water from the skin (sweating) helps maintain body temperature. When water is low and sweating decreases, the body temperature can rise to a harmful level.

Every day, humans lose about 3 liters of water as sweat, in urine, and in feces. To avoid dehydration, we must replace this water. We can replace some of it by consuming food that contains water. A typical adult obtains about 1.5 liters of water per day from food consumption, leaving a deficit of about 1.5 liters that must be replaced.

While it works quite well to replace this water with tap water, many people drink bottled water, in part because of concerns about the quality of tap water. However, the U.S. Food and Drug Administration (FDA), the government agency that sets the standards for bottled water, uses the same standards applied by the Environmental Protection Agency (EPA) to tap water. In other words, water from both sources should be equally clean. In fact, nearly 40% of bottled waters are actually derived directly from municipal tap water. Bottlers may, however, distill or filter this water to remove molecules that impart a smell or taste that consumers find objectionable.

Many people also chose to hydrate with sports drinks, most of which are high in simple sugars and calories. In fact, some of these drinks have so many additives that their solute concentration is higher than that of blood. When this happens, as you will learn more about in this chapter, water will leave the tissues, actually increasing the risk of dehydration.

Bottled waters and sports drinks are also chosen for convenience and portability, but this does come at a cost. The billions of bottles used each year require 1.5 million barrels of oil to produce. Although the bottles are recyclable, 86% of the bottles go to landfills each year. In addition, the energy required to transport water far exceeds the energy required to clean an equivalent amount of tap water.

Besides obtaining a healthy dose of water every day, people must consume foods that contain carbohydrates, proteins, and fats. (In Chapter 2, we explored the structure of these macromolecules, and now we focus on how they function in the body.)

Carbohydrates as Nutrients. Foods such as bread, cereal, rice, and pasta, as well as fruits and vegetables, are rich in sugars called carbohydrates. Carbohydrates are the major source of energy for cells. Energy is stored in the chemical bonds between the carbon, hydrogen, and oxygen atoms that comprise carbohydrate molecules (Chapter 2). Carbohydrates can exist as single-unit monosaccharides (individual sugar molecules) or can be bonded to each other to produce polysaccharides (long chains of sugar molecules).

The single-unit simple sugars are digested and enter the bloodstream quickly after ingestion. Sugars found in milk, juice, honey, and most refined foods are simple sugars. A sugar found in corn syrup, is shown in Figure 3.1a.

When multisubunit sugars are composed of many different branching chains of sugar monomers, they are called **complex carbohydrates.** Complex carbohydrates are found in vegetables, breads, legumes, and pasta.

When more sugar is present than needed, it can be stored for later use. Plants, such as potatoes, store their excess carbohydrates as the complex carbohydrate starch (Figure 3.1b). Animals store their excess carbohydrates as the complex carbohydrate glycogen in muscles and the liver (Figure 3.1c). Both starch and glycogen are polymers of glucose.

The body digests complex carbohydrates more slowly than it does simpler sugars because complex carbohydrates have more chemical bonds to break. Endurance athletes will load up on complex carbohydrates for several days before a race to increase the amount of stored energy they can draw on during competition.

Nutritionists agree that most of the carbohydrates in a healthful diet should be in the form of complex carbohydrates, and that it is best to consume only minimal amounts of processed sugars. A **processed food** is one that has undergone extensive refinement and, in doing so, has been stripped of much of its nutritive value. For example, unrefined raw brown sugar is made from the juice of the sugar cane plant and contains the minerals and nutrients found in the plant. Processing brown sugar to produce refined brown sugar results in the loss of these vitamins and minerals.

Fiber is an important part of a healthy diet. Also called *roughage,* fiber is composed mainly of those complex carbohydrates that humans cannot digest. For this reason, dietary fiber is passed into the large intestine, where some

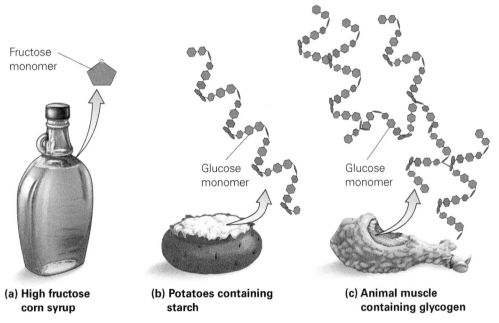

Fructose monomer

Glucose monomer

Glucose monomer

(a) High fructose corn syrup

(b) Potatoes containing starch

(c) Animal muscle containing glycogen

Figure 3.1 Dietary carbohydrates. (a) The simple sugar fructose is found in corn syrup. Complex carbohydrates include (b) starch in potatoes and other vegetables and (c) glycogen found in animal muscle.

of it is digested by bacteria, and the remainder gives bulk to the feces. Foods that have not been stripped of their nutrition by processing are called **whole foods.** Whole grains, beans, and many fruits and vegetables are good sources of dietary fiber.

Although fiber is not a nutrient because it is not absorbed by the body, it is still an important part of a healthful diet. Fiber helps maintain healthy cholesterol levels. Fiber may also decrease your risk of various cancers.

One type of dietary supplement you can find on the shelves of grocery stores is fiber bars. While fiber bars can be rich in fiber, they are also often rich in processed sugars. In addition, many of these bars contain chocolate chips or other sugary sweeteners. Nutritionists recommend that such highly processed, sugar-rich foods occupy a very small portion of your diet because they contain more calories than nutrition.

Proteins as Nutrients. Protein-rich foods include beef, poultry, fish, beans, eggs, nuts, and dairy products such as milk, yogurt, and cheese.

Proteins are composed of amino acids, which differ from each other based on their side groups (Chapter 2). Amino acids are bonded to each other in an infinite variety of combinations to produce a diverse array of proteins with many different functions.

Your body is able to synthesize many of the commonly occurring amino acids. Those your body cannot synthesize are called **essential amino acids** and must be supplied by the foods you eat. **Complete proteins** contain all the essential amino acids your body needs. Proteins obtained by eating meat are more likely to be complete than are those obtained by eating plants; plant proteins can often be missing one or more essential amino acids (**Figure 3.2**).

In the past, there was some concern that vegetarians might be at risk for deficiencies in certain amino acids. However, scientific studies have shown that there is little cause for concern. If a vegetarian's diet is rich in a wide variety of plant-based foods, the body will have little trouble obtaining all the amino acids it needs to build proteins.

Stop & Stretch Many cultures have characteristic meals that include protein from two different sources. For example, beans and rice, eggs and toast, and cereal and milk are common food pairings. Why might this practice of pairing different foods have evolved?

People looking to gain muscle mass will sometimes supplement their diet with protein powders that can be mixed with milk or water to help rebuild protein-rich muscle after a strenuous workout. However, extra protein does not

Figure 3.2 Essential amino acids. Essential amino acids, such as lysine and valine, cannot be synthesized by the body and must be obtained from the diet. Some proteins are high in one amino acid but low in another. Eating a wide variety of proteins thus ensures that all the necessary amino acids are available for growth, development, and maintenance.

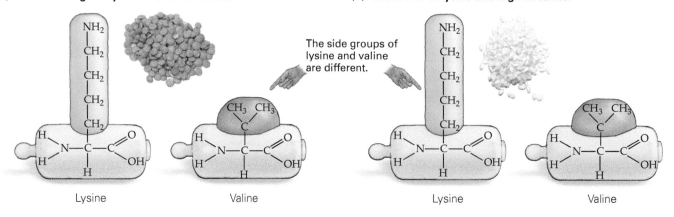

(a) Lentils are high in lysine and low in valine.

(b) Rice is low in lysine and high in valine.

The side groups of lysine and valine are different.

Lysine   Valine   Lysine   Valine

always lead to extra muscle. Like any nutrient, excess protein is stored as fat. So if you eat too much protein, you will end up increasing body fat. In addition, diets that are too rich in protein can lead to health problems such as bone loss and kidney damage. If you are looking to build muscle, the most sensible thing to do is to combine your workout regimen with a healthful diet. Nutritionists recommend that a 100-pound person eat around 36 grams of protein a day, a 150-pound person around 54 grams of protein, and a 200-pound person around 72 grams of protein. Since the average chicken breast contains close to 50 grams of protein, a slice of cheese around 7 grams of protein, and one cup of broccoli around 5—you can see that it is easy to obtain all the protein you need through your diet. Even the most active people, endurance athletes during training, need a maximum of 50% more protein, still easily obtained through whole foods.

Fats as Nutrients. The body uses fat, a type of lipid, as a source of energy. Gram for gram, fat contains a little more than twice as much energy as carbohydrates or protein. This energy is stored in the chemical bonds of the fat molecule.

Foods that are rich in fat include meat, milk, cheese, vegetable oils, and nuts. Muscle is often surrounded by stored fat, but some animals store fat throughout muscle, leading to the marbled appearance of some red meat. Other animals—chickens, for example—store fat on the surface of the muscle, making it easy to remove for cooking (Figure 3.3). Most mammals, including humans, store fat just below the skin to help cushion and protect vital organs, to insulate the body from cold weather, and to store energy in case of famine. Some scientists believe that prehistoric humans often faced times of famine and may have evolved to crave fat.

Fats consist of a glycerol molecule with hydrogen- and carbon-rich fatty acid tails attached (Chapter 2). Your body can synthesize most of the fatty acids it requires. Those that cannot be synthesized are called **essential fatty acids.** Like essential amino acids, essential fatty acids must be obtained from the diet. Omega-3 and omega-6 fatty acids are essential fatty acids that can be obtained by eating fish. These fatty acids are thought to help protect against heart disease. Nutritionists recommend that people eat about 12 ounces of fish every week. Some people who aren't eating that much fish supplement their diets with fish oil capsules. However, fish are a good source of vitamin D and the mineral selenium, while fish oil capsules are not.

The fatty acid tails of a fat molecule can differ in the number and placement of double bonds (Figure 3.4a on the next page). When the carbons of a fatty acid are bound to as many hydrogens as possible, the fat is said to be a **saturated fat** (saturated in hydrogens). When there are carbon-to-carbon double bonds, the fat is not saturated in hydrogens, and it is therefore an **unsaturated fat** (Figure 3.4b). The more double bonds there are, the higher the degree of unsaturation will be. When it contains many unsaturated carbons, the fat is referred to as **polyunsaturated.** The double bonds in unsaturated fats make the structures kink instead of lying flat. This form prevents the adjacent fat molecules from packing tightly together, so unsaturated fat tends to be liquid at room temperature. Cooking oil is an example of an unsaturated fat. Unsaturated fats are more likely to come from plant sources, while the fats found in animals are typically saturated. Saturated fats, with their absence of carbon-to-carbon double bonds, pack tightly together to make a solid structure. This is why saturated fats, such as butter, are solid at room temperature.

Commercial food manufacturers sometimes add hydrogen atoms to unsaturated fats by combining hydrogen gas with vegetable oils under pressure. This process, called **hydrogenation,** increases the level of saturation of a fat. This process solidifies liquid oils, thereby making food seem less greasy and extending their shelf life. Margarine is vegetable oil that has undergone hydrogenation.

(a) Fat within muscle

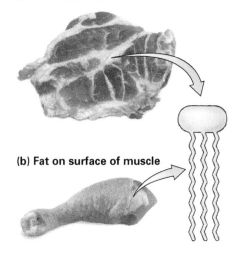

(b) Fat on surface of muscle

Figure 3.3 **Fat storage.** Fat can be intertwined with muscle tissue, as seen in this marbled piece of beef (a), or it can lie on the surface, as seen on this chicken leg (b).

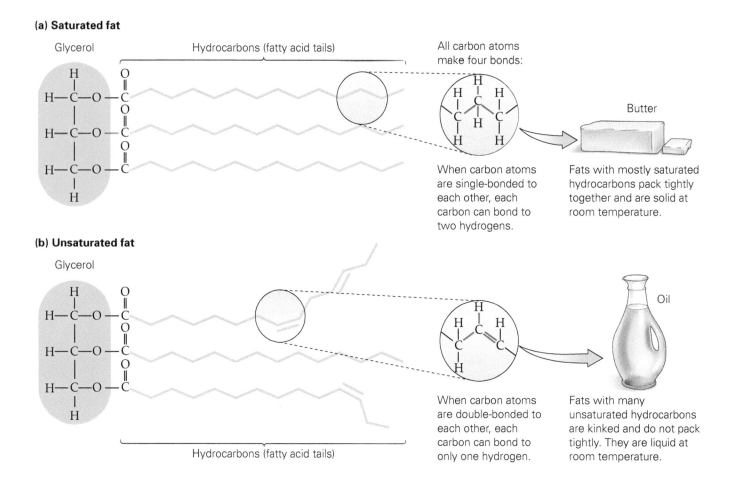

(a) Saturated fat

(b) Unsaturated fat

**Figure 3.4 Saturated and unsatu-
rated fats.** The types of bonds formed
determine whether a fat will be (a) solid or
(b) liquid at room temperature.
Visualize This: Which of the fatty acid
tails in this figure is polyunsaturated?

Trans fats are produced by incomplete hydrogenation, which also changes
the structure of the fatty acid tails in the fat so that, even though there are
carbon-carbon double bonds, the fatty acids are flat and not kinked.

In contrast to most fats, trans fats are not required or beneficial. Fast food
along with many of the energy bars, protein bars, fiber bars, and other nutrition bars on the market contain trans fats.

While definitive studies are currently under way, the potential health risks
of consuming foods rich in trans-fatty acids, common in fast foods, include
increased risk of clogged arteries, heart disease, and diabetes. Because fat
contains more stored energy per gram than carbohydrate and protein do, and
because excess fat intake is associated with several diseases, nutritionists recommend that you limit the amount of fat in your diet.

Micronutrients

Nutrients that are essential in minute amounts, such as vitamins and minerals,
are called **micronutrients**. They are neither destroyed by the body during use
nor burned for energy.

Vitamins. Vitamins are organic substances, most of which the body cannot
synthesize. Most vitamins function as **coenzymes**, or molecules that help enzymes, and thus speed up the body's chemical reactions. When a vitamin is not
present in sufficient quantities, deficiencies can affect every cell in the body because many different enzymes, all requiring the same vitamin, are involved in
numerous different bodily functions. (The structure and function of enzymes
will be covered in detail in Chapter 4.) Vitamins also help with the absorption

of other nutrients; for example, vitamin C increases the absorption of iron from the intestine. Some vitamins may even help protect the body against cancer and heart disease and slow the aging process.

Vitamin D, also called calcitriol, is the only vitamin your cells can synthesize. Because sunlight is required for synthesis, people living in cold climates can develop deficiencies in vitamin D, and many health-care providers recommend vitamin D supplementation.

All other vitamins must be supplied by the foods you eat. Many vitamins, such as the B vitamins and vitamin C, are water soluble, so boiling causes them to leach out into the water—this is why fresh vegetables are more nutritious than cooked ones. Steaming vegetables or using the vitamin-rich broth of canned vegetables when making soup helps preserve the vitamin content. Because water-soluble vitamins are not stored by the body, a lack of water-soluble vitamins is more likely than a lack of fat-soluble vitamins to be the cause of dietary deficiencies. Vitamins A, D, E, and K are fat soluble and build up in stored fat; allowing an excess of these vitamins to accumulate in the body can be toxic. **Table 3.1** lists some vitamins and their roles in the body.

Many people take a daily vitamin supplement, and while this does not hurt, it is better, when possible, to obtain vitamins through eating whole foods.

TABLE 3.1

Vitamins. Water-soluble and fat-soluble vitamins.			
Water-Soluble Vitamins	• Small organic molecules • Will dissolve in water • Cannot be synthesized by body • Supplements packaged as pressed tablets • Excesses usually not a problem because water-soluble vitamins are excreted in urine, not stored 		
Vitamin	**Sources**	**Functions**	**Effects of Deficiency**
Thiamin (B$_1$)	Pork, whole grains, leafy green vegetables	Required component of many enzymes	Water retention and heart failure
Riboflavin (B$_2$)	Milk, whole grains, leafy green vegetables	Required component of many enzymes	Skin lesions
Folic acid	Dark green vegetables, nuts, legumes (dried beans, peas, and lentils), whole grains	Required component of many enzymes	Neural tube defects, anemia, and gastrointestinal problems
B$_{12}$	Chicken, fish, red meat, dairy	Required component of many enzymes	Anemia and impaired nerve function
B$_6$	Red meat, poultry, fish, spinach, potatoes, and tomatoes	Required component of many enzymes	Anemia, nerve disorders, and muscular disorders
Pantothenic acid	Meat, vegetables, grains	Required component of many enzymes	Fatigue, numbness, headaches, and nausea
Biotin	Legumes, egg yolk	Required component of many enzymes	Dermatitis, sore tongue, and anemia
C	Citrus fruits, strawberries, tomatoes, broccoli, cabbage, green pepper	Collagen synthesis; improves iron absorption	Scurvy and poor wound healing
Niacin (B$_3$)	Nuts, leafy green vegetables, potatoes	Required component of many enzymes	Skin and nervous system damage

(continued)

TABLE 3.1 *(continued)*

Vitamins. Water-soluble and fat-soluble vitamins.

| Fat-Soluble Vitamins | • Small organic molecules
• Will not dissolve in water
• Cannot be synthesized by body (except vitamin D)
• Supplements packaged as oily gel caps
• Excesses can cause problems since fat-soluble vitamins are not excreted readily |  |

Vitamin	Sources	Functions	Effects of Deficiency	Effects of Excess
A	Leafy green and yellow vegetables, liver, egg yolk	Component of eye pigment	Night blindness, scaly skin, skin sores, and blindness	Drowsiness, headache, hair loss, abdominal pain, and bone pain
D	Milk, egg yolk	Helps calcium be absorbed and increases bone growth	Bone deformities	Kidney damage, diarrhea, and vomiting
E	Dark green vegetables, nuts, legumes, whole grains	Required component of many enzymes	Neural tube defects, anemia, and gastrointestinal problems	Fatigue, weakness, nausea, headache, blurred vision, and diarrhea
K	Leafy green vegetables, cabbage, cauliflower	Helps blood clot	Bruising, abnormal clotting, and severe bleeding	Liver damage and anemia

Instead of taking a vitamin C supplement, it would be better to eat an orange because the orange, in addition to providing vitamin C, provides beta carotene (an antioxidant), calcium, fiber, and other nutrients. Vitamin C tablet supplements lack these other healthy components.

Minerals. Minerals are substances that do not contain carbon but are essential for many cell functions. Because they lack carbon, minerals are said to be inorganic. They are important for proper fluid balance, in muscle contraction and conduction of nerve impulses, and for building bones and teeth. Calcium, chloride, magnesium, phosphorus, potassium, sodium, and sulfur are all minerals. Like some vitamins, minerals are water soluble and can leach out into the water during boiling. Also like vitamins, minerals are not synthesized in the body and must be supplied through your diet. **Table 3.2** lists the various functions of minerals that your body requires and describes what happens when there is a deficiency or an excess in certain minerals.

Calcium is one of the more commonly supplemented minerals. Your body needs calcium to help blood clot, muscles contract, nerves fire, and maintain healthy bone structure. When dietary calcium is low, calcium is removed from the bones, weakening them. If you are not getting enough calcium in your diet, around 1000 mg/day, many health-care providers will recommend calcium supplements. Eight ounces of yogurt, 1.5 ounces of cheese, and one glass of milk together contain around 1000 mg of calcium.

Antioxidants

In addition to containing vitamins and minerals, many whole foods contain molecules called **antioxidants** that are thought to play a role in the prevention of many diseases, including cancer. Biologists are currently investigating antioxidants to see whether these substances can slow the aging process. Antioxidants protect cells from damage caused by highly reactive molecules that are

TABLE 3.2

Minerals. The minerals we require and their roles in the body.				
Minerals	• Will dissolve in water • Inorganic elements (do not contain carbon) • Cannot be synthesized by body • Supplements packaged as pressed tablets			

Mineral	Sources	Functions	Effects of Deficiency	Effects of Excess
Calcium	Milk, cheese, dark green vegetables, legumes	Bone strength, blood clotting	Stunted growth, osteoporosis	Kidney stones
Chloride	Table salt, processed foods	Formation of stomach acid	Muscle cramps, reduced appetite, poor growth	High blood pressure
Magnesium	Whole grains, leafy green vegetables, legumes, dairy, nuts	Required component of many enzymes	Muscle cramps	Neurological disturbances
Phosphorus	Dairy, red meat, poultry, grains	Bone and tooth formation	Weakness, bone damage	Impaired ability to absorb nutrients
Potassium	Meats, fruits, vegetables, whole grains	Water balance, muscle function	Muscle weakness	Muscle weakness, paralysis, and heart failure
Sodium	Table salt, processed foods	Water balance, nerve function	Muscle cramps, reduced appetite	High blood pressure
Sulfur	Meat, legumes, milk, eggs	Components of many proteins	None known	None known

generated by normal cell processes. These highly reactive molecules, called free radicals, have an incomplete electron shell, which makes them more chemically reactive than molecules with complete electron shells. Free radicals can damage cell membranes and DNA. Antioxidants inhibit the chemical reactions that involve free radicals and decrease the damage they can do in cells. Antioxidants are abundant in fruits and vegetables, nuts, grains, and some meats. **Table 3.3** (on the next page) describes food sources of common antioxidants.

Many different companies manufacture supplements claiming to be rich in antioxidants, and they even advertise that antioxidants have been shown to decrease risk of heart disease and stroke. While this may be true for people who eat diets rich in antioxidants, those findings have not yet been shown to hold true for people who take antioxidant supplements.

3.2 Transport Across Membranes

Once nutrients arrive at cells, they must traverse the membrane that surrounds cells, the plasma membrane. Molecules must cross the plasma membrane to gain access to the inside of the cell, where they can be used to synthesize cell components or be metabolized to provide energy for the cell. The chemistry of the membrane facilitates the transport of some substances and prevents the transport of others.

An Overview: Membrane Transport

The plasma membrane that surrounds cells is composed of a phospholipid bilayer (Chapter 2). The interior of the bilayer is hydrophobic. Hydrophobic substances can dissolve in the membrane and pass through it more easily than hydrophilic

TABLE 3.3

Antioxidants. Antioxidants are being investigated for their disease-preventing abilities.

Antioxidants	• Present in whole foods • Protect cells from damage caused by free radicals • Thought to have a role in disease prevention

Antioxidant	Source
Beta-carotene	Foods rich in beta-carotene are orange in color; they include carrots, cantaloupe, squash, mangoes, pumpkin, and apricots. Beta-carotene is also found in some leafy green vegetables, such as collard greens, kale, and spinach.
Flavenoid	Cocoa and dark chocolate contain flavenoids.
Lutein	Lutein, which is known to help keep eyes healthy, is also found in leafy green vegetables such as collard greens, kale, and spinach.
Lycopene	Lycopene is a powerful antioxidant found in watermelon, papaya, apricots, guava, and tomatoes.
Selenium	Selenium is a mineral (not an antioxidant) that serves as a coenzyme for many antioxidant enzymes, thereby increasing their effectiveness. Rice, wheat, meats, bread, and Brazil nuts are major sources of dietary selenium.
Vitamin A 	Foods rich in vitamin A include sweet potatoes, liver, milk, carrots, egg yolks, and mozzarella cheese.
Vitamin C	Foods rich in vitamin C include most fruits, vegetables, and meats.
Vitamin E	Vitamin E is found in almonds, many cooking oils, mangoes, broccoli, and nuts.

ones. In this sense, the membrane of the cell is differentially permeable to the transport of molecules, allowing some to pass and blocking others from passing.

Substances that can cross the membrane will do so until the concentration is equal on both sides of the membrane, a condition called equilibrium. Carbon dioxide, water, and oxygen move freely across the membrane. Larger molecules, charged molecules, and ions cannot cross the lipid bilayer on their own. If these substances need to be moved across the membrane, they must move through proteins embedded in the membrane. Proteins in the membrane

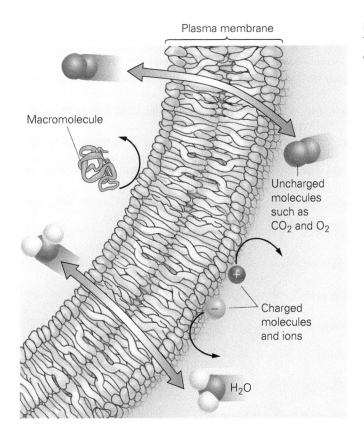

Plasma membrane

Macromolecule

Uncharged molecules such as CO_2 and O_2

Charged molecules and ions

H_2O

Figure 3.5 Transport of substances across membranes. The ability of a substance to cross a membrane is, in part, a function of its size and charge.

can serve as channels to allow such molecules to cross until their concentration is equal on both sides of the membrane. Proteins in the membrane can also move substances across the membrane when more of a substance is required on one side of the membrane than the other (Figure 3.5). *See the next section for A Closer Look at membrane transport.*

A Closer Look:
Membrane Transport

The manner in which a substance is moved across a membrane depends on its particular chemistry and concentration in the cell. Each type of transport has its own characteristics and energy requirements.

Passive Transport: Diffusion, Facilitated Diffusion, and Osmosis

All molecules contain energy that makes them vibrate and bounce against each other, scattering around like billiard balls during a game of pool. In fact, molecules will bounce against each other until they are spread out over all the available area. In other words, molecules will move from their own high concentration to their own low concentration. This movement of molecules from where they are in high concentration to where they are in low concentration is called **diffusion**. During diffusion, the net movement of molecules is from their own high to their own low concentration or *down*

a concentration gradient. This movement does not require an input of outside energy; it is spontaneous. Diffusion will continue until there is equilibrium, at which time no concentration gradient exists, and no net movement of molecules.

Diffusion also occurs in living organisms. When substances diffuse across the plasma membrane, we call the movement **passive transport**. Passive transport is so named because it does not require an input of energy. The structure of the phospholipid bilayer that comprises the plasma membrane prevents many substances from diffusing across it. Only very small, hydrophobic molecules are able to cross the membrane by diffusion. In effect, these molecules dissolve in the membrane to slip from one side of the membrane to the other (Figure 3.6a).

Hydrophilic molecules and charged molecules such as ions are unable simply to diffuse across the hydrophobic core of the membrane. For example, when you have a meal of chicken, rich in charged amino acids, and a green salad, rich

(continued on the next page)

(A Closer Look continued)

in hydrophilic carbohydrates and ions such as calcium (Ca^+), these amino acids, sugars, and ions cannot gain access to the inside of the cell on their own. Instead, these molecules are transported across membranes by proteins embedded in the lipid bilayer. This type of passive transport does not require an input of energy and is called **facilitated diffusion**.

Facilitated diffusion is so named because the specific membrane transport proteins are helping or "facilitating" the diffusion of substances across the plasma membrane

(**Figure 3.6b**). Although transport proteins are used to help the substance move across the plasma membrane, this form of transport is still considered to be diffusion because substances are traveling from high to low concentration.

The movement of water across a membrane is a type of passive transport called **osmosis**. Like other substances, water moves from its own high concentration to its own low concentration. Water can move through protein pores in the membrane, called aquaporins, but even without these, water can still cross the membrane. When an animal cell is placed in a solution of salt water, water leaves the cell, causing the cell

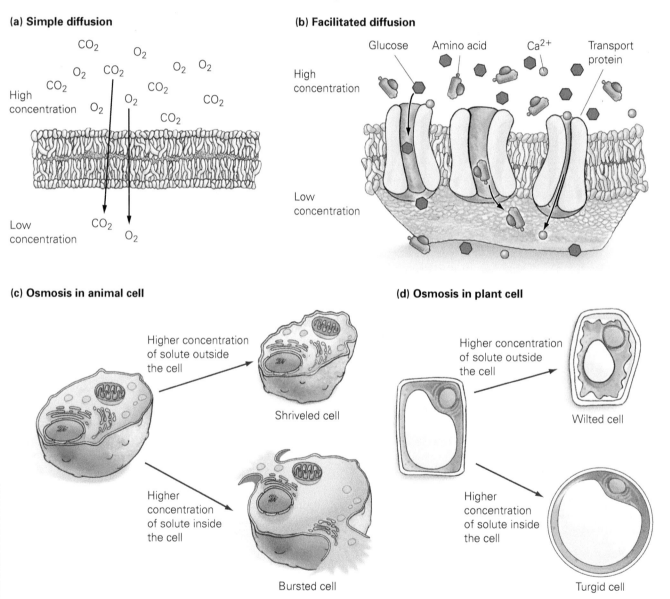

(a) Simple diffusion

(b) Facilitated diffusion

(c) Osmosis in animal cell

(d) Osmosis in plant cell

Figure 3.6 Three types of passive transport. (a) Simple diffusion of molecules across the plasma membrane occurs with the concentration gradient and does not require energy. Small hydrophobic molecules, carbon dioxide, and oxygen can diffuse across the membrane. (b) Facilitated diffusion is the diffusion of molecules assisted by substrate-specific proteins. Molecules move with their concentration gradient, which does not require energy. (c) Osmosis is a special type of diffusion that involves the movement of water in response to a concentration gradient. Water moves toward a region that has more dissolved solute. When water leaves an animal cell, it shrinks. (d) When water leaves a plant cell, the plant wilts instead of shrinks due to the presence of the cell wall.

to shrivel (Figure 3.6c). When an animal cell is placed in a solution with less dissolved solute than the cell, water will enter the cell, and it will expand and may even break open. Likewise, plants that are overfertilized or exposed to road salt wilt because water leaves the cells to equilibrate the concentration of water on either side of the plasma membrane (Figure 3.6d).

Stop & Stretch Suppose an endurance athlete was drinking a sports drink rich in substances that dissociate into ions called electrolytes, sugars, and other dissolved solutes. After drinking this drink, the athlete experiences dehydration. What might be happening?

Active Transport: Pumping Substances Across the Membrane

In some situations, a cell will need to maintain a concentration gradient. For example, nerve cells require a high concentration of certain ions inside the cell to transmit nerve impulses. To maintain this difference in concentration across the membrane requires the input of energy. Think of a hill with a steep incline or grade. Riding your

bike down the hill requires no energy, but riding your bike up the grade requires energy. At the cellular level, that energy is in the form of ATP (adenosine triphosphate, the main source of energy in cell reactions). **Active transport** is transport that uses proteins, powered by ATP, to move substances up a concentration gradient (Figure 3.7).

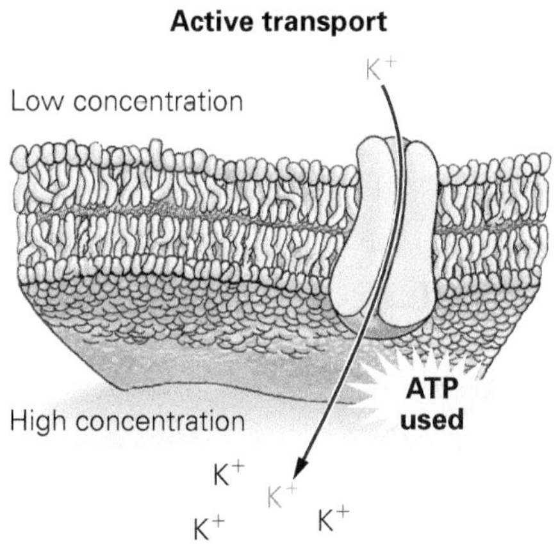

Stop & Stretch Muscle cells require a high internal concentration of potassium (K^+). Suppose you took a potassium supplement that needed to get inside a muscle cell. What type of transport would allow this ion to exist in a higher concentration inside than outside muscle cells?

Exocytosis and Endocytosis: Movement of Large Molecules Across the Membrane

Larger molecules are often too big to diffuse across the membrane or to be transported through a protein, regardless of whether they are hydrophobic or hydrophilic. Instead, they must be moved around inside membrane-bound vesicles (small sacs) that can fuse with membranes. **Exocytosis** (Figure 3.8a) occurs when a membrane-bound vesicle, carrying some substance, fuses with the plasma membrane and releases its contents into the exterior of the cell. **Endocytosis** (Figure 3.8b) occurs when a substance is brought into the cell by a vesicle pinching the plasma membrane inward.

Active transport

K⁺

Low concentration

ATP used

High concentration

K⁺ K⁺ K⁺ K⁺

Figure 3.7
Active transport.
Active transport moves substances against their concentration gradient and requires ATP energy to do so.

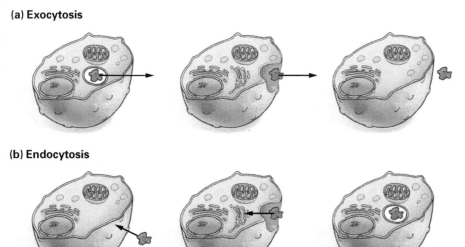

(a) Exocytosis

(b) Endocytosis

Figure 3.8 Movement of large substances. (a) Exocytosis is the movement of substances out of the cell. (b) Endocytosis is the movement of substances into the cell

(a) Protein breakdown

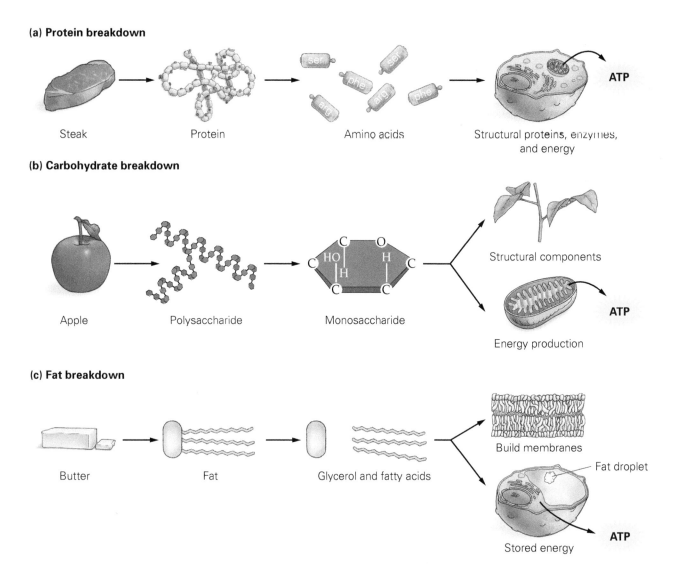

Steak   Protein   Amino acids   Structural proteins, enzymes, and energy   ATP

(b) Carbohydrate breakdown

Apple   Polysaccharide   Monosaccharide   Structural components   Energy production   ATP

(c) Fat breakdown

Butter   Fat   Glycerol and fatty acids   Build membranes   Fat droplet   Stored energy   ATP

Figure 3.9 You are what you eat. Food is digested into component molecules that are used to build cellular structures and generate ATP. (a) Proteins are broken down into amino acids and reassembled into proteins the cell requires and can be sources of energy. (b) Carbohydrates are broken down into sugars that can be reassembled into other sugars the cell requires or uses to produce energy. (c) Fats are broken down and reassembled to be used in membranes, burned to produce energy, or stored as fat.

You Are What You Eat

As you have seen, all the nutrients you consume and dismantle must find some way into your cells so that they can be used for energy and to build cellular components. Knowing all this should give heightened meaning to the phrase, "You are what you eat" (Figure 3.9). Because you are what you eat, it is important to eat a wide variety of healthy whole foods (Figure 3.10). If you do so, there is usually no need to supplement. In rare cases, it might make sense to supplement. For example, some vegetarians are deficient in vitamin B$_{12}$ or iron, pregnant women are advised to take folic acid to help prevent spinal cord anomalies in their babies, and people living in northern climates might benefit from a vitamin D supplement.

Before a prescription or nonprescription drug is released to the public, the FDA requires scientific testing to prove its safety and effectiveness. However, the same is not true of dietary supplements. Therefore, the claims made on the bottles and packages of such products have not necessarily been proven. This is why you will find the following asterisked footnote: "These statements have not been evaluated by the Food and Drug Administration. This product is not intended to diagnose, treat, cure, or prevent any disease."

As with any product that makes claims about its benefits, you should carefully evaluate the claims before deciding to use the product. Use what you know about the process of science and the importance of well-controlled studies in

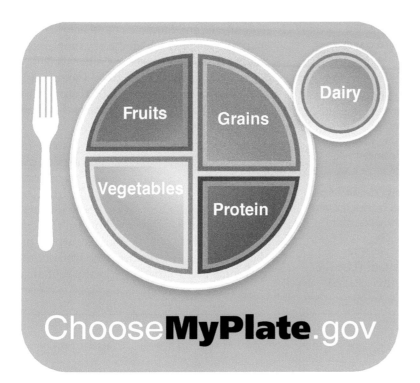

Figure 3.10 USDA MyPlate. The USDA recently replaced the MyPyramid icon with the MyPlate icon. MyPlate illustrates the proportions of each food group that a healthy meal should consist of. Along with the new icon comes advice to make half of your plate fruits and vegetables and to make at least half of your grains whole grains.

your evaluation. If you don't have the time or desire to do the research yourself, consult your physician or a trustworthy website. (Use the guidelines presented in Chapter 1.) The FDA also keeps a record on their website of dietary supplements that are under review for causing adverse effects.

Why pop high doses of expensive pills, powders, and potions when a healthy diet will get you all the nutrients you need, without the processing and costs?

SAVVY READER

The Açaí Bandwagon

It is often difficult to find trustworthy, research-based information about dietary supplements. This text was excerpted from a June 2009 Wellness Alert from the U.C. Berkeley Guide to Dietary Supplements (http://www.wellnessletter.com/html/ds/dsAcai.php).

In the nutraceutical or nutritional supplements market, there is never any shortage of bandwagons. One of the loudest and largest these days is the açaí bandwagon. Harvested from a Brazilian palm, açaí (ah-SAH-ee) berries are a dietary staple in Brazil and have also been used medicinally by Amazonian tribes. Açaí juice was introduced in the U.S. in 2001, and there are now more than 50 new food and drink products containing açaí. As a juice, pulp, powder, or capsule, it is marketed as a magic path to weight loss, a wrinkle remover, a way to cleanse the body of "toxins," and indeed just a plain old miracle cure. It is often combined with other

ingredients, such as glucosamine, so that the claims for benefits multiply exponentially.

Since açaí came on the market there have been a few studies pointing to potential benefits. Like many other fruits, açaí berries are high in antioxidants (molecules that quell cell-damaging free radicals) and other interesting compounds. But these were lab studies, and the results may not apply to humans. There is no scientific basis for weight-loss claims or any other health claims for açaí. The term "antioxidant" has become a sales tool.

Consumer protection groups such as the Center for Science in the Public Interest (CSPI) and the Better Business Bureau (BBB) have now come out against açaí marketers. "If Bernard Madoff were in the food business," said a CSPI nutritionist, "he'd be offering 'free' trials of açaí-based weight-loss products." On-line ads regularly promise a free trial, saying that all

(continued on the next page)

you have to pay is shipping and handling. The catch is that you must supply your credit card number, and you'll automatically be signed up for $50 monthly shipments that will prove hard to cancel.

We urge you not to give your credit card number to anybody selling açaí products. Hundreds of complaints have been registered, and you may never get your money back. Beware of web-sites warning you of açaí scams—far from helping you get your money back, most turn out to be just sales pitches for more açaí.

There is no magic berry for weight loss or good health. Açaí berries are no doubt a good food, like other berries, but why pay a fortune for them or supplements containing them?

1. What are some indicators that this website might be a good source of information? (Use the checklist in Chapter 1 to evaluate this article.)

2. This article, and hundreds of other articles summarizing the research on dietary supplements, is put out by the University of California Berkeley. Should you be more skeptical of claims made by a university putting together a database of resources for the public or a private company that manufactures an herbal supplement? Justify your answer.

3. Many product websites will include some data that seem to support their claims. Private companies can hire their own scientists to perform studies that often have results that differ from those of government and university-sponsored scientists. Would you be more skeptical of results produced by scientists hired by the company whose product they are testing or scientists who work for the government or a university? Justify your answer.

Chapter Review

Learning Outcomes

MasteringBIOLOGY®

Go to the Study Area at www.masteringbiology.com for practice quizzes, myeBook, BioFlix™ 3-D animations, MP3Tutor sessions, videos, current events, and more.

LO1 Describe the role of nutrients in the body (Section 3.1).

- Nutrients provide structural units and energy for cells (p. 56).

LO2 Describe the function of water in the body (Section 3.1).

- Water is an important dietary constituent that helps dissolve and eliminate wastes and maintain blood pressure and body temperature (p. 56).

LO3 Describe the major dietary macronutrients and discuss the functions of each (Section 3.1).

- Macronutrients are required in large amounts for proper growth and development. Macronutrients include carbohydrates, proteins, and fats. All of these molecules are composed of subunits that can be broken down for use by the cell (pp. 56–60).

LO4 List the major dietary micronutrients and describe their functions (Section 3.1).

- Micronutrients are dietary substances required in minute amounts for proper growth and development; they include vitamins and minerals (p. 60).

- Vitamins are organic substances, most of which the body cannot synthesize. Many vitamins serve as coenzymes to help enzymes function properly (pp. 60–61).
- Minerals are inorganic substances essential for many cell functions (p. 62).

LO5 Describe the structure and function of the plasma membrane (Section 3.2).

- To gain access to cells, nutrients move across the plasma membrane, which functions as a semipermeable barrier that allows some substances to pass and prevents others from crossing (p. 63).
- The plasma membrane is composed of two layers of phospholipids, in which are embedded proteins and cholesterol (pp. 63–65).

LO6 Distinguish between passive transport and active transport (Section 3.2).

- Passive transport mechanisms include simple diffusion and facilitated diffusion (diffusion through proteins). Passive transport always moves substances with their concentration gradient and does not require energy (pp. 65–67).

- Osmosis, the diffusion of water across a membrane, can involve the movement of water through protein pores in the membrane (pp. 66–67).
- Active transport is an energy-requiring process that requires proteins in cell membranes to move substances against their concentration gradients (p. 67).

L07 **Describe the processes of endocytosis and exocytosis (Section 3.2).**

- Larger molecules move into (endocytosis) and out (exocytosis) of cells enclosed in membrane-bound vesicles (p. 67).

Roots to Remember

The following roots of words come mainly from Latin and Greek and will help you to decipher terms:

endo-   means inside. Chapter term: endocytosis

exo-   means outside. Chapter term: exocytosis

osmo-   refers to water. Chapter term: osmosis

Learning the Basics

1. **LO1** What are the two main functions of nutrients?
2. **LO6** List three common cellular substances that can pass through cell membranes unaided.
3. **LO3** Macronutrients _____.

    A. include carbohydrates and vitamins; B. should comprise a small percentage of a healthful diet; C. are essential in minute amounts to help enzymes function; D. include carbohydrates, fats, and proteins; E. are synthesized by cells and not necessary to obtain from the diet.

4. **LO2** Which of the following is not a function of water?

    A. dispersing nutrients throughout the body; B. helping the body eliminate digestive wastes; C. helping to regulate body temperature; D. helping to regulate blood pressure.

5. **LO4** Micronutrients _____.

    A. include vitamins and carbohydrates; B. are not metabolized to produce energy; C. contain more energy than fatty acids; D. can be synthesized by most cells.

6. **LO5** The main constituents of the plasma membrane are _____.

    A. carbohydrates and lipids; B. proteins and phospholipids; C. fats and carbohydrates; D. fatty acids and nucleic acids.

7. **LO6** A substance moving across a membrane against a concentration gradient is moving by _____.

    A. passive transport; B. osmosis; C. facilitated diffusion; D. active transport; E. diffusion.

8. **LO6** A cell that is placed in salty seawater will _____.

    A. take sodium and chloride ions in by diffusion; B. move water out of the cell by active transport; C. use facilitated diffusion to break apart the sodium and chloride ions; D. lose water to the outside of the cell via osmosis.

9. **LO6 LO7** Which of the following forms of membrane transport require specific membrane proteins?

    A. diffusion; B. exocytosis; C. facilitated diffusion; D. active transport; E. facilitated diffusion and active transport.

10. **LO6** Water crosses cell membranes _____.

    A. by active transport; B. through protein pores called aquaporins; C. against its concentration gradient; D. in plant cells but not in animal cells.

Analyzing and Applying the Basics

1. **LO1 LO3** A friend of yours does not want to eat meat, so instead she consumes protein shakes that she buys at a nutrition store. Can you think of a dietary strategy that would allow her to be a vegetarian while not consuming protein shakes?

2. **LO1 LO2** The energy drink Red Bull is banned in several European countries due in part to concerns over its high caffeine content. Excess caffeine content can cause anxiety, heart palpitations, irritability, and difficulty sleeping. Many people use high-caffeine energy drinks as a mixer for alcohol, a practice that worries many health care practitioners. Why might mixing alcohol with energy drinks have even more negative health consequences than mixing alcohol with more conventional mixes such as pop, water, or juice?

3. **LO4** A vitamin water manufacturer claims that their product has all the vitamin C of an orange. What substances would you not be getting from the vitamin water that you would by eating an orange?

Connecting the Science

1. Select one fitness water or sports energy drink and research whether the claims made on its label are backed up by scientific evidence.

2. The New York City Board of Health recently adopted the nation's first major municipal ban on the use of trans fats in restaurant cooking. Do you think state health regulators should also ban trans fats in commercially prepared food like cookies and potato chips?

Answers to **Stop & Stretch, Visualize This, Savvy Reader,** and **Chapter Review** questions can be found in the **Answers** section at the back of the book

Fat: How Much Is Right for You?

Enzymes, Metabolism, and Cellular Respiration

Media images of attractive people tend to show the same body type.

The average college student spends a lot of time thinking about his or her body and ways to make it more attractive. While people come in all shapes and sizes, attractiveness tends to be more narrowly defined by the images of men and women we see in the popular media. Nearly all media images equate attractiveness and desirability with a limited range of body types.

For men, the ideal includes a tall, broad-shouldered, muscular physique with so little body fat that every muscle is visible. The standards for female beauty are equally unforgiving and include small hips, long and thin limbs, large breasts, and no body fat.

Because these ideals are difficult, if not impossible, to reach, we can end up feeling bad about our bodies and confused about what a healthy normal body looks like. How do you know if your body weight is a healthful one? Should you be dieting or trying to gain weight, or are you fine just as you are? This chapter will try to help you answer these questions.

Males are tall, thin, broad-shouldered, and muscular, while females are thin and large breasted.

For many people, trying to attain a body type that differs from their own can lead to anorexia ...

or obesity.

LEARNING OUTCOMES

LO1 Define the term metabolism.

LO2 Describe the structure and function of enzymes.

LO3 Explain how enzymes decrease a reaction's activation energy barrier.

LO4 List several reasons why metabolic rates differ between people.

LO5 Describe the structure and function of ATP.

LO6 Describe the process of cellular respiration from the breakdown of glucose through the production of ATP.

LO7 Explain how proteins and fats are broken down during cellular respiration.

LO8 Compare and contrast aerobic and anaerobic respiration.

LO9 Outline the health risks that occur with being overweight and underweight.

4.1 Enzymes and Metabolism

It is important to eat a well-balanced diet rich in unprocessed foods and the right amount of food (Chapter 3). All food, whether carbohydrate, protein, or fat, can be turned into fat when too much is consumed. In this manner, energy stored in the chemical bonds of food is converted into fat and stored for later use.

The amount of fat that a given individual will store depends partly on how quickly or slowly he or she breaks down food molecules into their component parts. **Metabolism** is a general term used to describe all of the chemical reactions occurring in the body.

Enzymes

All metabolic reactions are regulated by proteins called **enzymes** that speed up, or **catalyze,** the rate of biological reactions. Enzymes can help break down or build up substances or build more complex substances from simpler ones. The enzymes that help your body break down the foods you ingest liberate the energy stored in the food's chemical bonds. Enzymes are usually named for the reaction they catalyze and end in the suffix -*ase.* For example, sucrase is the enzyme that breaks down the table sugar sucrose.

To break chemical bonds, molecules must absorb energy from their surroundings, often by absorbing heat. This is why heating chemical reactants will speed up a reaction. However, heating cells to an excessively high temperature can damage or kill them, in part because proteins begin to break down. Enzymes do not require heat to catalyze the body's chemical reactions, so they break chemical bonds without damaging or killing cells. Furthermore, by eliminating the heat energy required to start a chemical reaction, enzymes allow the breakdown of chemical bonds to occur more quickly.

Activation Energy. The energy required to start the metabolic reaction serves as a barrier to catalysis and is called the **activation energy** (**Figure 4.1**). If not for the activation energy barrier, all of the chemical reactions in cells would occur relentlessly, whether the products of the reactions were needed or not. Because most metabolic reactions need to surpass the activation energy barrier before proceeding, they can be regulated by enzymes. In other words, a given chemical reaction will occur only if the correct enzyme is available and active. How do enzymes decrease the activation energy barrier?

Induced Fit. The chemicals that are metabolized by an enzyme-catalyzed reaction are called the enzyme's **substrate.** Enzymes decrease activation energy by binding to their substrate and placing stress on its chemical bonds, decreasing the amount of initial energy required to break the bonds. The region of the enzyme where the substrate binds is called the enzyme's **active site.** Each active site has its own shape and chemical climate. When the substrate binds to the active site, the enzyme changes shape slightly to envelop the substrate. This shape change by the enzyme in response to substrate binding results in stress being placed on the bonds of the substrate. This is called the **induced fit** model of enzyme catalysis. When the enzyme changes shape, it binds to the substrate more tightly, making it easier to break the substrate's chemical bonds. In this manner, the enzyme helps convert the substrate to a reaction product and then resumes its original shape so that it can perform the reaction again (**Figure 4.2**).

Different enzymes catalyze different reactions by a property called **specificity.** The specificity of an enzyme is the result of its shape and the shape of its active site. Different enzymes have unique shapes because they are composed of amino acids in varying sequences. The 20 amino acids, each with its own unique side group, are arranged in distinct numbers and orders for

(a) No enzyme present

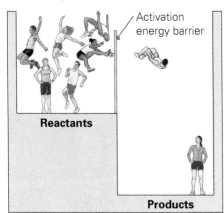

(b) Enzyme present

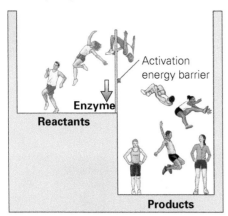

Figure 4.1 Activation energy.
(a) Few high jumpers can make it across the bar when the bar is high. Likewise, few reactants can break bonds to form products without surmounting an activation energy barrier. (b) Lowering the high-jump bar allows more jumpers to clear the bar. Enzymes lower the activation energy barrier in cells.

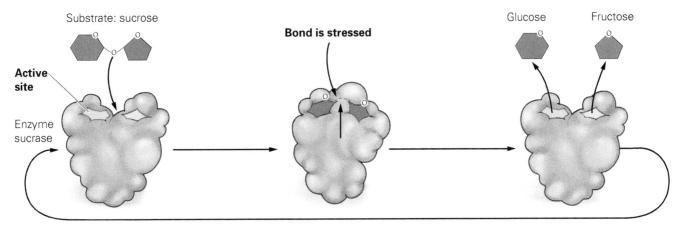

(1) The shape of the substrate matches the shape of the enzyme's active site.

(2) The induced fit model suggests that initial substrate binding to the active site changes the shape of the active site, inducing the substrate to fit even more snugly in the active site, stressing the bonds of the substrate.

(3) The shape change splits the substrate and releases the two subunits. The enzyme is able to perform the reaction again.

Figure 4.2 Enzymes. The enzyme sucrase is cleaving (splitting) the disaccharide sucrose into its monosaccharide subunits, fructose and glucose.
Visualize This: Is the enzyme itself permanently altered by the process of catalysis?

each enzyme, producing enzymes of all shapes and sizes, each with an active site that can bind with its particular substrate. Although an infinite variety of enzymes could be produced, it is quite often the case that different organisms will utilize very similar enzymes, likely due to their evolution from a common ancestor.

Enzymes mediate all of the metabolic reactions occurring in an organism's cells. They even affect the rate of a particular individual's metabolism. This means that two similar-size individuals might need to consume different amounts of food to meet their daily energy requirements.

Calories and Metabolic Rate

Energy is measured in units called calories. A **calorie** is the amount of energy required to raise the temperature of 1 gram of water by 1 degree Celsius (1°C). In scientific literature, energy is usually reported in kilocalories, and 1 kilocalorie equals 1000 calories of energy. However, in physiology—and on nutritional labels—the prefix *kilo-* is dropped, and a kilocalorie is referred to as a **Calorie** (with a capital C). Calories are consumed to supply the body with energy to do work, which includes maintaining body temperature.

Balancing energy intake versus energy output means eating the correct amount of food to maintain health. When foods are eaten, they are broken down into their component subunits. The energy stored in the chemical bonds of food can be used to make a form of energy that the cell can use. When the supply of Calories is greater than the demand, the excess Calories can be stored as fat.

The speed and efficiency of many different enzymes will lead to an overall increase or decrease in the rate at which a person can break down food. Thus, when you say that your metabolism is slow or fast, you are actually referring to the speed at which enzymes catalyze chemical reactions in your body.

A person's **metabolic rate** is a measure of his or her energy use. This rate changes according to the person's activity level. For example, we require less energy when asleep than we do when exercising. The **basal metabolic rate** represents the resting energy use of an awake, resting, but alert person. The average basal metabolic rate is 70 Calories per hour, or 1680 Calories per day. However, this rate varies widely among individuals because many factors

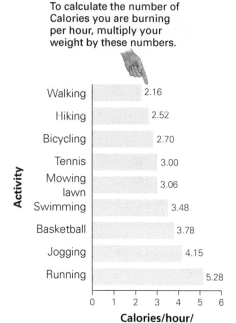

To calculate the number of Calories you are burning per hour, multiply your weight by these numbers.

Activity	Calories/hour/pound of body weight
Walking	2.16
Hiking	2.52
Bicycling	2.70
Tennis	3.00
Mowing lawn	3.06
Swimming	3.48
Basketball	3.78
Jogging	4.15
Running	5.28

Figure 4.3 Energy expenditures for various activities. This bar graph can help you determine how many Calories you burn during certain activities.
Visualize This: How many Calories would a 160-pound person burn in 30 minutes of swimming?

influence each person's basal metabolic rate: exercise habits, body weight, sex, age, and genetics. Overall nutritional status can also affect metabolism. For example, vitamins can function as coenzymes (Chapter 3). A diet lacking in a particular vitamin may lead to slowed metabolism because an enzyme that is missing its vitamin coenzyme may not be able to perform at its optimal rate.

Exercise requires energy, which allows you to consume more Calories without having to store them. As for body weight, a heavy person utilizes more Calories during exercise than a thin person does. Figure 4.3 shows the number of Calories used per hour for different activities, and allows calculation of the different rate of energy expenditure for different sizes of people. Males require more Calories per day than females do because testosterone, a hormone produced in larger quantities by males, increases the rate at which fat breaks down. Men also have more muscle than women, which requires more energy to maintain than fat does.

Age and genetics also play a role in metabolic rate. Two people of the same size and sex, who consume the same number of Calories and exercise the same amount, will not necessarily store the same amount of fat. The rate at which the foods you eat are metabolized slows as you age, and some people are simply born with lower basal metabolic rates. To obtain a rough measure of how many Calories you should consume per day, multiply the weight you wish to maintain by 11 and add the number of Calories you burn during exercise.

The properties of metabolic enzymes, like those of all proteins, are determined by the genes that encode them. Genes that influence a person's rate of fat storage and utilization are passed from parents to children. All of these variables help explain why some people seem to eat and eat and never gain an ounce, while others struggle with their weight for their entire lives.

Stop & Stretch Losing 1 pound of fat requires you to burn 3500 more Calories than you consume. If a person is trying to lose 1 pound per week and he decreases his caloric intake by 300 Calories per day, how many more Calories must he burn each day by exercising to reach his goal?

To be fully metabolized, food is broken down by the digestive system and then transported to individual cells via the bloodstream. Once inside cells, food energy can be converted into chemical energy by the process of cellular respiration.

4.2 Cellular Respiration

Cellular respiration is a series of metabolic reactions that converts the energy stored in chemical bonds of food into energy that cells can use while releasing waste products. Energy is stored in the electrons of chemical bonds, and when bonds are broken in a three-stage process, **adenosine triphosphate,** or **ATP,** is produced. ATP can supply energy to cells because it stores energy obtained from the movement of electrons that originated in food into its own bonds. Before trying to understand cellular respiration, it is important to have a better understanding of this chemical.

Structure and Function of ATP

Structurally, ATP is a nucleotide triphosphate. It contains the nitrogenous base adenine, a sugar, and three phosphates (Figure 4.4). Each phosphate in the series of three is negatively charged. These negative charges repel each other, which contributes to the stored energy in this molecule.

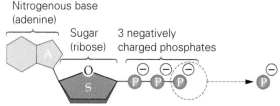

Nitrogenous base (adenine)
Sugar (ribose)
3 negatively charged phosphates

Figure 4.4 The structure of ATP. ATP is a nucleotide (sugar + phosphate + nitrogenous base) with a total of 3 negatively charged phosphates. Releasing a phosphate group from ATP releases energy.

Removal of the terminal phosphate group of ATP releases energy that can be used to perform cellular work. In this manner, ATP behaves much like a coiled spring. To think about this, imagine loading a dart gun. Pushing the dart into the gun requires energy from your arm muscles, and the energy you exert will be stored in the coiled spring inside the dart gun (**Figure 4.5**). When you shoot the dart gun, the energy is released from the gun and used to perform some work—in this case, sending a dart through the air.

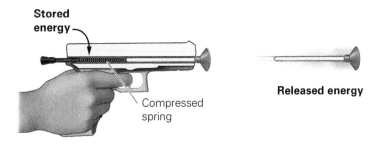

Figure 4.5 Stored energy. A dart gun uses energy stored in the coiled spring and supplied by the arm muscle to perform the work of propelling a dart.

The phosphate group that is removed from ATP can be transferred to another molecule. Thus, ATP can energize other compounds through **phosphorylation,** which means that it transfers a phosphate to another molecule. When a molecule, say an enzyme, needs energy, the phosphate group is transferred from ATP to the enzyme, and the enzyme undergoes a change in shape that allows the enzyme to perform its job. After removal of a phosphate group, ATP becomes adenosine diphosphate (ADP) (**Figure 4.6**). The energy released by the removal of the outermost phosphate of ATP can be used to help cells perform many different kinds of work. ATP helps power mechanical work such as the movement of cells, transport work such as the movement of substances across membranes during active transport, and chemical work such as the making of complex molecules from simpler ones (**Figure 4.7**).

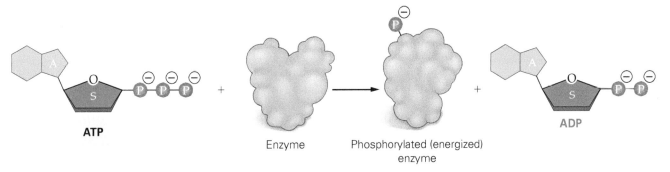

Figure 4.6 Phosphorylation. The terminal phosphate group of an ATP molecule can be transferred to another molecule, in this case an enzyme, to energize it. When ATP loses a phosphate, it becomes ADP.

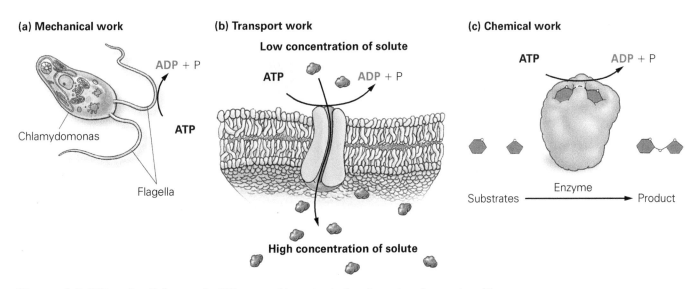

Figure 4.7 ATP and cellular work. ATP powers (a) mechanical work, such as the moving of the flagella of this single-celled green algae; (b) transport work, such as the active transport of a substance across a membrane from its own low to high concentration; and (c) chemical work, such as the enzymatic conversion of substrates to a product.

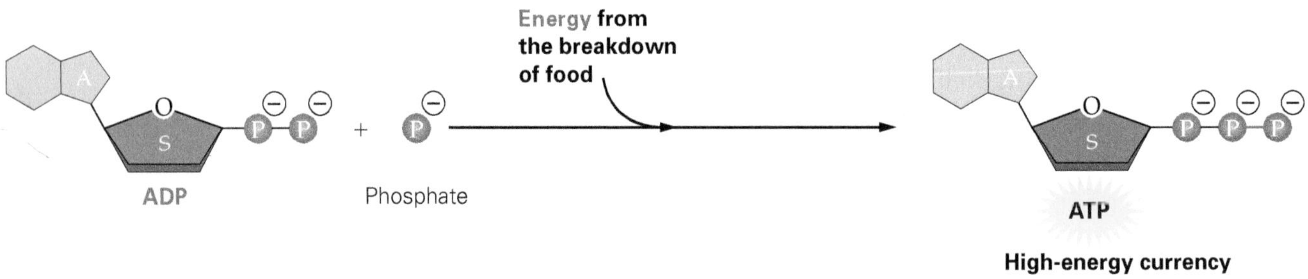

Energy from the breakdown of food

ADP   Phosphate   +   ATP

High-energy currency

Figure 4.8 Regenerating ATP. ATP is regenerated from ADP and phosphate during the process of cellular respiration.

Cells are continuously using ATP. Exhausting the supply of ATP means that more ATP must be regenerated. ATP is synthesized by adding back a phosphate group to ADP during the process of cellular respiration (**Figure 4.8**). During this process, oxygen is consumed, and water and CO_2 are produced. Because some of the steps in cellular respiration require oxygen, they are said to be **aerobic** reactions, and this type of cellular respiration is called **aerobic respiration.**

An Overview: Cellular Respiration

The word *respiration* can also be used to describe breathing. When we breathe, we take oxygen in through our lungs and expel carbon dioxide. The oxygen we breathe in is delivered to cells, which undergo cellular respiration and release carbon dioxide (**Figure 4.9**).

Most foods can be broken down to produce ATP as they are routed through this process. Carbohydrate metabolism begins earliest in the pathway, while proteins and fats enter at later points.

The equation for carbohydrate breakdown is

$$C_6H_{12}O_6 + 6O_2 \rightarrow 6CO_2 + 6H_2O$$

Glucose + Oxygen → Carbon dioxide + Water

Glucose is an energy-rich sugar, but the products of its digestion—carbon dioxide and water—are energy poor. So where does the energy go? The energy released during the conversion of glucose to carbon dioxide and water is used to synthesize ATP. Many of the chemical reactions in this process occur in the mitochondria, organelles that are found in both plant and animal cells. Mitochondria are surrounded by an inner and an outer membrane. The space between the two membranes is called the **intermembrane space.** The semifluid medium inside the mitochondrion is called the matrix (**Figure 4.10a**).

Through a series of complex reactions in the mitochondrion, a glucose molecule breaks apart, and carbon and oxygen are released from the cell as carbon dioxide. Hydrogens from glucose combine with oxygen to produce water (**Figure 4.10b**). Gaining an appreciation for *how* this happens requires a more in-depth look. *See the next section for **A Closer Look** at glycolysis, the citric acid cycle, and electron transport and ATP synthesis.*

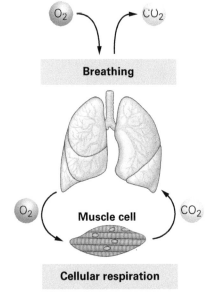

Figure 4.9 Breathing and cellular respiration. When you inhale, you bring oxygen from the atmosphere into your lungs. This oxygen is delivered through the bloodstream to tissues, such as the muscle tissue shown at the bottom of this figure, which use it to drive cellular respiration. The carbon dioxide produced by cellular respiration is released from cells and diffuses into the blood and to the lungs. Carbon dioxide is released from the lungs when you exhale.

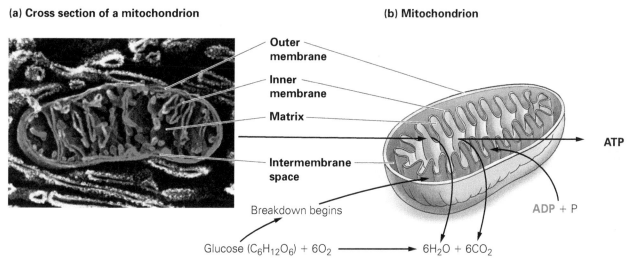

(a) Cross section of a mitochondrion

(b) Mitochondrion

Outer membrane

Inner membrane

Matrix

Intermembrane space

ATP

ADP + P

Breakdown begins

Glucose ($C_6H_{12}O_6$) + $6O_2$ ⟶ $6H_2O$ + $6CO_2$

Figure 4.10 Overview of cellular respiration. (a) Much of this process of cellular respiration takes place within the mitochondrion. (b) The breakdown of glucose by cellular respiration requires oxygen and ADP plus phosphate. The energy stored in the bonds of glucose is harvested to produce ATP (from ADP and P), releasing carbon dioxide and water.

A Closer Look:
Glycolysis, the Citric Acid Cycle, and Electron Transport and ATP Synthesis

Cellular respiration occurs in three stages: glycolysis, the citric acid cycle, and electron transport and ATP synthesis (**Figure 4.11**).

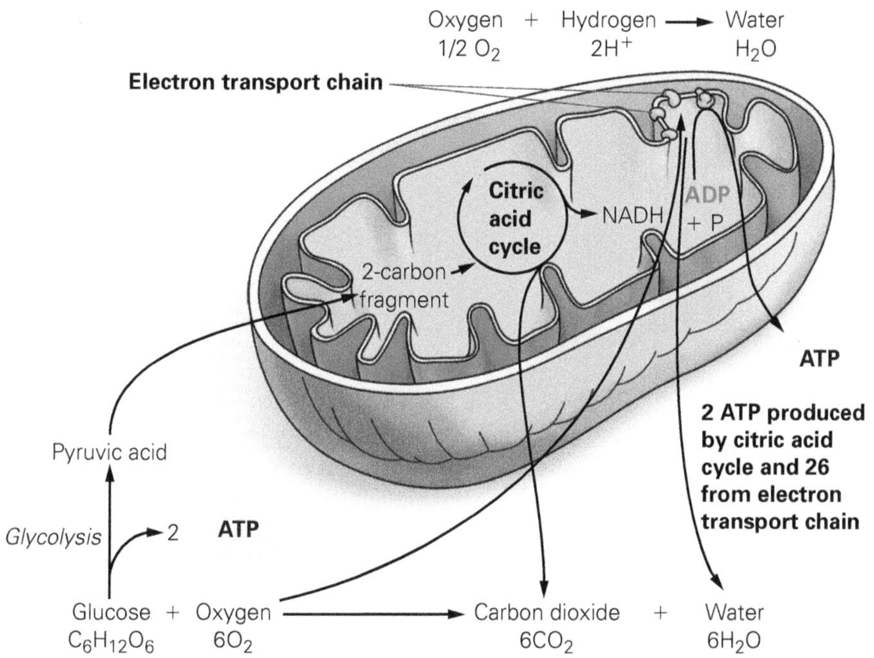

Stage 1: Glycolysis. To harvest energy from glucose, the 6-carbon glucose molecule is first broken down into two 3-carbon **pyruvic acid** molecules (**Figure 4.12** on the next page). This part of the process of cellular respiration actually occurs outside any organelle, in the fluid cytosol. Glycolysis does not require oxygen and produces 2 molecules of ATP.

In addition to producing ATP, glycolysis removes electrons from glucose for use in producing ATP during the final stage of cellular respiration. These electrons do not simply float around in a cell; this would damage the cell. Instead, they are carried by molecules called *electron carriers*. One of the electron carriers utilized by cellular respiration is a chemical called **nicotinamide adenine dinucleotide (NAD⁺).** NAD⁺ picks up 2 hydrogen atoms (along with their electrons) and releases

Oxygen + Hydrogen ⟶ Water
$1/2\,O_2$ $2H^+$ H_2O

Electron transport chain

Citric acid cycle

NADH

ADP + P

2-carbon fragment

ATP

2 ATP produced by citric acid cycle and 26 from electron transport chain

Pyruvic acid

Glycolysis ⟶ 2 **ATP**

Glucose + Oxygen ⟶ Carbon dioxide + Water
$C_6H_{12}O_6$ $6O_2$ $6CO_2$ $6H_2O$

Figure 4.11 A more detailed look at cellular respiration. This figure diagrams the inputs and outputs of cellular respiration.

(continued on the next page)

(A Closer Look continued)

Figure 4.12 Glycolysis. Glycolysis is the enzymatic conversion of glucose into 2 pyruvic acid molecules. The pyruvic acid molecules are further broken down in the mitochondrion. Two ATP and 2 NADH are made during glycolysis.
Visualize This: Why is glycolysis an appropriate name for this process?

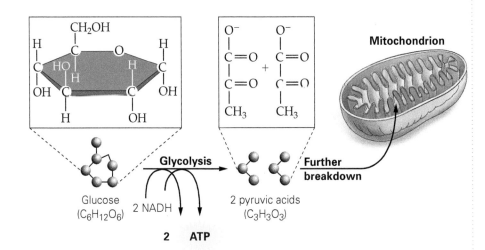

Glucose ($C_6H_{12}O_6$) — Glycolysis — 2 pyruvic acids ($C_3H_3O_3$) — Further breakdown — Mitochondrion

2 NADH   2 ATP

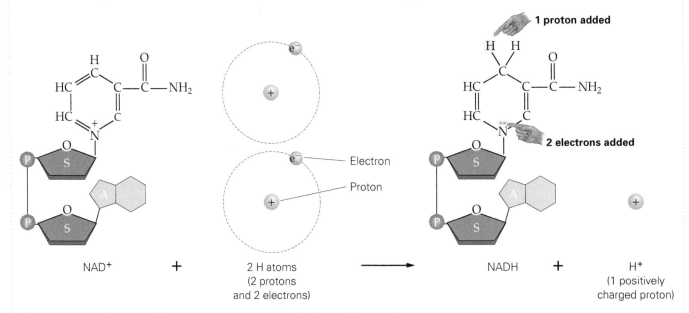

NAD$^+$ +   2 H atoms (2 protons and 2 electrons)   ⟶   NADH   +   H$^+$ (1 positively charged proton)

Electron — Proton — 1 proton added — 2 electrons added

Figure 4.13 Nicotinamide adenine dinucleotide (NAD$^+$). NAD$^+$ can pick up a hydrogen atom along with its electron. Hydrogen atoms are composed of 1 negatively charged electron that circles around the 1 positively charged proton. When the electron carrier NAD$^+$ encounters 2 hydrogen atoms (from food), it utilizes each hydrogen atom's electron and only 1 proton, thus releasing 1 proton.
Visualize This: Why does the number of double bonds in the nitrogenous base change when a proton is added to NAD$^+$ to produce NADH?

1 positively charged hydrogen ion (H$^+$), becoming **NADH** (**Figure 4.13**).

NADH serves as a sort of taxicab for electrons. The empty taxicab (NAD$^+$) picks up electrons. The full taxicab (NADH) carries electrons to their destination, where they are dropped off, and the empty taxicab returns for more electrons (**Figure 4.14**). NADH will deposit its electrons for use in the final step of cellular respiration.

Stage 2: The Citric Acid cycle. After glycolysis, the pyruvic acid is decarboxylated (loses a carbon dioxide

molecule), and the 2-carbon fragment that is left is further metabolized inside the mitochondria.

Once inside the mitochondrion, the energy stored in the remains of glucose is converted into the energy stored in the bonds of ATP. The first stage of this conversion is called the citric acid cycle.

The **citric acid cycle** is a series of reactions catalyzed by 8 different enzymes, located in the matrix of each mitochondrion. This cycle breaks down the remains of a carbohydrate, harvesting its electrons and releasing carbon dioxide into the atmosphere (**Figure 4.15**). These reactions are a cycle because every trip around the pathway regenerates the first reactant, a 4-carbon molecule called oxaloacetate (OAA). OAA is always available to react with carbohydrate fragments entering the citric acid cycle.

Stage 3: Electron Transport and ATP Synthesis. Electrons harvested during the citric acid cycle are also carried by NADH to the final stage in cellular respiration. The **electron transport chain** is a series of proteins embedded in the inner mitochondrial membrane that functions as a sort of conveyer belt for electrons, moving them from one protein to another. The electrons, dropped off by NADH molecules generated during glycolysis and the citric acid cycle, move toward the bottom of the electron transport chain toward the matrix of the mitochondrion, where they combine with oxygen to produce water.

Each time an electron is picked up by a protein or handed off to another protein, the protein moving it changes shape. This shape change allows the movement of hydrogen ions (H⁺) from the matrix of the mitochondrion to the intermembrane space. So, while the proteins in the electron transport chain are moving electrons down the electron transport chain toward oxygen, they are also moving H⁺ ions across the inner mitochondrial membrane and into the intermembrane space. This decreases the concentration of H⁺ ions in the matrix and increases their concentration within the intermembrane space. Whenever a concentration gradient of a molecule exists, molecules will diffuse from an area of high concentration to an area of low concentration (Chapter 3). Because charged ions cannot diffuse across the hydrophobic core of the membrane, they escape through a protein channel in the membrane called **ATP synthase.** This enzyme uses the energy generated by the rushing H⁺ ions to synthesize 26 ATP from ADP and phosphate in the same manner that water rushing through a mechanical turbine can be used to generate electricity (**Figure 4.16** on the next page).

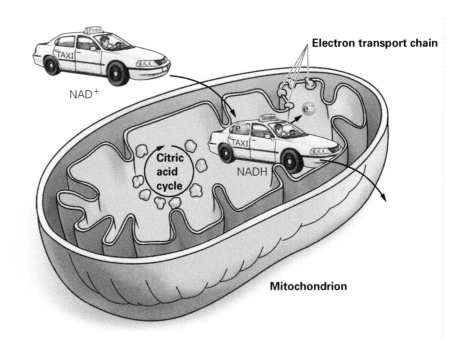

Figure 4.14 Electron carriers. NADH serves as an electron carrier, bringing electrons removed from the original glucose molecule to the electron transport chain. After dropping off its electrons, the electron carrier can be loaded up again and bring more electrons to the electron transport chain.

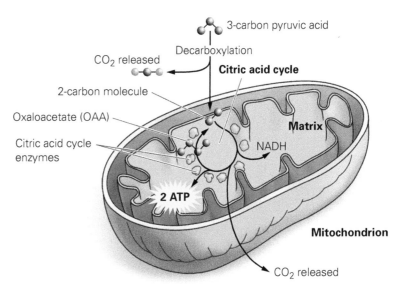

Figure 4.15 The citric acid cycle. The 3-carbon pyruvic acid molecules generated by glycolysis are decarboxylated, leaving a 2-carbon molecule that enters the citric acid cycle within the mitochondrial matrix. The 2-carbon fragment reacts with a 4-carbon OAA molecule and proceeds through a stepwise series of reactions that results in the production of more carbon dioxide and regenerates OAA. NADH and 2 ATP are also produced.

Overall, the two pyruvic acids produced by the breakdown of glucose during glycolysis are converted into carbon dioxide and water. Carbon dioxide is produced when it is removed from the pyruvic acid molecules during the citric acid cycle, and water is formed when oxygen combines with hydrogen ions at the bottom of the electron transport chain.

(continued on the next page)

(A Closer Look continued)

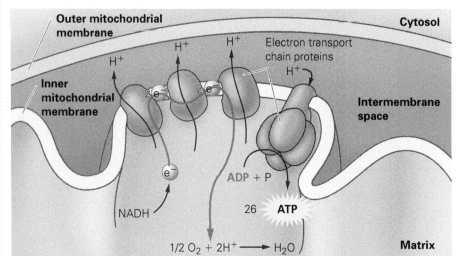

Figure 4.16 The electron transport chain of the inner mitochondrial membrane. NADH brings electrons from the citric acid cycle to the electron transport chain. As electrons move through the proteins of the electron transport chain, hydrogen ions are pumped into the intermembrane space. Hydrogen ions flow back through an ATP synthase protein, which converts ADP to ATP. In this manner, energy from electrons added to the electron transport chain is used to produce ATP.
Visualize This: What is meant by $1/2\ O_2$ in this figure?

Metabolism of Other Nutrients

Metabolism refers to all the chemical reactions that occur in an organism's cells. Most cells can break down not only carbohydrates but also proteins and fats. **Figure 4.17** shows the points of entry during cellular respiration for proteins and fats. Protein is broken down into component amino acids, which are then used to synthesize new proteins. Most organisms can also break down proteins to supply energy. However, this process takes place only when fats or carbohydrates are unavailable. In humans and other animals, the first step in producing energy from the amino acids of a protein is to remove the nitrogen-containing amino group of the amino acid. Amino groups are then converted to a compound called urea, which is excreted in the urine. The carbon, oxygen, and hydrogen remaining after the amino group is removed undergo further breakdown and eventually enter the mitochondria, where they are fed through the citric acid cycle and produce carbon dioxide, water, and ATP. The subunits of fats (glycerol and fatty acids) also go through the citric acid cycle and produce carbon dioxide, water, and ATP. Some cells will break down fat only when carbohydrate supplies are depleted.

Figure 4.17 Metabolism of other macromolecules. Carbohydrates, proteins, and fats can all undergo cellular respiration; they just feed into different parts of the metabolic pathway.

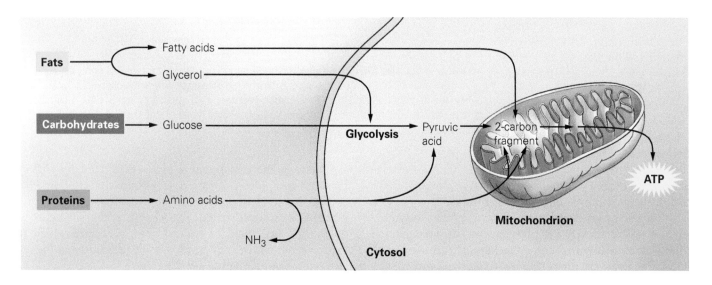

Metabolism Without Oxygen: Anaerobic Respiration and Fermentation

Aerobic cellular respiration is one way for organisms to generate energy. It is also possible for some cells to generate energy in the absence of oxygen, by a metabolic process called **anaerobic respiration.**

Muscle cells normally produce ATP by aerobic respiration. However, oxygen supplies diminish with intense exercise. When muscle cells run low on oxygen, they must get most of their ATP from glycolysis, the only stage in the cellular respiration process that does not require oxygen. When glycolysis happens without aerobic respiration, cells can run low on NAD⁺ because it is converted to NADH during glycolysis. When this happens, cells use a process called **fermentation** to regenerate NAD⁺.

Fermentation cannot, however, be used for very long because one of the by-products of this reaction leads to the buildup of a compound called lactic acid. Lactic acid is produced by the actions of the electron acceptor NADH, which has no place to dump its electrons during fermentation because there is no electron transport chain and no oxygen to accept the electrons. Instead, NADH deposits its electrons by giving them to the pyruvic acid produced by glycolysis (**Figure 4.18a**). Lactic acid is transported to the liver, where liver cells use oxygen to convert it back to pyruvic acid.

This requirement for oxygen, to convert lactic acid to pyruvic acid, explains why you continue to breathe heavily even after you have stopped working out. Your body needs to supply oxygen to your liver for this conversion, sometimes referred to as "paying back your oxygen debt." The accumulation of lactic acid also explains the phenomenon called "hitting the wall." Anyone who has ever felt as though their legs were turning to wood while running or biking knows this feeling. When your muscles are producing lactic acid by fermentation for a long time, the oxygen debt becomes too large, and muscles shut down until the rate of oxygen supply outpaces the rate of oxygen utilization, restoring proper feeling to your legs.

> **Stop & Stretch** Aerobic exercise (such as running, swimming, and biking) strengthens the heart, allowing it to pump more blood per beat. Given the effects of anaerobic respiration on the body, how does aerobic exercise increase stamina?

Some fungi and bacteria also produce lactic acid during fermentation. Certain microbes placed in an anaerobic environment transform the sugars in milk into yogurt, sour cream, and cheese. It is the lactic acid present in these dairy products that gives them their sharp or sour flavor. Yeast in an anaerobic environment produces ethyl alcohol instead of lactic acid. Ethyl alcohol is formed when carbon dioxide is removed from pyruvic acid (**Figure 4.18b**). The yeast used to help make beer and wine converts sugars present in grains (beer) or grapes (wine) into ethyl alcohol and carbon dioxide. Carbon dioxide, produced by baker's yeast, helps bread to rise.

4.3 Body Fat and Health

You have seen how cells can use food to make energy. When more energy is consumed than is utilized, the excess is stored as fat. A clear understanding of how much body fat is healthful is hard to come by because cultural and biological definitions of the term "overweight" differ markedly. Cultural

(a) Human muscle

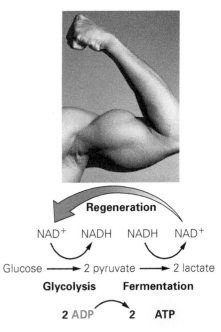

(b) Yeast

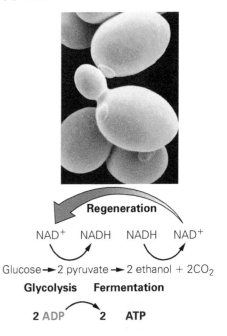

Figure 4.18 Metabolism without oxygen. Glycolysis can be followed by (a) lactate fermentation to regenerate NAD⁺. This pathway also produces 2 ATP during glycolysis. Glycolysis followed by (b) alcohol fermentation also regenerates NAD⁺ and produces 2 ATP.

(a)

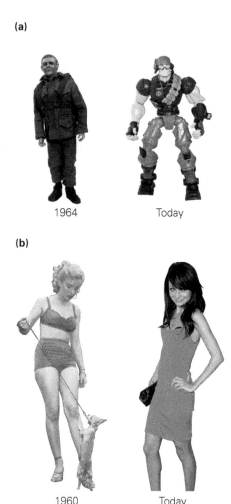

1964   Today

(b)

1960   Today

Figure 4.19 The perception of beauty. (a) GI Joe has become more muscular over time. (b) Female sex symbols have become thinner over time.

definitions of overweight have changed over the years. Men and women who were considered to be of normal weight in the past might not be seen as meeting today's standards. In the United States, the evolution of this trend has been paralleled by changes in children's action figures and dolls and in celebrities and movie stars over the last several decades. You need only compare the physiques of action figures from the 1960s and 1970s to the physiques seen on today's action figures to see how the standards have changed (**Figure 4.19a**). Standards for women have also changed. The 1960s sex symbol and movie star Marilyn Monroe was much curvier than more recent stars, such as Nicole Richie (**Figure 4.19b**).

The next time you read the newspaper, take note of advertisements for diets promoting diet products. It is often the case that "before" pictures show individuals of healthful weights, and "after" pictures show people who are too thin. Indeed, the average woman in the United States weighs 140 pounds and wears a size 12. The average model weighs 103 pounds and wears a size 4. With these distorted messages about body fat, it is difficult for average people to know how much body fat is right for them.

Evaluating How Much Body Fat Is Healthful

A person's sex, along with other factors, determines his or her ideal amount of body fat. Women need more body fat than men do to maintain their fertility. On average, healthy women have 22% body fat, and healthy men have 14%. To maintain essential body functions, women need at least 12% body fat but not more than 32%; for men, the range is between 3% and 29%. This difference between females and males, a so-called sex difference, exists because women store more fat on their breasts, hips, and thighs than men do. A difference in muscle mass leads to increased energy use by males because muscles use more energy than does fat.

Women also have an 8% thicker layer of tissue called the dermis under the outer epidermal layer of the skin as compared to men. This means that in a woman and a man of similar strength and body fat, the woman's muscles would look smoother and less defined than the man's muscles would.

A person's frame size also influences body fat—larger-boned people carry more fat. In addition, body fat tends to increase with age.

Unfortunately, it is a bit tricky to determine what any individual's ideal body weight should be. In the past, you simply weighed yourself and compared your weight to a chart showing a range of acceptable weights for given heights. The weight ranges on these tables were associated with the weights of a group of people who bought life insurance in the 1950s and whose health was monitored until they died. The problem with using these tables is that the subjects may not have been representative of the whole population. Generalizing results seen in one group to another group can lead to erroneous conclusions (Chapter 1). People who had the money to buy life insurance tended to have the other benefits of money as well, including easier access to health care, better nutrition, *and* lower body weight. Their longer lives may have had more to do with better health care and nutrition than with their weight.

To deal with some of the ambiguities associated with the insurance company's weight tables, a new measure of weight and health risk, the **body mass index (BMI),** has been developed. BMI is a calculation that uses both height and weight to determine a value that correlates an estimate of body fat with the risk of illness and death (**Table 4.1**).

Although the BMI measurement is a better approximation of ideal weight than are the insurance charts of the past, it is not perfect; BMI still does not account for differences in frame size, gender, or muscle mass. In fact, studies

TABLE 4.1

| Body mass index (BMI). A chart based on height and weight correlations. | | | | | | | | | | | |

Height	Weight										
4'10"	91	96	100	105	110	115	119	124	129	134	138	143
4'11"	94	99	104	109	114	119	124	128	133	138	143	148
5'0"	97	102	107	112	118	123	128	133	138	143	148	153
5'1"	100	106	111	116	122	127	132	137	143	148	153	158
5'2"	103	109	115	120	126	131	136	142	148	153	158	164
5'3"	107	113	118	124	130	135	141	146	152	158	163	169
5'4"	110	116	122	128	134	140	145	151	157	163	169	174
5'5"	114	120	126	132	138	144	150	156	162	168	174	180
5'6"	117	124	130	136	142	148	155	161	167	173	179	186
5'7"	121	127	134	140	146	153	159	166	172	178	185	191
5'8"	125	131	138	144	151	158	164	171	177	184	190	197
5'9"	129	135	142	149	155	162	169	176	183	189	196	203
5'10"	132	139	146	153	160	167	174	181	188	195	202	207
5'11"	136	143	150	157	165	172	179	186	193	200	208	215
6'0"	140	147	154	162	169	177	184	191	198	206	213	221
6'1"	144	151	159	166	174	182	189	197	205	212	219	227
6'2"	148	155	163	171	179	186	194	202	210	218	225	233
6'3"	151	160	168	176	184	192	200	208	216	224	232	240
6'4"	156	164	172	180	189	197	205	213	221	230	238	246
BMI	19	20	21	22	23	24	25	26	27	28	29	30

16 — Anorexic · Underweight and possibly anorexic · 20 · Healthy · 25 · ★ · Overweight · 30 · Obese

show that as many as 1 in 4 people may be misclassified by BMI tables because this measurement provides no means to distinguish between lean muscle mass and body fat. For example, an athlete with a lot of muscle will weigh more than a similar-size person with a lot of fat because muscle is heavier than fat.

If your BMI falls within the healthy range (BMI of 20–25), you probably have no reason to worry about health risks from excess weight. If your BMI is high, you may be at increased risk for diseases associated with obesity.

Obesity

Close to 1 in 3 Americans has a BMI of 30 or greater and is therefore considered to be obese. This crisis in **obesity** is the result you would expect when constant access to cheap, high-fat, energy-dense, unhealthful food is combined with lack of exercise. This relationship is clearly illustrated by the case of the Pima Indians.

Several hundred years ago, the ancestral population of Pima Indians split into 2 tribes. One branch moved to Arizona and adopted the American diet and lifestyle; the typical Pima of Arizona gets as much exercise as the average American and, like most Americans, eats a high-fat, low-fiber diet. Unfortunately, the health consequences for these people are more severe than they are for most other Americans—close to 60% of the Arizona Pima are obese and diabetic. In contrast, the Pima of Mexico maintained their ancestral farming life; their diet is rich in fruits, vegetables, and fiber. The Pima of Mexico also engage in physical labor for close to 22 hours per week and are on average 60 pounds lighter than their Arizona relatives. Consequently, diabetes is virtually unheard of in this group.

This example illustrates the impact of lifestyle on health: The Pima of Arizona share many genes with their Mexican relatives but have far less healthful lives due to their diet and lack of exercise. The example also shows that genes influence body weight because the Pima of Arizona have higher rates of obesity and diabetes than those of other Americans whose lifestyle they share.

Whether obesity is the result of genetics, diet, or lack of exercise, the health risks associated with obesity are the same. As your weight increases, so do your risks of diabetes, hypertension, heart disease, stroke, and joint problems.

Diabetes. **Diabetes** is a disorder of carbohydrate metabolism characterized by the impaired ability of the body to produce or respond to insulin. **Insulin** is a hormone secreted by beta cells, which are located within clusters of cells in the pancreas. Insulin's role in the body is to trigger cells to take up glucose so that they can convert the sugar into energy. People with diabetes are unable to metabolize glucose; as a result, the level of glucose in the blood rises.

There are two forms of the disease. Type 1, the insulin-dependent form of diabetes, usually arises in childhood. People with Type 1 diabetes cannot produce insulin because their immune systems mistakenly destroy their own beta cells. When the body is no longer able to produce insulin, daily injections of the hormone are required. Type 1 diabetes is not correlated with obesity.

Type 2, the non-insulin-dependent form of diabetes, usually occurs after 40 years of age and is more common in the obese. Type 2 diabetes arises either from decreased pancreatic secretion of insulin or from reduced responsiveness to secreted insulin in target cells. People with Type 2 diabetes are able to control blood glucose levels through diet and exercise and, if necessary, by insulin injections.

Hypertension. **Hypertension,** or high blood pressure, places increased stress on the circulatory system and causes the heart to work too hard. Compared to a person with normal blood pressure, a hypertensive person is six times more likely to have a heart attack.

Blood pressure is the force, originated by the pumping action of the heart, exerted by the blood against the walls of the blood vessels. Blood vessels expand and contract in response to this force. Blood pressure is reported as 2 numbers: The higher number, called the **systolic blood pressure,** represents the pressure exerted by the blood against the walls of the blood vessels as the heart contracts; the lower number, called the **diastolic blood pressure,** is the pressure that exists between contractions of the heart when the heart is relaxing. Normal blood pressure is around 120 over 80 (symbolized as 120/80). Blood pressure is considered to be high when it is persistently above 140/90.

Problematic weight gain is typically the result of increases in the amount of fatty tissue versus increases in muscle mass. Fat, like all tissues, relies on oxygen and other nutrients from food to produce energy. As the amount of fat on your body increases, so does the demand for these substances. Therefore, the amount of blood required to carry oxygen and nutrients also increases. Increased blood volume means that the heart has to work harder to keep the blood moving through the vessels, thus placing more pressure on blood vessel walls and leading to increased heart rate and blood pressure.

Heart Attack, Stroke, and Cholesterol. A **heart attack** occurs when there is a sudden interruption in the supply of blood to the heart caused by the blockage of a vessel supplying the heart. A **stroke** is a sudden loss of brain function that results when blood vessels supplying the brain are blocked or ruptured. Heart attack and stroke are more likely in obese people because the elevated blood pressure caused by obesity also damages the lining of blood vessels and increases the likelihood that cholesterol will be deposited there. Cholesterol-lined vessels are said to be *atherosclerotic*.

Because lipids like cholesterol are not soluble in aqueous solutions, cholesterol is carried throughout the body, attached to proteins in structures called lipoproteins. **Low-density lipoproteins (LDLs)** have a high proportion of cholesterol (in other words, they are low in protein). LDLs distribute both the cholesterol synthesized by the liver and the cholesterol derived from diet throughout the body. LDLs are also important for carrying cholesterol to cells, where it is used to help make plasma membranes and hormones. **High-density lipoproteins (HDLs)** contain more protein than cholesterol. HDLs scavenge excess cholesterol from the body and return it to the liver, where it is used to make bile. The cholesterol-rich bile is then released into the small intestine, and from there much of it exits the body in the feces. The LDL/HDL ratio is an index of the rate at which cholesterol is leaving body cells and returning to the liver.

Your physician can measure your cholesterol level by determining the amounts of LDL and HDL in your blood. If your total cholesterol level is over 200 or your LDL level is above 100 or so, then your physician may recommend that you decrease the amount of cholesterol and saturated fat in your diet. This may mean eating more plant-based foods and less meat (because plants do not have cholesterol and meat does have cholesterol), as well as reducing the amount of saturated fats in your diet. Saturated fat is thought to raise cholesterol levels by stimulating the liver to step up its production of LDLs and slowing the rate at which LDLs are cleared from the blood.

Stop & Stretch A friend of yours has his cholesterol level checked and tells you that he is really relieved because his cholesterol is 195. What other factors should your friend be considering before he decides his cholesterol level gives him no cause to be concerned?

Cholesterol is not all bad; in fact, some cholesterol is necessary—it is present in cell membranes to help maintain their fluidity, and it is the building block for steroid hormones such as estrogen and testosterone. You do, however, synthesize enough cholesterol so that you do not need to obtain much from your diet.

For some people, those with a genetic predisposition to high cholesterol, controlling cholesterol levels through diet is difficult because dietary cholesterol makes up only a fraction of the body's total cholesterol. People with high cholesterol who do not respond to dietary changes may have inherited genes that increase the liver's production of cholesterol. These people may require prescription medications to control their cholesterol levels.

Cholesterol-laden, atherosclerotic vessels increase your risk of heart disease and stroke. Fat deposits narrow your heart's arteries, so less blood can flow to your heart. Diminished blood flow to your heart can cause chest pain, or angina. A complete blockage can lead to a heart attack. Lack of blood flow to the heart during a heart attack can cause the oxygen-starved heart tissue to die, leading to irreversible heart damage.

The same buildup of fatty deposits also occurs in the arteries of the brain. If a blood clot forms in a narrowed artery in the brain, it can completely block blood flow to an area of the brain, resulting in a stroke. If oxygen-starved brain tissue dies, permanent brain damage can result.

Anorexia and Bulimia

Eating disorders that make you underweight cause health problems that are as severe as those caused by too much weight (**Table 4.2**). **Anorexia,** or self-starvation, is rampant on college campuses. Estimates suggest that 1 in 5 college women and 1 in 20 college men restrict their intake of Calories so severely that they are essentially starving themselves to death. Others allow themselves to eat—sometimes very large amounts of food (called binge eating)—but prevent the nutrients from being turned into fat by purging themselves, by vomiting or using laxatives. Binge eating followed by purging is called **bulimia.**

Anorexia has serious long-term health consequences. Anorexia can starve heart muscles to the point that altered rhythms develop. Blood flow is reduced, and blood pressure drops so much that the little nourishment present cannot get to the cells. The lack of body fat accompanying anorexia can also lead to the cessation of menstruation, a condition known as amenorrhea. Amenorrhea occurs when a protein called leptin, which is secreted by fat cells, signals the brain that there is not enough body fat to support a pregnancy. Hormones (such as estrogen) that regulate menstruation are blocked, and menstruation ceases. Amenorrhea can be permanent and causes sterility in a substantial percentage of people with anorexia.

The damage done by the lack of estrogen is not limited to the reproductive system; bones are affected as well. Estrogen secreted by the ovaries during the menstrual cycle acts on bone cells to help them maintain their strength and size. Anorexics reduce the development of dense bone and put themselves at a much higher risk of breaking their weakened bones, in a condition called **osteoporosis.**

Besides experiencing the same health problems that anorexics face, people with bulimia can rupture their stomachs through forced vomiting. They often have dental and gum problems caused by stomach acid being forced into their mouths during vomiting, and they can become fatally dehydrated.

Achieving Ideal Weight

As you have seen, the health problems associated with obesity, anorexia, and bulimia are severe. To avoid these problems it is best to focus more on fitness and healthy eating and less on body weight.

Working slowly toward being fit and eating healthfully rather than trying the latest fad diet are more realistic and attainable ways to achieve the positive health outcomes that we all desire. In fact, fitness may be more important than body weight in terms of health. Studies show that fit but overweight people have better health outcomes than unfit slender people. In other words, lack of fitness is associated with higher health risks than excess body weight. Therefore, it makes more sense to focus on eating right and exercising than it does to focus on the number on the scale.

TABLE 4.2

Obesity and anorexia or bulimia. Health problems result from being either overweight or underweight.	
Health Problems Resulting from Obesity	**Health Problems Resulting from Anorexia and Bulimia**
• Adult-onset diabetes	• Altered heart rhythms
• Hypertension (high blood pressure)	• Amenorrhea (cessation of menstruation)
• Heart attack	• Osteoporosis (weakened bones)
• Stroke	• Ruptured stomach
• Joint problems	• Dental/gum problems
	• Dehydration

SAVVY READER

Hoodia for Weight Loss

Several websites claim that celebrities, including Angelina Jolie, used a substance derived from the Hoodia Gordonii plant to help them lose weight. Summarized below are some of the claims made by these sites:

- Hoodia plants require over five years to grow to maturity. This cactus plant is also a fairly uncommon plant. These two factors limit its availability. Because the supply of Hoodia cannot meet the demand, Hoodia pills are more expensive than one would expect.

- Hoodia diet pills are natural. They contain no additives or preservatives, just parts of the plant. This natural substance acts on your brain to make you eat less.

- Hoodia has been used by indigenous people to help them traverse the Kalahari Desert.

- Drug companies have spent millions of dollars performing tests of Hoodia diet pills. No side effects have been found. Hoodia must be safe for everyone.

- If you don't believe Hoodia works, just look at these photos of Angelina Jolie (**Figure 4.20**). This is proof of the effectiveness of this natural appetite suppressant.

1. List several claims presented on this website. Is any evidence presented for these claims?

2. Many products are marketed as natural and many consumers let their guard down when they hear that a product is natural. Are all

Figure 4.20 Angelina Jolie

natural substances good for humans? Give an example of something that is natural that is not good for humans.

3. One site claims that a drug company performed tests on Hoodia and found that there were no side effects and that Hoodia appears to be safe for everyone. Is this the way scientists report data? What kinds of information should be here to substantiate this claim?

4. Look at the before (left) and after (right) photos of Angelina Jolie. Did she need to lose weight in the before picture? Is there any evidence presented that she actually took Hoodia?

Chapter Review

Learning Outcomes

LO1 **Define the term metabolism (Section 4.1).**
- Metabolic reactions include all the chemical reactions that occur in cells to build up or break down macromolecules (p. 74).

LO2 **Describe the structure and function of enzymes (Section 4.1).**
- Enzymes are proteins that catalyze specific cellular reactions. The active site of an enzyme is composed of amino acids that affect its ability to bind to its substrate (pp. 74–75).

LO3 **Explain how enzymes decrease a reaction's activation energy barrier (Section 4.1).**
- The binding of a substrate to the enzyme's active site causes the enzyme to change shape (induced fit), placing more stress on the bonds of the substrate and thereby lowering the activation energy (pp. 74–75).

LO4 **List several reasons why metabolic rates differ between people (Section 4.1).**
- An individual's metabolic rate is affected by many factors, including age, sex, exercise level, body weight, and genetics (pp. 75–76).

LO5 **Describe the structure and function of ATP (Section 4.2).**
- ATP is a nucleotide triphosphate. The nucleotide found in ATP contains a sugar and the nitrogenous base adenine (p. 76).
- The energy stored in the chemical bonds of food can be released by metabolic reactions and stored in the bonds of ATP. Cells use ATP to power energy-requiring processes (pp. 76–77).
- Breaking the terminal phosphate bond of ATP releases energy that can be used to perform cellular work and produces ADP plus a phosphate (p. 77).
- ATP is generated in most organisms by the process of cellular respiration, which consumes carbohydrates and releases water and carbon dioxide as waste products (p. 78).

LO6 **Describe the process of cellular respiration from the breakdown of glucose through the production of ATP (Section 4.2).**
- Cellular respiration begins in the cytosol, where a 6-carbon sugar is broken down into two 3-carbon pyruvic acid molecules during the anaerobic process of glycolysis (pp. 79–80).

Mastering**BIOLOGY**

Go to the Study Area at www.masteringbiology.com for practice quizzes, myeBook, BioFlix™ 3-D animations, MP3Tutor sessions, videos, current events, and more.

- The pyruvic acid molecules then move across the two mitochondrial membranes, where they are decarboxylated. The remaining 2-carbon fragment then moves into the matrix of the mitochondrion, where the citric acid cycle strips them of carbon dioxide and electrons (p. 80).
- Electrons removed from chemicals that are part of glycolysis and the citric acid cycle are carried by electron carriers, such as NADH, to the inner mitochondrial membrane; there they are added to a series of proteins called the electron transport chain. At the bottom of the electron transport chain, oxygen pulls the electrons toward itself. As the electrons move down the electron transport chain, the energy that they release is used to drive protons (H⁺) into the intermembrane space. Once there, the protons rush through the enzyme ATP synthase and produce ATP from ADP and phosphate (pp. 81–82).
- When electrons reach the oxygen at the bottom of the electron transport chain, they combine with the oxygen and hydrogen ions to produce water (pp. 81–82).

LO7 **Explain how proteins and fats are broken down during cellular respiration (Section 4.2).**
- Proteins and fats are also broken down by cellular respiration, but they enter the pathway at later points than glucose (p. 82).

LO8 **Compare and contrast aerobic and anaerobic respiration (Section 4.2).**
- Anaerobic respiration is cellular respiration that does not use oxygen as the final electron acceptor (p. 83).

LO9 **Outline the health risks that occur with being overweight and underweight (Section 4.3).**
- Obesity is associated with many health problems, including hypertension, heart attack and stroke, diabetes, and joint problems (pp. 86–87).
- Anorexia and bulimia can cause altered heart rhythms, amenorrhea, osteoporosis, dehydration, and dental and stomach problems (pp. 88–89).

Roots to Remember

The following roots of words come mainly from Latin and Greek and will help you decipher terms:

- **an-** means absence of. Chapter term: anaerobic
- **-ase** is a common suffix in names of enzymes. Chapter term: sucrase

glyco- means sugar. Chapter term: glycolysis

lipo- refers to fat or lipid. Chapter term: lipoprotein

osteo- refers to bone. Chapter term: osteoporosis

Learning the Basics

1. LO1 Define the term metabolism.

2. LO3 What is meant by the term induced fit?

3. LO6 What are the reactants and products of cellular respiration?

4. LO2 Which of the following is a *false* statement regarding enzymes?

A. Enzymes are proteins that speed up metabolic reactions; B. Enzymes have specific substrates; C. Enzymes supply ATP to their substrates; D. An enzyme may be used many times.

5. LO3 Enzymes speed up chemical reactions by _____.

A. heating cells; B. binding to substrates and placing stress on their bonds; C. changing the shape of the cell; D. supplying energy to the substrate.

6. LO6 Cellular respiration involves _____.

A. the aerobic metabolism of sugars in the mitochondria by a process called glycolysis; B. an electron transport chain that releases carbon dioxide; C. the synthesis of ATP, which is driven by the rushing of protons through an ATP synthase; D. electron carriers that bring electrons to the citric acid cycle; E. the production of water during the citric acid cycle.

7. LO6 The electron transport chain _____.

A. is located in the matrix of the mitochondrion; B. has the electronegative carbon dioxide at its base; C. is a series of nucleotides located in the inner mitochondrial membrane; D. is a series of enzymes located in the intermembrane space; E. moves electrons from protein to protein and moves protons from the matrix into the intermembrane space.

8. LO5 Most of the energy in an ATP molecule is released _____.

A. during cellular respiration; B. when the terminal phosphate group is hydrolyzed; C. in the form of new nucleotides; D. when it is transferred to NADH.

9. LO9 The function of low-density lipoproteins (LDLs) is to _____.

A. break down proteins; B. digest starch; C. transport cholesterol from the liver; D. carry carbohydrates into the urine.

10. LO8 Anaerobic respiration _____.

A. generates proteins for muscles to use; B. occurs in yeast cells only; C. does not use oxygen as the final electron acceptor; D. utilizes glycolysis, the citric acid cycle, and the electron transport chain.

Analyzing and Applying the Basics

1. LO7 A friend decides he will eat only carbohydrates because carbohydrates are burned to make energy during cellular respiration. He reasons that he will generate more energy by eating carbohydrates than by eating fats and proteins. Is this true?

2. LO9 A friend thinks taking an oral enzyme supplement will speed up his metabolism. Do you think this is true? What will happen to the tablet supplement once it is in the stomach?

3. LO4 What would you say to a friend who qualifies as obese on a BMI chart but who exercises regularly and eats a well-balanced diet?

Connecting the Science

1. Early Earth was devoid of oxygen. How might bacteria on early Earth have generated energy?

2. Why do you think that anorexia and bulimia are more common among women than men?

Answers to **Stop & Stretch, Visualize This, Savvy Reader,** and **Chapter Review** questions can be found in the **Answers** section at the back of the book.

Life in the Greenhouse

Photosynthesis and Global Warming

Massive flooding
triggered by Hurricane
Katrina devastated 80%
of New Orleans.

LEARNING OUTCOMES

LO1 Describe the greenhouse effect.

LO2 Explain why water is slow to change temperature.

LO3 Compare and contrast the water cycle and the carbon cycle.

LO4 Discuss the origin of fossil fuels and their relationship to the carbon cycle.

LO5 State the basic equation of photosynthesis.

LO6 Describe the light reactions of photosynthesis.

LO7 Explain the events that occur during the Calvin cycle, and describe the relationship between the light reactions and the Calvin cycle.

LO8 Explain the role of stomata in balancing photosynthesis and water loss.

LO9 Define *photorespiration*, and explain why this process is detrimental to plants that experience it.

LO10 Discuss how our own activities contribute to or reduce the risk of global warming.

New Orleans. New York. Miami. Amsterdam. Alexandria. Mumbai. Ho Chi Minh City. Bangkok. Hong Kong. Shanghai. Tokyo. According to predictions of the effect of global warming, some of the great cities of the world may share a common future—portions of each may experience catastrophic flooding and become essentially uninhabitable. New Orleans provides a stark example. In August 2005, when Hurricane Katrina came onshore along the Gulf Coast of the United States, 80% of the city was inundated. A massive evacuation and relief effort still could not prevent the deaths of over 1500 residents or adequately attend to the needs of those impacted by the storm and flooding. Even today, significant portions of New Orleans—a major metropolis in one of the richest countries on Earth—remain essentially abandoned. Imagine the impact of such a disaster on cities and communities that lack the resources of the United States.

Does the same fate await many of the other great cities of the world?

Global warming is causing Earth's ice to melt, raising sea levels by a meter over this century.

This future catastrophic flooding is possible because rising temperatures brought about by global warming are causing ice all over Earth's surface to melt. As this stored water runs off the land and into the oceans, sea levels rise. According to the U.S. National Climate Data Center, the entire planet has warmed by 0.25°C (0.5°F) each decade during the twentieth century, which has contributed to a sea-level rise of 10 to 20 cm (4 to 8 inches). The rising tides have already caused the widespread erosion of beaches, and even submerged islands. If global warming continues as predicted, by the end of the twenty-first century the sea level will be between 50 and 100 centimeters (20 to 36 inches) higher than today—a much faster and more devastating change than what we have already witnessed. Compounding this rapid raise in level, as ocean temperatures increase, the amount of

Rising sea levels are already threatening millions of people along ocean coasts.

energy available to be released in storms will rise, resulting in more powerful hurricanes and causing even more catastrophic flooding.

The currently occurring melt of Earth's surface ice is dramatic and sobering. In Antarctica, rising temperatures have led to the collapse of massive portions of the continent's ice cap. In recent years, Rhode Island-sized shelves of ice have fallen into the ocean. The Greenland ice cap is becoming thinner at its margins every year, and melt water collecting underneath the ice threatens to cause large portions to slide off the island. Meanwhile, the dramatic glaciers in Glacier National Park, located in the northwest corner of Montana, are disappearing. Several of the park's glaciers have already shrunk to half their original size, and the total number of glaciers has decreased from approximately 150 in 1850 to around 35 today. If this trend continues, scientists predict that by the year 2030 not a single glacier will be left in Glacier National Park. All of this melt water must go somewhere—and like an overflowing bathtub, it spills over the edges of ocean basins, inundating low-lying cities and coasts.

Why is Earth's ice melting and threatening to swamp our coastlines, and can anything be done to stop or slow the global warming that is causing the melt?

5.1 The Greenhouse Effect

Global warming is the progressive increase of Earth's average temperature that has been occurring over the past century. Although the general public seems to believe that there is debate among scientists and government-appointed panels about global warming, that is not really the case. Scientists who publish in peer-reviewed journals and respected scientific panels and societies, such as the Intergovernmental Panel on Climate Control (IPCC), National Academy of Sciences, and the American Association for the Advancement of Science (AAAS), all agree that the climate is warming and that most of the warming observed in the last century is attributable to human activities.

Global warming is caused by recent increases in the concentrations of particular gases in the atmosphere, including water vapor, carbon dioxide (CO_2), methane (CH_4), and ozone (O_3). The accumulation of many of these *greenhouse* gases is a direct result of human activity, namely, coal, oil, and natural gas combustion. The most abundant gas emitted by combustion of these fuels is carbon dioxide; for this reason, carbon dioxide is considered the most important greenhouse gas to control.

The presence of carbon dioxide and the other greenhouse gases in the atmosphere leads to a phenomenon called the **greenhouse effect.** Despite this name, the phenomenon caused by these gases is not exactly like that of a greenhouse, where panes of glass allow radiation from the sun to penetrate inside and then trap the heat that radiates from warmed-up surfaces. On Earth, the greenhouse effect works like this: Warmth from the sun heats Earth's surface, which then radiates the heat energy outward. Most of this heat is radiated back into space, but some of the heat warms up the greenhouse gases in the atmosphere and then is reradiated to Earth's surface. In effect, greenhouse gases act like a blanket (**Figure 5.1**). When you sleep under

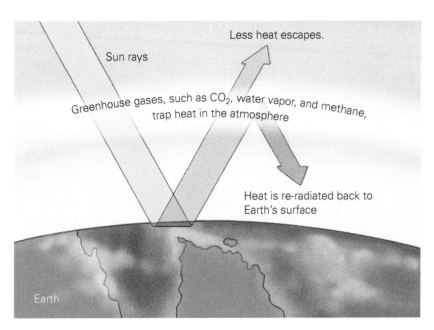

Figure 5.1 The greenhouse effect. Heat from the sun is absorbed in the atmosphere by water vapor, carbon dioxide, and other greenhouse gases and reradiated back to Earth.
Visualize This: What effect do you think increased levels of carbon dioxide have on the greenhouse effect?

a blanket at night, your body heat warms the blanket, which in turn keeps you warm. When the levels of greenhouse gases in the atmosphere increase, the effect is similar to sleeping under a thicker blanket—more heat is retained and reradiated, and therefore the temperature underneath the atmosphere is warmer.

The greenhouse effect is not in itself a dangerous phenomenon. If Earth's atmosphere did not have some greenhouse gases, too much heat would be lost to space, and Earth would be too cold to support life. At historical levels, greenhouse gases keep temperatures on Earth stable and hospitable to life. The danger posed by the greenhouse effect is in the *excess* warming caused by more and more carbon dioxide accumulation in the atmosphere as a result of coal, oil, and natural gas burning.

Stop & Stretch The moon's average daytime temperature is 107°C (225°F), and its average nighttime temperature is –153°C (–243°F). But the moon and Earth are the same distance from the sun. Think about the greenhouse gases on Earth to come up with a hypothesis for why temperatures on the moon might fluctuate so dramatically.

Water, Heat, and Temperature

Bodies of water absorb energy and help maintain stable temperatures on Earth. You have perhaps noticed that when you heat water on a stove, the metal pot becomes hot before the water. This is because water heats more slowly than metal and has a stronger resistance to temperature change than most substances.

Heat and temperature are measures of energy. **Heat** is the total amount of energy associated with the movement of atoms and molecules in a substance. **Temperature** is a measure of the intensity of heat—for example, how fast the molecules in the substance are moving. When you are swimming in a cool lake, your body has a higher temperature than the water; however, the lake contains more heat than your body because even though its molecules are moving more slowly, the sum total of molecular movement in its large volume is much greater than the sum total of molecular movements in your much smaller body.

The formation of hydrogen bonds between neighboring molecules of water makes it more cohesive than other liquids (Chapter 2). The hydrogen bonds also make water resistant to temperature change, even when a large amount of heat is added. This phenomenon occurs because when water is first heated, the heat energy disrupts the hydrogen bonds. Only after enough of the hydrogen bonds have been broken can heat cause individual water molecules to move faster, thus increasing the temperature. As the temperature continues to rise, individual water molecules can move fast enough to break free of all hydrogen bonds and rise into the air as water vapor. This is the basis for the water cycle that moves water from land, oceans, and lakes to clouds and then back again to Earth's surfaces (**Figure 5.2**). When water cools, hydrogen bonds re-form between adjacent molecules, releasing heat into the atmosphere. A body of water can release a large amount of heat into its surroundings while not decreasing its temperature much (**Figure 5.3**).

Water's high heat-absorbing capacity has important effects on Earth's climate. The vast amount of water contained in Earth's oceans and lakes keeps temperatures moderate by absorbing huge amounts of heat radiated by the sun and releasing that heat during less-sunny times, warming the air and preventing large temperature swings.

Figure 5.2 The water cycle. Water moves from the oceans and other surface water to the atmosphere and back, with stops in living organisms, underground pools and soil, and ice caps and glaciers on land.
Visualize This: What would happen to the amount of water in the ocean if all of the stored ice and snow melted? How would this melting affect the shoreline?

The water cycle

Water storage in the atmosphere

Condensation into clouds

Water storage in ice and snow

Precipitation (rain, snow, or fog)

Evaporation from plants

Evaporation

Snowmelt runoff to streams

Spring

Streamflow

Evaporation

Surface runoff

Freshwater storage

Infiltration into ground

Ground-water discharge

Ground-water storage

Water storage in oceans

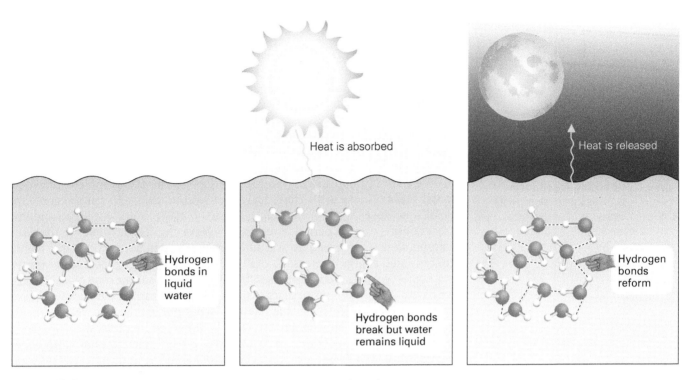

Figure 5.3 Hydrogen bonding in water. Hydrogen bonds break as they absorb heat and re-form as water releases heat.

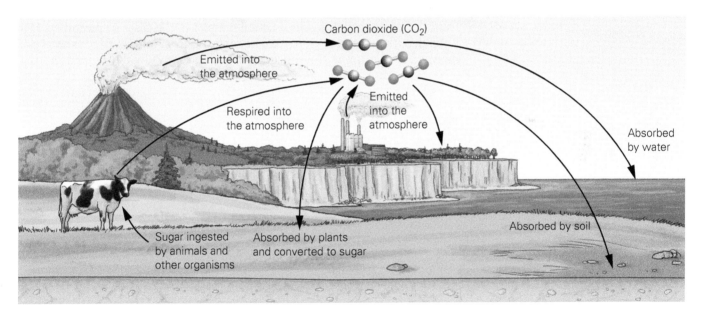

The balance between the release and storage of heat energy is vital to the maintenance of climate conditions on Earth. This balance can be disrupted when increasing levels of carbon dioxide cause more heat to be trapped in the atmosphere. Carbon dioxide in the atmosphere comes from many different sources, some of which are increasing in output.

5.2 The Flow of Carbon

The atoms that make up the complex molecules of living organisms move through the environment via biogeochemical cycles. **Figure 5.4** illustrates how carbon, like water, cycles back and forth between living organisms, the

Figure 5.4 The flow of carbon.
Living organisms, volcanoes, and fossil fuel emissions produce CO_2. Plants, oceans, and soil absorb CO_2 from the air.
Visualize This: Based on the observations that carbon dioxide levels have increased in the last 100 years and that volcanic activity has remained constant, which would you predict releases more carbon dioxide into the air: volcanic or human activity?

Figure 5.5 Burning fossil fuels. The burning of fossil fuels by industrial plants and automobiles adds more carbon dioxide to the environment.

atmosphere, bodies of water, and rock. The carbon dioxide you exhale enters the atmosphere, where it can absorb heat; these molecules can return to Earth's surface, where they can dissolve in water or be absorbed by plants. As you will learn in this chapter, carbon dioxide taken up by plants is converted into carbohydrates using the energy from sunlight. Other organisms use some of these carbohydrates for energy, rereleasing the carbon dioxide into the atmosphere in the process. Any unconsumed carbohydrates may become buried in the very rock of Earth for millennia; the carbon contained there can later be released through volcanic activity or by extraction and combustion by humans. It is the latter activity that is contributing to a buildup of carbon dioxide in the atmosphere.

The stored carbohydrates discussed in the previous paragraph are known as **fossil fuels** (Figure 5.5). These fuels—petroleum, coal, and natural gas—are "fossils" because they formed from the buried remains of ancient plants and microorganisms. Over a period of millions of years, the carbohydrates in these organisms were transformed by heat and pressure deep in Earth's crust into highly concentrated energy sources. Humans now tap these energy sources to power our homes, vehicles, and businesses, but as a result of our burning of these fuels, we have released millions of years of stored carbon as carbon dioxide.

Human use of fossil fuels is having a measurable effect; increases in carbon dioxide in the atmosphere are well documented by direct measurements over the past 50 years (Figure 5.6). Using "fossil air," scientists have also documented that the current level of carbon dioxide in the atmosphere greatly exceeds the levels present on Earth at any time in human history. Carbon dioxide concentrations in the past are measured by examining ice sheets that have existed for thousands of years. This measurement is possible because snow falling on an ice sheet surface traps air. As it accumulates, underlying snow is compressed into ice, and the trapped air becomes tiny ice-encased air bubbles. Thus, these bubbles are fossils—actual samples of the gases in the atmosphere at the time they formed. Cores can be removed from long-lived ice sheets and analyzed to determine the concentration of carbon dioxide trapped in fossil

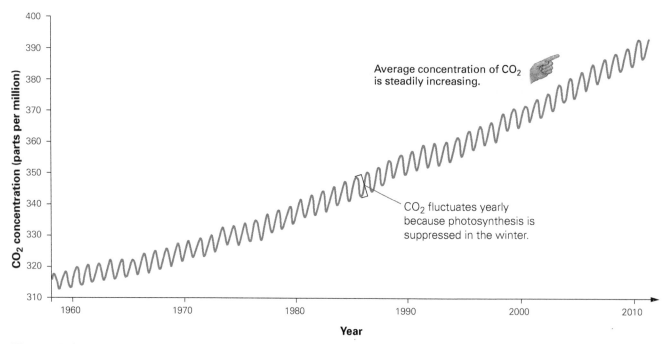

Figure 5.6 Increases in atmospheric carbon dioxide. Carbon dioxide levels from 1960 to present as measured by instruments at Mauna Loa observatory in Hawaii.

Visualize This: What evidence in the graph demonstrates the increased rate of carbon dioxide accumulation from 2000 to 2010 compared to the 1960s?

air (**Figure 5.7**). Other gases in the bubbles can provide indirect information about temperatures at the time the bubbles formed.

Ice core data from Antarctica (**Figure 5.8**) indicate that although Earth has gone through cycles of high carbon dioxide, the concentration of carbon dioxide in the atmosphere today is higher than at any other time in the last 400,000 years. The ice core data also demonstrate that increased levels of carbon dioxide occur at the same time as increased temperatures, suggesting that the carbon dioxide measurably warmed Earth. Taken together, these data are quite worrisome; they tell us that Earth may soon be facing temperatures well above those we experience in the current climate and warmer than Earth has seen in millennia.

5.3 Can Photosynthesis Reduce the Risk of Global Warming?

Modern plants and microorganisms do as their ancient predecessors did—absorb carbon dioxide from the atmosphere. Is it possible that we could depend on modern plants and microorganisms to reduce the amount of greenhouse gases released by fossil fuel burning and thus the threat of global warming?

Photosynthesis is the process by which plants and other microorganisms trap light energy from the sun and use it to convert carbon dioxide and water into sugar. In other words, photosynthesis transforms solar energy into the chemical energy required by all living things.

Chloroplasts: The Site of Photosynthesis

Chloroplasts are the specialized organelles in plant cells where photosynthesis takes place. Chloroplasts are surrounded by two membranes (**Figure 5.9** on the next page). The inner and outer membranes together are called the *chloroplast envelope* (Chapter 2). The chloroplast envelope encloses a compartment filled with **stroma,** the thick fluid that houses some of the enzymes of photosynthesis. Suspended in the stroma are disk-like membranous structures called **thylakoids,** which are typically stacked in piles like pancakes. The large amount of

Figure 5.7 Ice core. By analyzing ice cores, scientists can measure past atmospheric concentrations of carbon dioxide.

Figure 5.8 Records of temperature and atmospheric carbon dioxide concentration from Antarctic ice cores. These data indicate that increases in carbon dioxide levels are correlated with higher temperatures.
Visualize This: This graph has axes on both the left and right sides. Which line is indicated by the left axis and which by the right axis? Why are both lines included on the same graph?

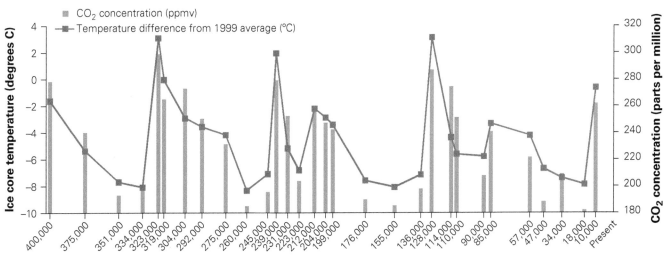

(a)

(b)

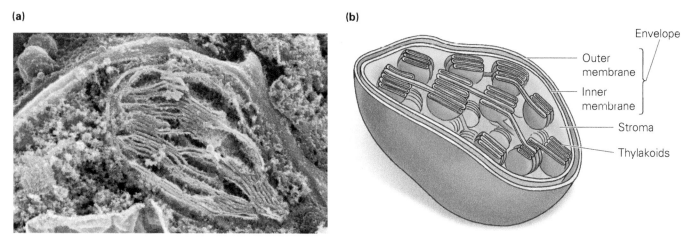

Envelope

Outer membrane

Inner membrane

Stroma

Thylakoids

Figure 5.9 Chloroplasts. The cross section (a) and drawing (b) of a chloroplast show the structures involved in photosynthesis.

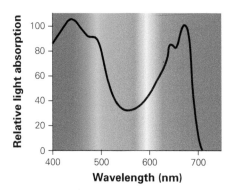

Figure 5.10 The absorption spectra of leaf pigments. Visible "white" light is actually made up of a series of wavelengths that appear as different colors to our eyes. The wavelengths absorbed by chlorophyll and other pigments in green plants are mainly red and blue and are thus removed from the spectrum of light that is reflected back to us from a leaf surface.
Visualize This: How would this graph appear for the pigments in the skin of a ripe red apple?

membrane inside the chloroplast provides abundant surface area on which some of the reactions of photosynthesis can occur.

On the surface of the thylakoid membrane are millions of molecules of **chlorophyll,** a pigment that absorbs energy from the sun. It is chlorophyll that gives leaves and other plant structures their green color. Like all pigments, chlorophyll absorbs light. Light is made up of rays with different colors, and each color corresponds to a different wavelength—to the human eye, shorter and middle wavelengths appear violet to green, and longer wavelengths appear yellow to red (**Figure 5.10**). Chlorophyll looks green to human eyes because it absorbs the shorter (blue) and longer (red) wavelengths of visible light and reflects the middle (green) range of wavelengths.

When a pigment such as chlorophyll absorbs sunlight, electrons associated with the pigment become excited, that is, increase in energy level. In effect, light energy is transferred to the chlorophyll and becomes chemical energy. For most pigments, the molecule remains excited for a very brief amount of time before this chemical energy is lost as heat. This is why a surface that looks black (that is, one composed of pigments that absorb all visible light wavelengths) heats quickly in the sun. (In contrast, a white surface, which does not absorb any light energy, remains relatively cool.) Inside a chloroplast, however, the chemical energy of the excited chlorophyll molecules is not released as heat; instead, the energy is captured.

Stop & Stretch In temperate areas, leaves change color in the fall. This occurs because chlorophyll is reabsorbed by the plant, making other pigments in the leaf visible. Why do you suppose the chlorophyll is reabsorbed?

An Overview: Photosynthesis

In plants and other photosynthetic organisms, solar energy is used to rearrange the atoms of carbon dioxide and water absorbed from the environment into carbohydrates (initially the sugar glucose). Photosynthesis produces oxygen as a waste product. The equation summarizing photosynthesis is as follows:

$$6\ CO_2 + 6\ H_2O + \text{Light energy} \rightarrow C_6H_{12}O_6 + 6\ O_2$$
Carbon dioxide + Water + Light energy → Glucose + Oxygen

Photosynthetic organisms use the carbohydrates that they produce to grow and supply energy to their cells. They, along with the organisms that eat them, lib-

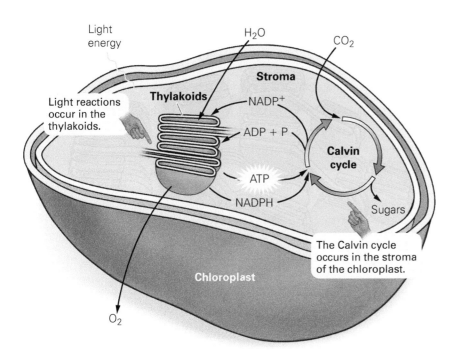

Figure 5.11 Overview of photosynthesis. Sunlight drives the synthesis of glucose and oxygen from carbon dioxide and water in plants and other photosynthetic organisms. The reaction requires two steps: the first that generates cellular energy (ATP) and harvests electrons from water (carried by NADPH); and the second that uses those products to convert carbon dioxide into sugar.

erate the energy stored in the chemical bonds of sugars via the process of cellular respiration. Any excess carbohydrates are stored in the body of the plant or other photosynthesizer and could become the raw material for fossil fuel production.

The process of photosynthesis can be broken down into two steps (**Figure 5.11**). The first, or "photo," step harvests energy from the sun during a series of reactions called the **light reactions,** which occur when there is sunlight. These reactions occur within the thylakoids of the chloroplast and produce the energy-carrying molecule ATP and energized electrons, which are transported around the cell by the molecule NADPH. The second, or "synthesis," step, called the **Calvin cycle,** uses the ATP and NADPH generated by the light reactions to synthesize sugars from carbon dioxide. This step can occur in either the presence or the absence of sunlight, and for this reason, the Calvin cycle is also sometimes referred to as the light-independent reactions. *See the next section for **A Closer Look** at photosynthesis.*

Stop & Stretch In what ways are photosynthesis and cellular respiration (Chapter 4) related?

A Closer Look:
The Two Steps of Photosynthesis

The Light Reactions

When sunlight strikes chlorophyll molecules in a thylakoid membrane, the excited electrons are trapped by another molecule (**Figure 5.12**, step 1, on the next page). The energized electrons are then transferred along a specified path to other molecules in an electron transport chain located in the thylakoid membrane—a process very similar to the electron transport chain in mitochondria (Chapter 4). The electrons

lose some of their energy with every step along the transport chain, but the chloroplast uses this energy to do work (**Figure 5.12**, step 2, on the next page). In particular, some of the proteins in the electron transport chain use the energy of the electrons to pump protons from the exterior to the interior of the thylakoid, setting up a gradient between the inside and outside of this structure. The stockpiled protons diffuse back out of the thylakoid through an enzyme in the thylakoid membrane, producing ATP in the same way that mitochondria

(continued on the next page)

(A Closer Look continued)

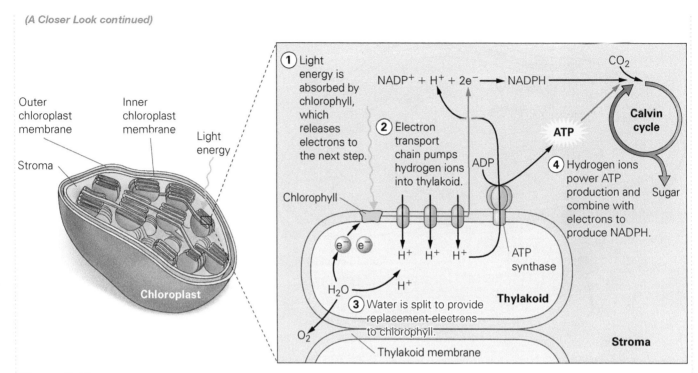

Figure 5.12 The light reactions of photosynthesis. (1) Sunlight strikes chlorophyll molecules located in the thylakoid membrane, exciting electrons, which then move to a higher energy level. (2) The electrons are captured by an electron transport chain, and their energy is used to pump hydrogen ions across the thylakoid membrane. (3) Water is split. Electrons removed from water are used to replace those lost from chlorophyll. Oxygen gas is released. (4) The movement of hydrogen ions out of the thylakoid power ATP production and generate NADPH. These molecules are produced in the stroma, where they will be available to the enzymes of the Calvin cycle.

Visualize This: How is this set of reactions similar to the electron transport chain in cellular respiration? How is it different? (To review cellular respiration, see Chapter 4.)

make ATP. The newly synthesized ATP is released into the stroma of the chloroplast, where it can be used by the enzymes of the Calvin cycle to produce sugars and other organic molecules.

Oxygen is produced during the light reactions when water (H_2O) is "split" into $2H^+$ ions and a single oxygen atom (O) (**Figure 5.12**, step 3). Two oxygen atoms combine to produce O_2, which is released as a waste product from the chloroplast. Since the hydrogen atom contains a single proton around which orbits a single electron, the splitting of water to produce two H^+ ions also releases two electrons. These two electrons are transferred back to chlorophyll molecules in the thylakoid membrane to replace those passed along the electron transport chain.

At the end of the electron transport chain, electrons are transferred to the electron carrier for plants, nicotinamide adenine dinucleotide phosphate, or NADP. Just like the NAD^+ involved in cellular respiration, $NADP^+$ functions as an electron taxicab. The $NADP^+$ used during photosynthesis picks up one H^+ ion and two electrons from the electron transport chain (**Figure 5.12**, step 4). The resulting NADPH ferries electrons to the stroma, where the enzymes of the Calvin cycle will use the electrons to assemble sugars. Thus, the

light reactions produce ATP and a source of electrons, both of which are used in the synthesis step, and release oxygen as a by-product.

Stop & Stretch   ATP is used to power all activities in a cell that require energy input. Why can't plants just use the ATP produced during the light reactions of photosynthesis for all of their activities, instead of going through the step of producing glucose, which the plants will later break down to produce ATP?

The Calvin Cycle

The Calvin cycle consists of a series of reactions occurring in the stroma that use the ATP and NADPH produced during photosynthesis to convert carbon dioxide into sugars. CH_2O is the general formula for sugars. For example, glucose is $C_6H_{12}O_6$ or $6(CH_2O)$. A quick comparison of the formulas of these molecules makes it obvious that converting CO_2 into CH_2O requires the incorporation of hydrogen atoms and their associated electrons. These components are provided by the NADPH produced during the light reactions.

During the Calvin cycle, carbon dioxide from the environment reacts with 5-carbon molecules called ribulose bisphosphate, or RuBP (Figure 5.13). The enzyme that

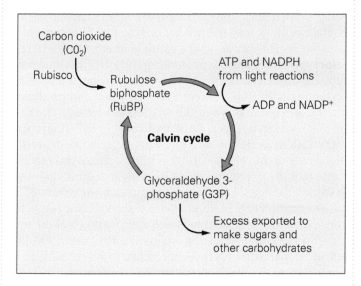

Figure 5.13 Reactions of the Calvin cycle. Carbon dioxide is incorporated into sugar in plants through a series of reactions that regenerate the initial carbon-containing starting product, the sugar ribulose bisphosphate (RuBP). The enzyme that attaches carbon dioxide to RuBP in the first step of the Calvin cycle is called rubisco. Glyceraldehyde 3-phosphate (G3P) is the simple sugar produced by these reactions. Excess G3P is exported to other pathways to produce organic molecules the plant needs.
Visualize This: How would an absence of sunlight affect the Calvin cycle?

catalyzes this reaction is ribulose bisphosphate carboxylase oxygenase, or **rubisco,** the most abundant protein on the planet. Immediately, the resulting 6-carbon molecules break down into pairs of 3-carbon molecules. These 3-carbon molecules go through several rearrangements, with the help of energy released from ATP and electrons from NADPH, to produce the sugar glyceraldehyde 3-phosphate (or G3P). In the last step of the cycle, five G3P molecules are rearranged into three RuBP molecules. Excess G3P produced by this process is used by the cell to make glucose and other carbohydrate compounds. Because the first stable product of the Calvin cycle is a 3-carbon compound, this pathway is often called C_3 photosynthesis.

Stop & Stretch Plants can make all of the macromolecules they need to survive from simple, inorganic (non-carbon-containing) components. They accumulate carbon from the atmosphere and hydrogen and oxygen from water to make the overwhelming majority of their mass. The remaining elements they require are from the soil. Consider the atoms found in proteins and nucleic acids and list two or three soil nutrients that you think are most important to plant growth.

While photosynthesis does take carbon dioxide out of the atmosphere, it is important to realize that the fossil fuels that we have been using only for the last century or so took over 100 million years to form. In other words, right now, carbon dioxide is being released into the atmosphere many times faster than it can be absorbed via natural photosynthesis. We cannot simply rely on photosynthesis to remove excess greenhouse gases as a way to prevent global warming. In fact, rising temperatures may actually slow photosynthesis and reduce its effectiveness.

5.4 How Global Warming Might Reduce Photosynthesis

The high temperatures caused by global warming can reduce photosynthesis in a number of ways, but primarily because higher temperatures generally lead to drier conditions.

Land plants bring carbon dioxide for photosynthesis into their leaves through openings on their surface called **stomata** (Figure 5.14). Stomatal openings are surrounded by two kidney-bean-shaped cells called **guard cells.** When the guard cells are compressed against each other, the stomata are closed, thus restricting the flow of gases into or out of the plant. When the guard cells change shape to create a gap between them, the stomata are open, and carbon dioxide and oxygen gases can be exchanged. In addition to the exchange of gases, water moves out of the plant through the stomatal opening via a process called

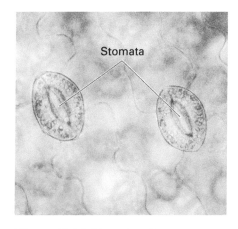

Figure 5.14 Stomata. Stomata are adjustable microscopic pores found on the surface of leaves that allow for gas exchange.

(a) Open

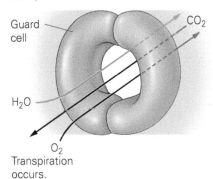

Guard cell

CO_2

H_2O

O_2

Transpiration occurs.

(b) Closed

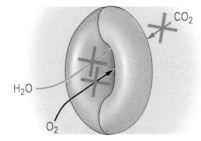

CO_2

H_2O

O_2

Transpiration does not occur.

Figure 5.15 Gas exchange and water loss. (a) When stomata are open, oxygen gases and carbon dioxide can be exchanged, but water can be lost from a plant through a process called transpiration. (b) When the guard cells change shape to block the opening, gas exchange and transpiration do not occur.

transpiration (Figure 5.15). The transpired water is replaced in the plant by water from the soil. On hot, dry days, plants cannot absorb enough water from the soil and will close their stomata to reduce the rate of water lost to transpiration. Closing stomatal openings also prevents carbon dioxide from entering the plant. Thus, with rising temperatures, the rate of photosynthesis declines.

Closing the stomatal openings may not just reduce the rate of photosynthesis; it may even counteract it. This occurs as the result of another series of reactions known as **photorespiration.** During photorespiration, the first enzyme in the Calvin cycle, rubisco, uses oxygen instead of carbon dioxide as a substrate for the reaction it catalyzes with RuBP. Oxygen is used when carbon dioxide levels are low inside the leaf, such as when the stomata are closed. The compound produced by the reaction between RuBP and oxygen, called glycolate, cannot be used in the Calvin cycle. In fact, glycolate must be destroyed by the plant because high levels of this acid will further inhibit photosynthesis. The breakdown of glycolate requires energy provided by cellular respiration—thus, instead of taking up carbon dioxide, plants that are photorespiring release it.

In warmer and drier environments, natural selection should favor plants that can minimize photorespiration despite having closed stomata much of the time. Two mechanisms for reducing photorespiration are known as C_4 and CAM photosynthesis, in which additional pathways that concentrate carbon dioxide in the plant (C_4) or capture it during the night when temperatures are cooler (CAM) occur before the C_3 photosynthetic pathway. C_4 and CAM, along with the C_3 process that occurs during the Calvin cycle, are briefly summarized in **Table 5.1**.

As Earth warms, it is possible that the abundance of C_4 and CAM plants will increase while the number of C_3 plants declines as a result of the increased burden of photorespiration. However, replacing C_3 plants, many of which are large trees, with C_4 plants, mostly grasses, will reduce the total amount of photosynthesis and the amount of carbon that can be taken out of the atmosphere. Because net rates of photosynthesis (as measured by grams of carbon dioxide removed from the atmosphere per acre per year) in grasslands are 30% to 60% less than rates in forests, the loss of trees significantly decreases the removal of carbon dioxide from the atmosphere.

The replacement of trees with grasses, **deforestation,** may happen naturally on a warming Earth as C_4 plants outperform C_3 types, but it is already happening at high rates thanks to human activities. The process occurs when forests are cleared for logging, farming, and ever-expanding human settlements. Deforestation also contributes directly to the increase in carbon dioxide within the atmosphere; current estimates are that up to 25% of the carbon dioxide introduced into the atmosphere originates from the cutting and burning of forests in the tropics alone. Clearly, one effective way to reduce global warming is to reduce deforestation and promote photosynthesis by planting trees through reforestation projects.

In the worst cases, rising temperatures due to global warming will cause vegetated land to become desert, completely eliminating photosynthesis in certain areas. Warming temperatures will expose more snow- and ice-covered regions, allowing additional photosynthesis there, but this additional carbon dioxide uptake is likely to be offset by carbon dioxide released from the more rapid decay of carbohydrates in the soil in these formerly frozen regions. As we have seen, photosynthesis is not likely to soak up much of the excess carbon dioxide released by fossil fuel combustion and, in fact, it may even decline on a warmer planet.

5.5 Decreasing the Effects of Global Warming

The devastation Hurricane Katrina caused in New Orleans is paralleled by environmental damage seen worldwide as a result of climate change. Sea-level rise is not the only effect; global warming will change weather patterns, resulting in droughts in some areas and torrential rains in others. It will

TABLE 5.1

C₃, C₄, and CAM plant photosynthesis. Plants have evolved adaptations that prevent water loss.		
Type of plant and example	**Stomata status**	**Description**
C₃ Soybean 	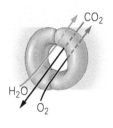	The Calvin cycle converts two 3-carbon sugars into glucose.
C₄ Corn 	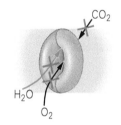	Enzymes scavenge CO_2 to produce 4-carbon sugars, even when stomata are closed. The 4-carbon sugars "pump" carbon dioxide molecules to the Calvin cycle when they break down.
CAM Jade 	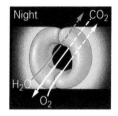 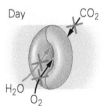	Water loss is slowed by opening stomata only at night. Carbon dioxide is stored as an organic acid in the vacuole. The acid's breakdown releases this carbon dioxide to the Calvin cycle during the day.

cause the extinction of animals and plants that have narrow temperature requirements. It may contribute to the spread of infectious disease and dangerous pests that are now limited by temperature to the tropics. And the carbon dioxide absorbed by the ocean is causing it to acidify, threatening coral reefs, algae in the open ocean, and the organisms that depend on them. Clearly, excess carbon dioxide in the atmosphere is imposing a cost on people.

The United States has a disproportionate impact on the rate of carbon dioxide emissions into the atmosphere. Home to only 4% of the world's population, the United States produces close to 25% of the carbon dioxide emitted by fossil fuel burning. The emissions rate of carbon dioxide for an average American is twice that of a Japanese or German individual, three times that of the global average, four times that of a Swede, and 20 times that of the average Indian.

Most of the emissions for an individual country come from industry, followed by transportation and then by commercial, residential, and agricultural emissions. All of us can work to reduce our personal contribution to global warming by decreasing residential and transportation emissions. Most residential emissions are from energy used to heat and cool homes and to power electrical appliances. Transportation emissions are affected by our choice of the vehicles we use for transport, the fuel economy of cars, and the number of miles traveled. **Table 5.2**

TABLE 5.2

Decreasing your greenhouse gas emissions. Here are some ideas that you can use to help slow the rate of global warming.

Action		Annual decrease in carbon dioxide production
Drive an energy-efficient vehicle. SUVs average 16 miles per gallon, while smaller cars average 25 miles per gallon.		13,000 pounds
Carpool 2 days per week.		1,590 pounds
Recycle glass bottles, aluminum cans, plastic, newspapers, and cardboard.		850 pounds
Walk 10 miles per week instead of driving.		590 pounds
Buy high-efficiency appliances.		400 pounds per appliance
Buy food and other products with reusable or recyclable packaging, or reduced packaging, to save the energy required to manufacture new containers.		230 pounds
Use a push mower instead of a power mower.		80 pounds
Plant shade trees around your home to decrease energy consumption and to remove carbon dioxide by photosynthesis.		50 pounds

describes many ways that you can decrease your greenhouse gas emissions and indicates the number of pounds of carbon dioxide that each action would save annually. These reductions may seem trivial in comparison to the scope of the problem, but when they are multiplied by the more than 300 million people in the United States, the savings become significant.

Having an effect on industrial, commercial, and agricultural sectors is difficult for any one individual. Changes in these areas will instead take leadership from the policy makers who are committed to reducing emissions. Making these changes requires that our leaders, and all of us, understand that even though the implications of and solutions to global warming may be open to debate, the fact that it is occurring at an unprecedented rate is not.

SAVVY READER

Refugees from Global Warming

by Sean Mattson

CARTI SUGDUB, Panama—Rising seas from global warming, coming after years of coral reef destruction, are forcing thousands of indigenous Panamanians to leave their ancestral homes on low-lying Caribbean islands.

Seasonal winds, storms and high tides combine to submerge the tiny islands, crowded with huts of yellow cane and faded palm fronds, leaving them ankle-deep in emerald water for days on end.

Pablo Preciado, leader of the island of Carti Sugdub, remembers that in his childhood floods were rare, brief and barely wetted his toes. "Now it's something else. It's serious," he said.

The increase of a few inches in flood depth is consistent with a global sea level rise over Preciado's 64 years of life and has been made worse by coral mining by the islanders that reduced a buffer against the waves.

If the islanders abandon their homes as planned, the exodus will be one of the first blamed on rising sea levels and global warming.

"This is no longer about a scientist saying that climate change and the change in sea level will flood (a people) and affect them," said Hector Guzman, a marine biologist and coral specialist at the Smithsonian Tropical Research Institute in Panama. "This is happening now in the real world."

The fiercely independent Kuna, famed for rebellions against Spanish conquistadors, French pirates and Panamanian overlords, have accelerated their fate by mining coral, which they use to expand islands and build artificial islets and breakwaters.

Guzman, based at a Pacific island research center on the edge of Panama City, has warned of the risks of coral mining for a decade but says speaking out against a legally permitted traditional activity is "taboo."

"(The Kuna) have increased their vulnerability to storms, wave action, and above all, the action of the rise in sea level," he told Reuters.

1. What is the hypothesis of this excerpted article?
2. Did the author explore other hypotheses about the increased flooding in Carti Sugdub, even if only to refute them?
3. Is the islanders' plight definitively caused by global warming?

Source: Thomson Reuters, July 12, 2010. http://www .reuters.com/article/idUSTRE66B0PL20100712

Chapter Review

Learning Outcomes

MasteringBIOLOGY

Go to the Study Area at www.masteringbiology.com for practice quizzes, myeBook, BioFlix™ 3-D animations, MP3Tutor sessions, videos, current events, and more.

LO1 **Describe the greenhouse effect (Section 5.1).**

- Greenhouse gases, particularly carbon dioxide, increase the amount of heat retained in Earth's atmosphere, which then leads to increased surface temperatures (pp. 94–95).

LO2 **Explain why water is slow to change temperature (Section 5.1).**

- Water can absorb large amounts of heat without undergoing rapid or drastic changes in temperature because heat must first be used to break hydrogen bonds between adjacent water molecules. A high heat-absorbing capacity is a characteristic of water (pp. 95–97).

LO3 **Compare and contrast the water cycle and the carbon cycle (Section 5.1, Section 5.2).**

- Both water and carbon cycle between animals, plants, soil, oceans, and the atmosphere. Most of the movement of water occurs outside living organisms, while carbon is mostly cycled due to biological activity (pp. 96–98).

LO4 **Discuss the origin of fossil fuels and their relationship to the carbon cycle (Section 5.2).**

- Fossil fuels are the buried remains of ancient plants, which took carbon dioxide out of the atmosphere and transformed it into carbohydrates. The burning of fossil fuels is returning carbon to the atmosphere, leading to rising concentrations of carbon dioxide and global warming (pp. 98–99).

LO5 **State the basic equation of photosynthesis (Section 5.3).**

- Photosynthesis utilizes carbon dioxide from the atmosphere to make sugars and other substances. During photosynthesis, energy from sunlight is used to rearrange the atoms of carbon dioxide and water to produce sugars and oxygen (p. 100).

LO6 **Describe the light reactions of photosynthesis (Section 5.3).**

- Photosynthesis occurs in chloroplasts. Sunlight strikes the chlorophyll molecule within chloroplasts, boosting electrons to a higher energy level. These excited electrons are dropped down an electron transport chain located in the thylakoid membrane, and ATP is made (pp. 101–102).
- Electrons are also passed to electron carriers (NADPH). The electrons that are lost from chlorophyll become replaced by electrons acquired during the splitting of water, and oxygen is released (p. 102).

LO7 **Explain the events that occur during the Calvin cycle, and describe the relationship between the light reactions and the Calvin cycle (Section 5.3).**

- The Calvin cycle utilizes the products of the light reactions (ATP and the electron carrier NADPH) to incorporate carbon dioxide into sugars, regenerating the starting products of the cycle and exporting excess sugars to be used as chemical building blocks for plant compounds (pp. 102–103).

LO8 **Explain the role of stomata in balancing photosynthesis and water loss (Section 5.3).**

- Stomata on a plant's surface not only allow in carbon dioxide for photosynthesis but also allow water to escape from the plant. Guard cells surrounding the stomata can change shape to close the stomata and restrict water loss (pp. 103–104).

LO9 **Define *photorespiration*, and explain why this process is detrimental to plants that experience it (Section 5.4).**

- Photorespiration occurs when stomata are closed, carbon dioxide declines in the plant, and oxygen is incorporated into the first step of the Calvin cycle. The resulting product is poisonous to the plant, and energy must be expended to reduce it (p. 104).
- C_4 and CAM plants have evolved to perform photosynthesis while reducing the risk of photorespiration in dry conditions (pp. 104–105).

LO10 **Discuss how our own activities contribute to or reduce the risk of global warming (Section 5.5).**

- Humans are deforesting Earth's land surface, reducing the rate of photosynthesis and thus the uptake of atmospheric carbon dioxide, so reforestation is one strategy to reduce global warming (p. 104).
- Humans can reduce carbon dioxide emissions by increasing energy efficiency and reducing energy use (pp. 106–107).

Roots to Remember

The following roots of words come mainly from Latin and Greek and will help you decipher terms:

chloro- means green. Chapter terms: chloroplast, chlorophyll

photo- means light. Chapter terms: photosynthesis, photorespiration

phyll- means leaf. Chapter term: chlorophyll

trans- means across or to the other side. Chapter term: transpiration

Learning the Basics

1. **LO5** What are the reactants and products of photosynthesis?

2. **LO1** Carbon dioxide functions as a greenhouse gas by _____.

   A. interfering with water's ability to absorb heat; B. increasing the random molecular motions of oxygen; C. allowing radiation from the sun to reach Earth and absorbing the reradiated heat; D. splitting into carbon and oxygen and increasing the rate of cellular respiration

3. **LO2** Water has a high heat-absorbing capacity because _____.

   A. the sun's rays penetrate to the bottom of bodies of water, mainly heating the bottom surface; B. the strong covalent bonds that hold individual water molecules together require large inputs of heat to break; C. it has the ability to dissolve many heat-resistant solutes; D. initial energy inputs are first used to break hydrogen bonds between water molecules and only after these are broken, to raise the temperature; E. all of the above are true

4. **LO4** The burning of fossil fuels _____.

   A. releases carbon dioxide to the atmosphere; B. primarily occurs as a result of human activity. C. is contributing to global warming; D. is possible thanks to photosynthesis that occurred millions of years ago; E. all of the above are correct.

5. **LO8** Stomata on a plant's surface _____.

   A. prevent oxygen from escaping; B. produce water as a result of photosynthesis; C. cannot be regulated by the plant; D. allow carbon dioxide uptake into leaves; E. are found in stacks called thylakoids.

6. **LO6** Which of the following **does not** occur during the light reactions of photosynthesis?

   A. Oxygen is split, releasing water; B. Electrons from chlorophyll are added to an electron transport chain; C. An electron transport chain drives the synthesis of ATP for use by the Calvin cycle; D. NADPH is produced and will carry electrons to the Calvin cycle; E. Oxygen is produced when water is split.

7. **LO6 LO7** Which of the following is a **false** statement about photosynthesis?

   A. During the Calvin cycle, electrons and ATP from the light reactions are combined with atmospheric carbon dioxide to produce sugars; B. The enzymes of the Calvin cycle are located in the chloroplast stroma; C. Oxygen produced during the Calvin cycle is released into the atmosphere; D. Sunlight drives photosynthesis by boosting electrons found in chlorophyll to a higher energy level; E. Electrons released when sunlight strikes chlorophyll are replaced by electrons from water.

8. **LO10** Which human activity generates the most carbon dioxide?

   A. driving; B. cooking; C. bathing; D. using aerosol sprays.

9. **LO9** Photorespiration occurs _____.

   A. under hot and dry conditions; B. when oxygen is incorporated in the first step of the Calvin cycle; C. when carbon dioxide levels are high inside the plant; D. A and B are correct; E. A, B, and C are correct.

10. **LO5** Select the **true** statement regarding metabolism in plant and animal cells.

    A. Plant and animal cells both perform photosynthesis and aerobic respiration; B. Animal cells perform aerobic respiration only, and plant cells perform photosynthesis only; C. Plant cells perform aerobic respiration only, and animal cells perform photosynthesis only; D. Plant cells perform cellular respiration and photosynthesis, and animal cells perform aerobic respiration only.

Analyzing and Applying the Basics

1. **LO3** During the last Ice Age, global temperatures were 4 to 5°C lower than they are today. How would this temperature difference affect the water cycle?

2. **LO3** Imagine an Earth without living organisms on it. How does this difference change the water cycle and the carbon cycle?

3. **LO5** Before life evolved on Earth, there was very little oxygen in the air, mostly because oxygen is a very reactive molecule and tends to combine with other compounds. Explain why oxygen can be maintained at 16% in our atmosphere today.

Connecting the Science

1. A "carbon footprint" is the amount of carbon dioxide you produce as a result of your lifestyle. List five realistic actions you can take to reduce your carbon footprint. Which of these actions will have the greatest impact on your footprint? How difficult for you would it be to take this action?

2. The impact of global warming is not uniformly negative—some regions will benefit from a change in the climate. In addition, technological fixes exist that can help humans deal with some of the consequences of sea-level rise, and people can move from threatened coastlines and areas experiencing the negative effects of global warming. Given these facts, should humans commit to trying to stop global warming, or should we instead focus on mitigating its consequences?

Answers to **Stop & Stretch, Visualize This, Savvy Reader,** and **Chapter Review** questions can be found in the **Answers** section at the back of the book.

CHAPTER 6

Cancer

DNA Synthesis, Mitosis, and Meiosis

Cancer cells divide when they should not.

Nicole's early college career was similar to that of most students. She enjoyed her independence and the wide variety of courses required for her double major in biology and psychology. She worried about her grades and finding ways to balance her coursework with her social life. She also tried to find time for lifting weights in the school's athletic center and snowboarding at a local ski hill. Some weekends, to take a break from school, she would ride the bus home to see her family.

Nicole got sick during her junior year of college.

Managing to get schoolwork done, see friends and family, and still have time left to exercise had been difficult, but possible, for Nicole during her first two years at school. That changed drastically during her third year of school.

One morning in October of her junior year, Nicole began having severe pains in her abdomen. The first time this happened, she was just beginning an experiment in her cell biology laboratory. Hunched over and sweating, she barely managed to make it through her two-hour biology lab. Over the next few days, the pain intensified so much that she was unable to walk from her apartment to her classes without stopping several times to rest.

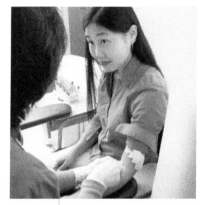

She had to undergo some procedures to see if she had cancer.

Later that week, as she was preparing to leave for class, the pain was so severe that she had to lie down in the hallway of her apartment. When her roommate got home a few minutes later, she took Nicole to the student health center for an emergency visit. The physician at the health center first determined that Nicole's appendix had not burst and then made an appointment for Nicole to see a gynecologist the next day.

After hearing Nicole's symptoms, the gynecologist pressed on her abdomen and felt what he

She wants to understand why she got cancer.

thought was a mass on her right ovary. He used a noninvasive procedure called ultrasound to try to get an image of her ovary. This procedure requires the use of high-frequency sound waves. These waves, which cannot be heard by humans, bounce off tissues and produce a pattern of echoes that can be used to create a picture called a sonogram. Healthy tissues, fluid-filled cysts, and tumors all look different on a sonogram.

Nicole's sonogram convinced her gynecologist that she had a large growth on her ovary. He told her that he suspected this growth was a *cyst,* or fluid-filled sac. Her gynecologist told her that cysts often go away without treatment, but this one seemed to be quite large and should be surgically removed.

Even though the idea of having an operation was scary for Nicole, she was relieved to know that the pain would stop. Her gynecologist also assured her that she had nothing to worry about because cysts are not cancerous. A week after the abdominal pain began, Nicole's gynecologist removed the cyst and her completely engulfed right ovary through an incision just below her navel.

After the operation, Nicole's gynecologist assured her that the remaining ovary would compensate for the missing ovary by ovulating (producing an egg cell) every month. He added that he would have to monitor her remaining ovary carefully to see that it did not become cystic or, even worse, cancerous. She could not afford to lose another ovary if she wanted to have children someday.

Monitoring her remaining ovary involved monthly visits to her gynecologist's office, where Nicole had her blood drawn and analyzed. The blood was tested for the level of a protein called CA125, which is produced by ovarian cells. Higher-than-normal CA125 levels usually indicate that the ovarian cells have increased in size or number and are thus associated with the presence of an ovarian tumor.

Nicole went to her scheduled checkups for five months after the original surgery. The day after her March checkup, Nicole received a message from her doctor asking that she come to see him the next day. Because she needed to study for an upcoming exam, Nicole tried to push aside her concerns about the appointment. By the time she arrived at her gynecologist's office, she had convinced herself that nothing serious could be wrong. She thought a mistake had probably been made and that he just wanted to perform another blood test.

The minute her gynecologist entered the exam room, Nicole could tell by his demeanor that something was wrong. As he started speaking to her, she began to feel very anxious—when he said that she might have a tumor on her remaining ovary, she could not believe her ears. When she heard the words *cancer* and *biopsy,* Nicole felt as though she was being pulled underwater. She could see that her doctor was still talking, but she could not hear or understand him. She felt too nauseated to think clearly, so she excused herself from the exam room, took the bus home, and immediately called her parents.

After speaking with her parents, Nicole realized that she had many questions to ask her doctor. She did not understand how it was possible for such a young woman to have lost one ovary to a cyst and then possibly to have a tumor on the other ovary. She wondered how this tumor would be treated and what her prognosis would be. Before seeing her gynecologist again, Nicole decided to do some research in order to make a list of questions for her doctor.

6.1 What Is Cancer?

Cancer is a disease that begins when a single cell replicates itself although it should not. **Cell division** is the process a cell undergoes to make copies of itself. This process is normally regulated so that a cell divides only when more cells are required and when conditions are favorable for division. A cancerous cell is a rebellious cell that divides without being given the go-ahead.

Tumors Can Be Cancerous

Unregulated cell division leads to a pileup of cells that form a lump or **tumor.** A tumor is a mass of cells; it has no apparent function in the body. Tumors that stay in one place and do not affect surrounding structures are said to be **benign.** Some benign tumors remain harmless; others become cancerous. Tumors that invade surrounding tissues are **malignant** or cancerous. The cells of a malignant tumor can break away and start new cancers at distant locations through a process called **metastasis** (Figure 6.1).

Cancer cells can travel virtually anywhere in the body via the lymphatic and circulatory systems. The lymphatic system collects fluids, or lymph, lost from blood vessels. The lymph is then returned to the blood vessels, thus allowing

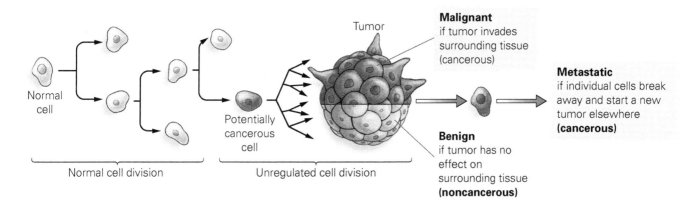

Figure 6.1 What is cancer? A tumor is a clump of cells with no function. Tumors may remain benign, or they can invade surrounding tissues and become malignant. Tumor cells may move, or metastasize, to other locations in the body. Malignant and metastatic tumors are cancerous.

cancer cells access to the bloodstream. Lymph nodes are structures that filter the lost fluids. When a cancer patient is undergoing surgery, the surgeon will often remove a few lymph nodes to see if any cancer cells are in the nodes. If cancer cells appear in the nodes, then some cells have left the original tumor and are moving through the bloodstream. If this has happened, cancerous cells likely have metastasized to other locations in the body.

When cancer cells metastasize, they can gain access not only to the **circulatory system,** which includes blood vessels to transport the blood, but also to the heart, which pumps the blood. Once inside a blood vessel, cancer cells can drift virtually anywhere in the body.

Cancer cells differ from normal cells in three ways: (1) They divide when they should not; (2) they invade surrounding tissues; and (3) they can move to other locations in the body. All tissues that undergo cell division are susceptible to becoming cancerous. However, there are ways to increase or decrease the probability of getting cancer.

Risk Factors for Cancer

Certain exposures and behaviors, called **risk factors,** increase a person's risk of obtaining a disease. General risk factors for virtually all cancers include tobacco use, a high-fat and low-fiber diet, lack of exercise, obesity, excess alcohol consumption, and increasing age. **Table 6.1** outlines other risk factors that are linked to particular cancers.

Tobacco Use. The use of tobacco of any type, whether via cigarettes, cigars, pipes, or chewing tobacco, increases your risk of many cancers. While smoking is the cause of 90% of all lung cancers, it is also the cause of about one-third of all cancer deaths. Cigar smokers have increased rates of lung, larynx, esophagus, and mouth cancers. Chewing tobacco increases the risk of cancers of the mouth, gums, and cheeks. People who do not smoke but who are exposed to secondhand smoke have increased lung cancer rates.

Tobacco smoke contains more than 20 known cancer-causing substances called **carcinogens.** For a substance to be considered carcinogenic, exposure to the substance must be correlated with an increased risk of cancer. Examples of carcinogens include cigarette smoke, radiation, ultraviolet light, asbestos, and some viruses.

The carcinogens that are inhaled during smoking come into contact with cells deep inside the lungs. Chemicals present in cigarettes and cigarette smoke have been shown to increase cell division, inhibit a cell's ability to repair damaged DNA, and prevent cells from dying when they should.

Stop & Stretch In addition to dividing uncontrollably, cancer cells also fail to undergo a type of programmed cell death called *apoptosis,* during which a cell uses specialized chemicals to kill itself. Why do you think it is useful that our own cells can "commit suicide" in certain situations?

, Chemicals in cigarette smoke also disrupt the transport of substances across cell membranes and alter many of the enzyme reactions that occur within cells. They have also been shown to increase the generation of *free radicals,* which remove electrons from other molecules. The removal of electrons from DNA or other molecules causes damage to these molecules—damage that, over time, may lead to cancer. Cigarette smoking provides so many different opportunities for DNA damage and cell damage that tumor formation and metastasis are quite likely for smokers. In fact, people who smoke cigarettes increase their odds of developing almost every cancer.

TABLE 6.1

Cancer risk. Risk factors and detection methods for particular cancers are given.

Cancer Location	Risk Factors	Detection	Comments
Ovary Oviduct / Ovary	• Smoking • Mutation to *BRCA2* gene • Advanced age • Oral contraceptive use and pregnancy decrease risk	• Blood test for elevated CA125 level • Rectovaginal exam	• Fifth leading cause of death among women in the United States
Breast	• Smoking • Mutation to *BRCA1* gene • High-fat, low-fiber diet • Use of oral contraceptives may slightly increase risk	• Monthly self-exams, look and feel for lumps or changes in contour • Mammogram	• Only 5% of breast cancers are due to *BRCA1* mutations • Second-highest cause of cancer-related deaths • 1% of breast cancer occurs in males
Cervix Uterus / Cervix / Vagina	• Smoking • Exposure to sexually transmitted human papilloma virus (HPV)	• Annual Pap smear tests for the presence of precancerous cells	• Precancerous cells can be removed by laser surgery or cryotherapy (freezing) before they become cancerous
Skin	• Smoking • Fair skin • Exposure to ultraviolet light from the sun or tanning beds	• Monthly self-exams, look for growths that change in size or shape	• Skin cancer is the most common of all cancers; usually curable if caught early
Blood (leukemia)	• Exposure to high-energy radiation such as that produced by atomic bomb explosions in Japan during World War II	• A sample of blood is examined under a microscope	• Cancerous white blood cells cannot fight infection efficiently; people with leukemia often succumb to infections

(continued)

TABLE 6.1 *(continued)*

Cancer risk. Risk factors and detection methods for particular cancers are given.

Cancer Location	Risk Factors	Detection	Comments
Lung	• Smoking • Exposure to secondhand smoke • Asbestos inhalation	• X-ray	• Lung cancer is the most common cause of death from cancer, and the best prevention is to quit, or never start, smoking
Colon and rectum Small intestine Colon	• Smoking • Polyps in the colon • Advanced age • High-fat, low-fiber diet	• Change in bowel habits • Colonoscopy is an examination of the rectum and colon using a lighted instrument	• Benign buds called polyps can grow in the colon; removal prevents them from mutating and becoming cancerous
Prostate Bladder Prostate Rectum	• Smoking • Advanced age • High-fat, low-fiber diet	• Blood test for elevated level of prostate-specific antigen (PSA) • Physical exam by physician, via rectum	• More common in African American men than Asian, white, or Native American men
Testicle Testicle Scrotum	• Abnormal testicular development	• Monthly self-exam, inspect for lumps and changes in contour	• Testicular cancer accounts for only 1% of all cancers in men but is the most common form of cancer found in males between the ages of 15 and 35

A High-Fat, Low-Fiber Diet. Cancer risk may also be influenced by diet. The American Cancer Society recommends eating at least 5 servings of fruits and vegetables every day as well as 6 servings of food from other plant sources, such as breads, cereals, grains, rice, pasta, or beans. Plant foods are low in fat and high in fiber. A diet high in fat and low in fiber is associated with increased risk of cancer. Fruits and vegetables are also rich in *antioxidants*. These substances help to neutralize the electrical charge on free radicals and thereby prevent the free radicals from taking electrons from other molecules, including DNA. There is some evidence that antioxidants may help prevent certain cancers by minimizing the number of free radicals that may damage the DNA in our cells.

Lack of Exercise. Regular exercise decreases the risk of most cancers, partly because exercise keeps the immune system functioning effectively. The immune system helps destroy cancer cells when it can recognize them as foreign to the host body. Unfortunately, since cancer cells are actually your own body's cells run amok, the immune system cannot always differentiate between normal cells and cancer cells.

Obesity. Exercise also helps prevent obesity, which is associated with increased risk for many cancers, including cancers of the breast, uterus, ovary, colon, gallbladder, and prostate. Because fatty tissues can store hormones, the abundance of fatty tissue has been hypothesized to increase the odds of hormone-sensitive cancers such as breast, uterine, ovarian, and prostate cancer.

Excess Alcohol Consumption. Drinking alcohol is associated with increased risk of some types of cancer. Men who want to decrease their cancer risk should have no more than two alcoholic drinks a day, and women one or none. People who both drink and smoke increase their odds of cancer in a multiplicative rather than additive manner. In other words, if one type of cancer occurs in 10% of smokers and in 2% of drinkers, someone who smokes and drinks multiplies chances of developing cancer to a rate that is closer to 20% than 12%. The risk factor percentages are multiplied rather than added.

Increasing Age. As you age, your immune system weakens, and its ability to distinguish between cancer cells and normal cells decreases. This weakening is part of the reason many cancers are far more likely in elderly people. Additional factors that help explain the higher cancer risk with increasing age include cumulative damage. If we are all exposed to carcinogens during our lifetime, then the longer we are alive, the greater the probability that some of those carcinogens will mutate genes involved in regulating the cell cycle. Also, because multiple mutations are necessary for a cancer to develop, it often takes many years to progress from the initial mutation to a tumor and then to full-blown cancer. Scientists estimate that most cancers large enough to be detected have been growing for at least five years and are composed of close to one billion cells.

Nicole's cancer affected ovarian tissue. Why might ovarian cells be more likely to become cancerous than some other types of cells? Cells that divide frequently are more prone to cancer than those that don't divide often. When an egg cell is released from the ovary during ovulation, the tissue of the ovary becomes perforated. Cells near the perforation site undergo cell division to heal the damaged surface of the ovary. For Nicole, these cell divisions may have become uncontrolled, leading to the growth of a tumor.

6.2 Passing Genes and Chromosomes to Daughter Cells

Most cell division does not lead to cancer. Cell division produces new cells to heal wounds, replace damaged cells, and help organisms grow and reproduce themselves. Each of us begins life as a single fertilized egg cell that undergoes millions of rounds of cell division to produce all the cells that comprise the tissues and organs of our bodies.

(a) Amoeba

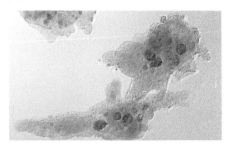

(b) English ivy

Figure 6.2 Asexual reproduction.
(a) This single-celled amoeba divides by copying its DNA and producing offspring that are genetically identical to the original, parent amoeba. (b) Some multicellular organisms, such as this English ivy plant, can reproduce asexually from cuttings.

(a) Uncondensed DNA

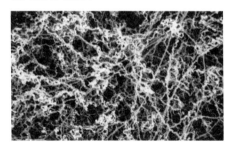

(b) DNA condensed into chromosomes

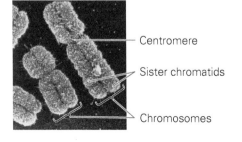

— Centromere

— Sister chromatids

— Chromosomes

Some organisms reproduce by producing exact copies of themselves via cell division. Reproduction of this type, called **asexual reproduction,** does not require genetic input from two parents and results in offspring that are genetically identical to the original parent cell. Single-celled organisms, such as bacteria and amoeba, reproduce in this manner (**Figure 6.2a**). Some multicellular organisms can reproduce asexually also. For example, most plants can grow from clippings of the stem, leaves, or roots and thereby reproduce asexually (**Figure 6.2b**). Organisms whose reproduction requires genetic information from two parents undergo **sexual reproduction.** Humans reproduce sexually when sperm and egg cells combine their genetic information at fertilization.

Genes and Chromosomes

Whether reproducing sexually or asexually, all dividing cells must first make a copy of their genetic material, the **DNA (deoxyribonucleic acid).** The DNA carries the instructions, called **genes,** for building all of the proteins that cells require. The DNA in the nucleus is wrapped around proteins to produce structures called **chromosomes.** Chromosomes can carry hundreds of genes along their length. Different organisms have different numbers of chromosomes in their cells. For example, dogs have 78 chromosomes in each cell, humans have 46, and dandelions have 24.

Chromosomes are in an uncondensed, string-like form prior to cell division (**Figure 6.3a**). Before cell division occurs, the DNA in each chromosome is condensed (compressed) in a short, linear form (**Figure 6.3b**). Condensed linear chromosomes are easier to maneuver during cell division and are less likely to become tangled or broken than are the uncondensed and string-like structures. When a chromosome is replicated, a copy is produced that carries the same genes. The copied chromosomes are called **sister chromatids,** and each sister chromatid is composed of one DNA molecule. Sister chromatids are attached to each other at a region toward the middle of the replicated chromosome, called the **centromere.**

The DNA molecule itself is double stranded and can be likened to a twisted rope ladder (Chapter 2). The backbone or "handrails" of each strand are composed of alternating sugar and phosphate groups. Across the width or "rungs" of the DNA helix are the nitrogenous bases, paired together via hydrogen bonds such that adenine (A) makes a base pair with thymine (T), and guanine (G) makes a base pair with cytosine (C).

You also learned that two of the people credited with determining DNA structure are James Watson and Francis Crick. Watson and Crick reported their hypothesis about the structure of the DNA molecule in a 1953 paper for the journal *Nature.* Although they did not go so far as to propose a detailed model for how the DNA molecule was replicated, they did say, "It has not escaped our notice that the specific pairing we have postulated immediately suggests a copying mechanism for the genetic material." The copying mechanism that Watson and Crick referred to is also called *DNA replication.*

Figure 6.3 DNA condenses during cell division. (a) DNA in its replicated but uncondensed form prior to cell division. (b) During cell division, each copy of DNA is wrapped neatly around many small proteins, forming the condensed structure of a chromosome. After DNA replication, two identical sister chromatids are produced and joined to each other at the centromere.

DNA Replication

During the process of **DNA replication** that precedes cell division, the double-stranded DNA molecule is copied, first by splitting the molecule in half up the middle of the helix. New nucleotides are added to each side of the original parent molecule, maintaining the A-to-T and G-to-C base pairings. This process results in two daughter DNA molecules, each composed of one strand of parental nucleotides and one newly synthesized strand (Figure 6.4a). Because each newly formed DNA molecule consists of one-half conserved parental DNA and one-half new daughter DNA, this method of DNA replication is referred to as *semiconservative replication.*

Replicating the DNA requires an enzyme that assists in DNA synthesis. This enzyme, called **DNA polymerase,** moves along the length of the unwound helix and helps bind incoming nucleotides to each other on the newly forming daughter strand (Figure 6.4b). When free nucleotides floating in the nucleus have an affinity for each other (A for T and G for C), they bind to each other across the width of the helix. Nucleotides that bind to each other are said to be *complementary* to each other.

(a) DNA replication

(b) The DNA polymerase enzyme facilitates replication.

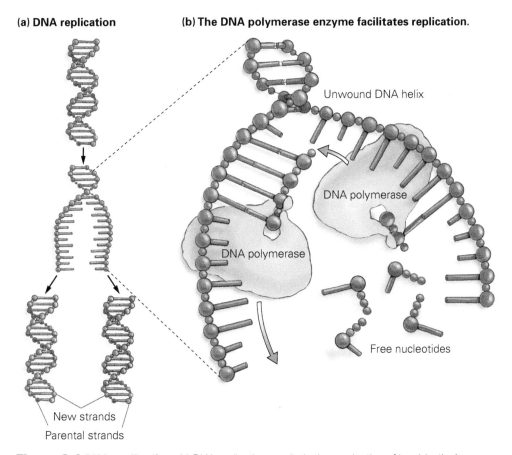

Unwound DNA helix

DNA polymerase

DNA polymerase

Free nucleotides

New strands

Parental strands

Figure 6.4 DNA replication. (a) DNA replication results in the production of two identical daughter DNA molecules from one parent molecule. Each daughter DNA molecule contains half of the parental DNA and half of the newly synthesized DNA.

Visualize This: Assume another round of replication were to occur with the incoming nucleotides still being purple in color. How many total DNA molecules would be produced, and what proportion of each DNA molecule would be purple?

(b) The DNA polymerase enzyme moves along the unwound helix, tying together adjacent nucleotides on the newly forming daughter DNA strand. Free nucleotides have three phosphate groups, two of which are cleaved to provide energy for this reaction before the nucleotide is added to the growing chain.

Figure 6.5 Unduplicated and duplicated chromosomes. An unreplicated chromosome is composed of one double-stranded DNA molecule. A replicated chromosome is X-shaped and composed of two identical double-stranded DNA molecules. Each DNA molecule of the duplicated chromosome is a copy of the original chromosome and is called a sister chromatid. In this illustration, the letters A, b, and C represent different genes along the length of the chromosome.

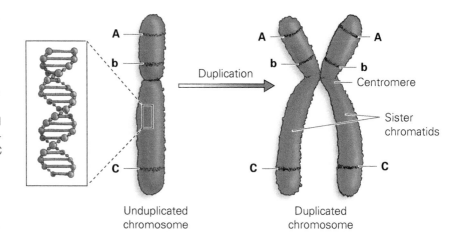

Unduplicated chromosome   Duplicated chromosome

The DNA polymerase enzyme catalyzes the formation of the covalent bond between nucleotides along the length of the helix. The paired nitrogenous bases are joined across the width of the backbone by hydrogen bonding, and the DNA polymerase advances along the parental DNA strand to the next unpaired nucleotide. When an entire chromosome has been replicated, the newly synthesized copies are identical to each other. They are attached at the centromere as sister chromatids (Figure 6.5).

Stop & Stretch DNA replication is not entirely analogous to photocopying. Explain how it differs.

6.3 The Cell Cycle and Mitosis

After a cell's DNA has been replicated, the cell is ready to divide. Mitosis is one way in which this division occurs. **Mitosis** is an asexual division that produces daughter nuclei that are exact copies of the parent nuclei. Mitosis is part of the cell cycle, or life cycle, of non-sex cells called **somatic cells.**

For cells that divide by mitosis, the cell cycle includes three steps: (1) **interphase,** when the DNA replicates; (2) *mitosis,* when the copied chromosomes split and move into the daughter nuclei; and (3) **cytokinesis,** when the cytoplasm of the parent cell splits (Figure 6.6a). As you will see, interphase and mitosis are further divided into steps as well.

Interphase

A normal cell spends most of its time in interphase (Figure 6.6b). During this phase of the cell cycle, the cell performs its typical functions and produces the proteins required for the cell to do its particular job. For example, during interphase, a muscle cell would be producing proteins required for muscle contraction, and a blood cell would be producing proteins required to transport oxygen. Different cell types spend varying amounts of time in interphase. Cells that frequently divide, like skin cells, spend less time in interphase than do those that seldom divide, such as some nerve cells. A cell that will divide also begins preparations for division during interphase. Interphase can be separated into three phases: G_1, S, and G_2.

During the G_1 (first gap or growth) phase, most of the cell's organelles duplicate. Consequently, the cell grows larger during this phase. During the

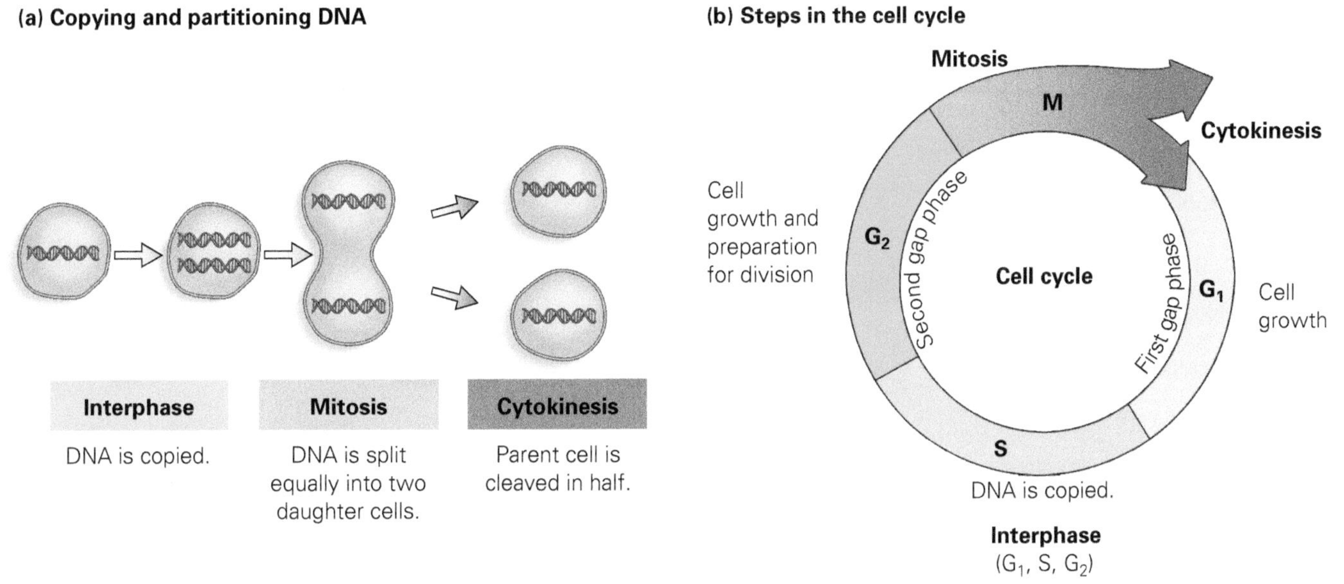

(a) Copying and partitioning DNA

(b) Steps in the cell cycle

Interphase	Mitosis	Cytokinesis
DNA is copied.	DNA is split equally into two daughter cells.	Parent cell is cleaved in half.

Mitosis

M

Cytokinesis

Cell growth and preparation for division

Second gap phase

G₂   **Cell cycle**   **G₁**

First gap phase

Cell growth

S

DNA is copied.

Interphase
(G₁, S, G₂)

Figure 6.6 The cell cycle. (a) During interphase, the DNA is copied. Separation of the DNA into two daughter nuclei occurs during mitosis. Cytokinesis is the division of the cytoplasm, creating 2 daughter cells. (b) During interphase, there are two stages when the cell grows in preparation for cell division, G₁ and G₂ stages, and one stage where the DNA replicates, the S stage. The chromosomes are separated and two daughter cells are formed during the M phase.

S (synthesis) phase, the DNA in the chromosomes replicates. During the G₂ (second gap) phase of the cell cycle, proteins are synthesized that will help drive mitosis to completion. The cell continues to grow and prepare for the division of chromosomes that will take place during mitosis.

Mitosis

The movement of chromosomes into new cells occurs during mitosis. Mitosis takes place in all cells with a nucleus, although some of the specifics of cell division differ among kingdoms. Whether these phases occur in an animal or a plant, the outcome of mitosis and the next phase, cytokinesis, is the same: the production of genetically identical daughter cells. To achieve this outcome, the sister chromatids of a replicated chromosome are pulled apart, and one copy of each is placed into each newly forming nucleus. Mitosis is accomplished during 4 stages: prophase, metaphase, anaphase, and telophase. **Figure 6.7** (on the next page) summarizes the cell cycle in animal cells. The four stages of mitosis are nearly identical in plant cells.

During **prophase,** the replicated chromosomes condense, allowing them to move around in the cell without becoming entangled. Protein structures called **microtubules** also form and grow, ultimately radiating out from opposite ends, or **poles,** of the dividing cell. The growth of microtubules helps the cell to expand. Motor proteins attached to microtubules also help pull the chromosomes around during cell division. The membrane that surrounds the nucleus, called the **nuclear envelope,** breaks down so that the microtubules can gain access to the replicated chromosomes. At the poles of each dividing animal cell, structures called **centrioles** physically anchor one end of each forming microtubule. Plant cells do not contain centrioles, but microtubules in these cells do remain anchored at a pole.

During **metaphase,** the replicated chromosomes are aligned across the middle, or equator, of each cell. To do this, the microtubules, which are

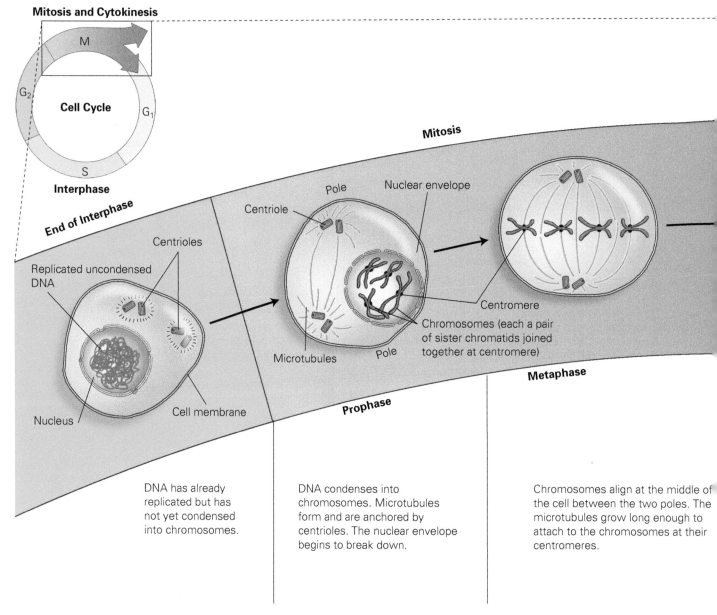

Figure 6.7 Cell division in animal cells. This diagram illustrates how cell division proceeds from interphase through mitosis and cytokinesis.

attached to each chromosome at the centromere, line up the chromosomes in single file across the middle of the cell.

During **anaphase,** the centromere splits, and the microtubules shorten to pull each sister chromatid of a chromosome to opposite poles of the cell.

In the last stage of mitosis, **telophase,** the nuclear envelopes re-form around the newly produced daughter nuclei, and the chromosomes revert to their uncondensed form.

Stop & Stretch How does the similarity in the process of mitosis between animals and plants support the idea that all organisms share a common ancestor?

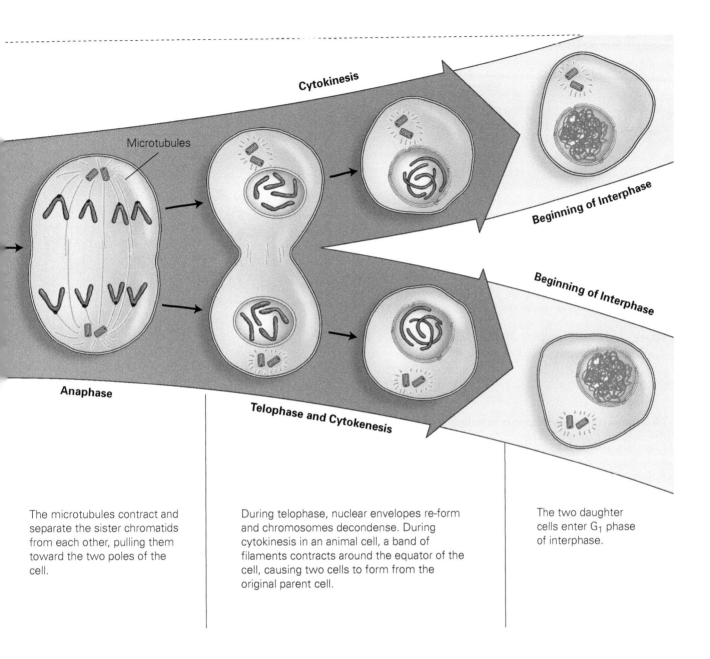

Microtubules

Cytokinesis

Beginning of Interphase

Beginning of Interphase

Anaphase

Telophase and Cytokenesis

The microtubules contract and separate the sister chromatids from each other, pulling them toward the two poles of the cell.

During telophase, nuclear envelopes re-form and chromosomes decondense. During cytokinesis in an animal cell, a band of filaments contracts around the equator of the cell, causing two cells to form from the original parent cell.

The two daughter cells enter G_1 phase of interphase.

Cytokinesis

Cytokinesis divides the cytoplasm, and daughter cells are produced. During cytokinesis in animal cells, a band of proteins encircles the cell at the equator and divides the cytoplasm. This band of proteins contracts to pinch apart the two nuclei and surrounding cytoplasm, creating two daughter cells from the original parent cell. Cytokinesis in plant cells requires that cells build a new **cell wall,** an inflexible structure surrounding the plant cells. **Figure 6.8** (on the next page) shows the difference between cytokinesis in animal and plant cells. During telophase of mitosis in a plant cell, membrane-bound vesicles from the Golgi apparatus (Chapter 2) deliver the materials required for building the cell wall to the center of the cell. The materials include a tough, fibrous carbohydrate called **cellulose** as well as some proteins. The membranes surrounding the vesicles gather in the center of the cell to form a structure called a **cell plate.** The cell plate and cell wall grow across the width of the cell

(a) Cytokinesis in an animal cell

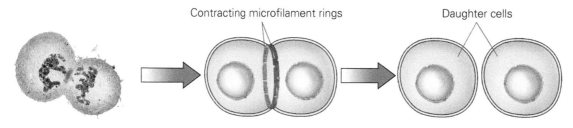

Contracting microfilament rings

Daughter cells

(b) Cytokinesis in a plant cell

Forming cell wall

Parent cell wall

Vesicles with cell wall material

Forming cell plate

New cell wall

Daughter cells

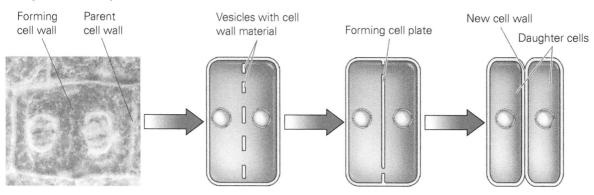

Figure 6.8 A comparison of cytokinesis in animal and plant cells. (a) Animal cells produce a band of filaments that divide the cell in half. (b) Plant cells undergoing mitosis must do so within the confines of a rigid cell wall. During cytokinesis, plant cells form a cell plate that grows down the middle of the parent cell and eventually forms a new cell wall.

and form a barrier that eventually separates the products of mitosis into two daughter cells.

After cytokinesis, the cell reenters interphase, and if the conditions are favorable, the cell cycle may repeat itself.

6.4 Cell-Cycle Control and Mutation

When cell division is working properly, it is a tightly controlled process. Cells are given signals for when and when not to divide. The normal cells in Nicole's ovary and the rest of her body were responding properly to the signals telling them when and how fast to divide. However, the cell that started her tumor was not responding properly to these signals.

An Overview: Controls in the Cell Cycle

Instead of proceeding in lockstep through the cell cycle, normal cells halt cell division at a series of **checkpoints.** At each checkpoint, proteins survey the cell to ensure that conditions for a favorable cellular division have been met. Three checkpoints must be passed before cell division can occur; one takes place during G_1, one during G_2, and the last during metaphase (Figure 6.9).

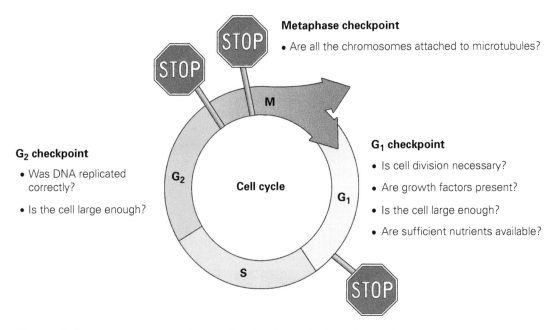

Figure 6.9 Controls of the cell cycle. Checkpoints at G₁, G₂, and metaphase determine whether a cell will continue to divide.
Visualize This: What would happen if a cell started M phase without having copied all of its chromosomes?

Proteins at the G₁ checkpoint determine whether it is necessary for a cell to divide. To do this, they survey the cell environment for the presence of other proteins called **growth factors** that stimulate cells to divide. Growth factors bind to cell membrane-bound proteins called **receptors** that elicit a response from the cell. If enough growth factors are present to trigger cell division, then other proteins check to see if the cell is large enough to divide and if all the nutrients required for cell division are available. When growth factors are limited in number, cell division does not occur. At the G₂ checkpoint, other proteins ensure that the DNA has replicated properly and double-check the cell size, again making sure that the cell is large enough to divide. The third and final checkpoint occurs during metaphase. Proteins present at metaphase verify that all the chromosomes have attached themselves to microtubules so that cell division can proceed properly.

If proteins surveying the cell at any of these three checkpoints determine that conditions are not favorable for cell division, the process is halted. When this happens, the cell may die.

Proteins that regulate the cell cycle, like all proteins, are coded by genes. When these proteins are normal, cell division is properly regulated. When these cycle-regulating proteins are unable to perform their jobs, unregulated cell division leads to large masses of cells called *tumors*. Mistakes in cell cycle regulation arise when the genes controlling the cell cycle are altered, or mutated, versions of the normal genes. A **mutation** is a change in the sequence of DNA. Changes to DNA can change a gene and in turn can alter the protein that the gene encodes, or provides instructions for. Mutant proteins do not perform their required cell functions in the same way that normal proteins do. If mutations occur to genes that encode the proteins regulating the cell cycle, cells can no longer regulate cell division properly. One or more cells in Nicole's ovary must have accumulated mutations in the cell-cycle control genes, leading to the development of cancer. *See the next section for A Closer Look at mutations to cell-cycle control genes.*

A Closer Look:
At Mutations to Cell-Cycle Control Genes

Those genes that encode the proteins regulating the cell cycle are called **proto-oncogenes** (*proto* means "before," and *onco* means "cancer"). Proto-oncogenes are normal genes located on many different chromosomes that enable organisms to regulate cell division. Cancer can develop when the normal proto-oncogenes undergo mutations and become **oncogenes.** A wide variety of organisms carry proto-oncogenes, which means that many different types of organisms can develop cancer (**Figure 6.10**).

Many proto-oncogenes provide the cell with instructions for building growth factors. A normal growth factor

stimulates cell division only when the cellular environment is favorable and all conditions for division have been met. Oncogenes can overstimulate cell division (**Figure 6.11a**).

One gene involved in many cases of ovarian cancer is called *HER2*. (Names of genes are italicized, while names of the proteins they produce are not.) The *HER2* gene carries instructions for building a **receptor** protein. When the shape of the receptor on the surface of the cell is normal, it signals the inside of the cell to allow division to occur. Mutations to the gene that encodes this receptor can result in a receptor protein with a different shape from that of the nor-

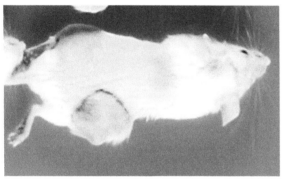

Figure 6.10 Cancer in nonhuman organisms. Many organisms carry proto-oncogenes, which can mutate into oncogenes and cause the development of tumors or cancer.

(a) Mutations to proto-oncogenes

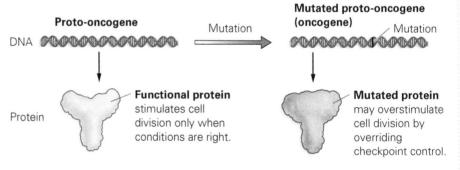

Proto-oncogene

DNA

Mutation

Mutated proto-oncogene (oncogene)

Mutation

Protein

Functional protein stimulates cell division only when conditions are right.

Mutated protein may overstimulate cell division by overriding checkpoint control.

(b) Mutations to tumor-suppressor genes

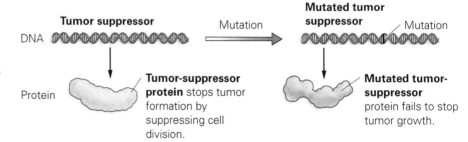

Figure 6.11 Mutations to proto-oncogenes and tumor-suppressor genes. (a) Mutations to proto-oncogenes and (b) tumor-suppressor genes can increase the likelihood of cancer developing.

Tumor suppressor

DNA

Mutation

Mutated tumor suppressor

Mutation

Protein

Tumor-suppressor protein stops tumor formation by suppressing cell division.

Mutated tumor-suppressor protein fails to stop tumor growth.

mal receptor protein. When mutated or misshapen, the receptor protein functions as if many growth factors were present, even when there are actually few-to-no growth factors.

Another class of genes involved in cancer are **tumor suppressors.** These genes, also present in all humans and many other organisms, carry the instructions for producing proteins that suppress or stop cell division if conditions are not favorable. These proteins can also detect and repair damage to the DNA. For this reason, normal tumor suppressors serve as backups in case the proto-oncogenes undergo mutation. If a growth factor overstimulates cell division, the normal tumor suppressor impedes tumor formation by preventing the mutant cell from moving through a checkpoint (Figure 6.11b).

When a tumor-suppressor protein is not functioning properly, it does not force the cell to stop dividing even though conditions are not favorable. Mutated tumor suppressors also allow cells to override cell-cycle checkpoints. One well-studied tumor suppressor, named p53, helps to determine whether cells will repair damaged DNA or commit cellular suicide if the damage is too severe. Mutations to the gene that encodes p53 result in damaged DNA being allowed to proceed through mitosis, thereby passing on even more mutations. Over half of all human cancers involve mutations to the gene that encodes p53.

Mutations to a tumor-suppressor gene are common in cells that have become cancerous. Researchers believe that a normal *BRCA2* gene encodes a protein that is involved in helping to repair damaged DNA. The misshapen, mutant version of the protein cannot help to repair damaged DNA. This means that damaged DNA will be allowed to undergo mitosis, thus passing new mutations on to their daughter cells. As more and more mutations occur, the probability that a cell will become cancerous increases. Figure 6.12 summarizes the roles of growth factors and tumor suppressors in the development of cancer.

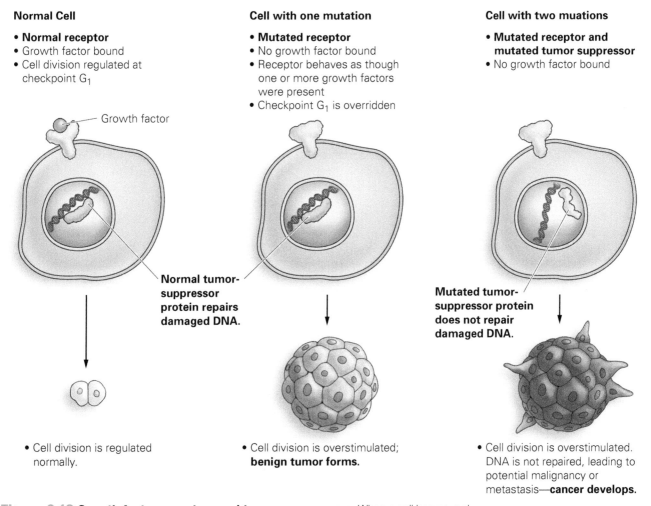

Normal Cell

- **Normal receptor**
- Growth factor bound
- Cell division regulated at checkpoint G_1

Growth factor

Normal tumor-suppressor protein repairs damaged DNA.

- Cell division is regulated normally.

Cell with one mutation

- **Mutated receptor**
- No growth factor bound
- Receptor behaves as though one or more growth factors were present
- Checkpoint G_1 is overridden

- Cell division is overstimulated; **benign tumor forms.**

Cell with two muations

- **Mutated receptor and mutated tumor suppressor**
- No growth factor bound

Mutated tumor-suppressor protein does not repair damaged DNA.

- Cell division is overstimulated. DNA is not repaired, leading to potential malignancy or metastasis—**cancer develops.**

Figure 6.12 Growth factor receptors and tumor suppressors. When a cell has normal growth factors and tumor suppressors, cell division is properly regulated. Mutations to growth factor receptors can cause cell division to be overstimulated, and a benign tumor can form. Additional mutations to tumor-suppressor genes increase the likelihood of malignancy and metastasis.

Cancer Development Requires Many Mutations

Some mutations that occur as a result of damaged DNA being allowed to undergo mitosis are responsible for the progression of a tumor from a benign state, to a malignant state, to metastasis. For example, some cancer cells can stimulate the growth of surrounding blood vessels through a process called **angiogenesis.** These cancer cells secrete a substance that attracts and reroutes blood vessels so that they supply a developing tumor with oxygen (necessary for cellular respiration) and other nutrients. When a tumor has its own blood supply, it can grow at the expense of other, noncancerous cells. Because the growth of rapidly dividing cancer cells occurs more quickly than the growth of normal cells in this process, entire organs can eventually become filled with cancerous cells. When this occurs, an organ can no longer work properly, leading to compromised functioning or organ failure. Damage to organs also explains some of the pain associated with cancer.

Normal cells also display a property called **contact inhibition,** which prevents them from dividing when doing so would require them to pile up on each other. Cancer cells, conversely, have undergone mutations that allow them to continue to divide and form a tumor (Figure 6.13a). In addition, normal cells do need some contact with an underlayer of cells to stay in place. This phenomenon is the result of a process called **anchorage dependence** (Figure 6.13b). Cancer cells override this requirement for some contact with other cells because cancer cells are dividing too quickly and do not expend enough energy to secrete adhesion molecules that glue the cells together. Once a cell loses its anchorage dependence, it may leave the original tumor and move to the blood, lymph, or surrounding tissues.

Most cells are programmed to divide a certain number of times—usually 50 to 70 times—and then they stop dividing. This limits most developing tumors to a small mole, cyst, or lump, all of which are benign. Cancer cells, however, have undergone mutations that allow them to be immortal. They achieve immortality by activating a gene that is usually turned off after early development. This gene produces an enzyme called **telomerase** that helps prevent the degradation of chromosomes. As chromosomes degrade with age, a cell loses its ability to divide. In cancer cells, telomerase is reactivated, allowing the cells to divide without limit.

Stop & Stretch   Telomerase is turned off early in development, causing chromosomes to shorten and eventually lose their ability to replicate. What does this fact imply about the life span of tissue or organ?

Figure 6.13 Contact inhibition and anchorage dependence. (a) When normal cells are grown on a solid support such as the bottom of a flask, they grow and divide until they cover the bottom of the flask. (b) Cancer cells lose the requirement that they adhere to other cells or a solid support.

(a) Contact inhibition

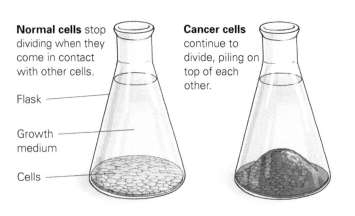

Normal cells stop dividing when they come in contact with other cells.

Flask

Growth medium

Cells

Cancer cells continue to divide, piling on top of each other.

(b) Anchorage dependence

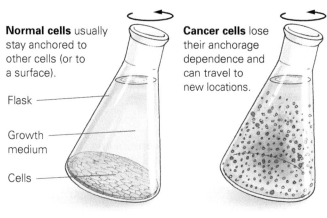

Normal cells usually stay anchored to other cells (or to a surface).

Flask

Growth medium

Cells

Cancer cells lose their anchorage dependence and can travel to new locations.

In Nicole's case, the progression from normal ovarian cells to cancerous cells may have occurred as follows: (1) One single cell in her ovary may have acquired a mutation to its *HER2* growth factor receptor gene. (2) The descendants of this cell would have been able to divide faster than neighboring cells, forming a small, benign tumor. (3) Next, a cell within the tumor may have undergone a mutation to its *BRCA2* tumor-suppressor gene, resulting in the inability of the BRCA2 protein to fix damaged DNA in the cancerous cells. (4) Cells produced by the mitosis of these doubly mutant cells would continue to divide even though their DNA is damaged, thereby enlarging the tumor and producing cells with more mutations. (5) Subsequent mutations could result in angiogenesis, lack of contact inhibition, reactivation of the telomerase enzyme, or overriding of anchorage dependence. If Nicole were very unlucky, the end result of these mutations could be that cells carrying many mutations would break away from the original ovarian tumor and set up a cancer at 1 or more new locations in her body.

Multiple-Hit Model. Because multiple mutations are required for the development and progression of cancer, scientists describe the process of cancer development using the phrase **multiple-hit model.** Nicole may have inherited some of these mutations, or they may have been induced by environmental exposures. Even though cancer is a disease caused by malfunctioning genes, most cancers are not caused only by the inheritance of mutant genes. In fact, scientists estimate that close to 70% of cancers are caused by mutations that occur during a person's lifetime.

Most of us will inherit few if any mutant cell-cycle control genes. Our level of exposure to risk factors will determine whether enough mutations will accumulate during our lifetime to cause cancer. Risk factors for ovarian cancer include smoking and uninterrupted ovulation. Ovarian cancer risk is thought to decrease for many years if ovulation is prevented for periods of time, as it is when a woman is pregnant, breastfeeding, or taking the birth control pill. When an egg cell is released from the ovary during ovulation, some tissue damage occurs. In the absence of ovulation, there is no damaged ovarian tissue and hence no need for extra cell divisions to repair damaged cells. Because mutations are most likely to occur when DNA is replicating before division, fewer divisions equal a lower likelihood of the appearance of cancer-causing mutations.

Regardless of its origin, once cancer is suspected, detection and treatment follow.

6.5 Cancer Detection and Treatment

Early detection and treatment of cancer increase the odds of survival dramatically. Being on the lookout for warning signs (**Figure 6.14**) can help alert individuals to developing cancers.

Detection Methods: Biopsy

Different cancers are detected using different methods. Some cancers are detected by the excess production of proteins that are normally produced by a particular cell type. Ovarian cancers often show high levels of CA125 protein in the blood. Nicole's level of CA125 led her gynecologist to think that a tumor might be forming on her remaining ovary. Once he suspected a tumor, Nicole's physician scheduled a biopsy. A **biopsy** is the surgical removal of cells, tissue, or fluid that will be analyzed to determine whether they are cancerous.

C hange in bowel or bladder habits

A sore that does not heal

U nusual bleeding or discharge

T hickening or lump

I ndigestion or difficulty swallowing

O bvious change in wart or mole

N agging cough or hoarseness

Figure 6.14 Warning signs of cancer. Self-screening for cancer can save your life. If you experience 1 or more of these warning signs, see your doctor.

When viewed under a microscope, benign tumors consist of orderly growths of cells that resemble the cells of the tissue from which thy were taken. Malignant or cancerous cells do not resemble other cells found in the same tissue; they are dividing so rapidly that they do not have time to produce all the proteins necessary to build normal cells. This leads to the often abnormal appearance of cancer cells as seen under a microscope.

A needle biopsy is usually performed if the cancer is located on or close to the surface of the patient's body. For example, breast lumps are often biopsied with a needle to determine whether the lump contains fluid and is a noncancerous cyst or whether it contains abnormal cells and is a tumor. When a cancer is diagnosed, surgery is often performed to remove as much of the cancerous growth as possible without damaging neighboring organs and tissues.

In Nicole's case, getting at the ovary to find tissue for a biopsy required the use of a surgical instrument called a **laparoscope.** For this operation, the surgeon inserted a small light and a scalpel-like instrument through a tiny incision above Nicole's navel.

Nicole's surgeon preferred to use the laparoscope because he knew Nicole would have a much easier recovery from laparoscopic surgery than she had from the surgery to remove her other, cystic ovary. Laparoscopy had not been possible when removing Nicole's other ovary—the cystic ovary had grown so large that her surgeon had to make a large abdominal incision to remove it.

A laparoscope has a small camera that projects images from the ovary onto a monitor that the surgeon views during surgery. These images showed that Nicole's tumor was a different shape, color, and texture from the rest of her ovary. They also showed that the tumor was not confined to the surface of the ovary; in fact, it appeared to have spread deeply into her ovary. Nicole's surgeon decided to shave off only the affected portion of the ovary and leave as much intact as possible, with the hope that the remaining ovarian tissue might still be able to produce egg cells. He then sent the tissue to a laboratory so that the pathologist could examine it. Unfortunately, when the pathologist looked through the microscope this time, she saw the disorderly appearance characteristic of cancer cells. Nicole's ovary was cancerous, and further treatment would be necessary.

Treatment Methods: Chemotherapy and Radiation

A treatment that works for one woman with ovarian cancer might not work for another ovarian cancer patient because a different suite of mutations may have led to the cancer in each woman's ovary. Luckily for Nicole, her ovarian cancer was diagnosed very early. Regrettably, this is not the case for most women with ovarian cancer, many of whom are diagnosed after the disease has progressed. The symptoms of ovarian cancer are often subtle and can be overlooked or ignored. They include abdominal swelling, pain, bloating, gas, constipation, indigestion, menstrual disorders, and fatigue. The difficulty of diagnosis is compounded because no routine screening tests are available. For instance, CA125 levels are checked only when ovarian cancer is suspected because (1) ovaries are not the only tissues that secrete this protein; (2) CA125 levels vary from individual to individual; and (3) these levels depend on the phase of the woman's menstrual cycle. Elevated CA125 levels usually mean that the cancer has been developing for a long time. Consequently, by the time the diagnosis is made, the cancer may have grown quite large and metastasized, making it much more difficult to treat.

Nicole's cancer was caught early. However, her physician was concerned that some of her cancerous ovarian cells may have spread through blood vessels or lymph ducts on or near the ovaries or spread into her abdominal cavity, so he started Nicole on chemotherapy after her surgery.

Chemotherapy. During **chemotherapy,** chemicals are injected into the bloodstream. These chemicals selectively kill dividing cells. A variety of chemotherapeutic agents act in different ways to interrupt cell division.

Chemotherapy involves many drugs because most chemotherapeutic agents affect only one type of cellular activity. Cancer cells are rapidly dividing and do not take the time to repair mistakes in replication that lead to mutations. These cells are allowed to proceed through the G_2 checkpoint with many mutations. Therefore, cancer cells can randomly undergo mutations, a few of which might allow them to evade the actions of a particular chemotherapeutic agent. Cells that are resistant to one drug proliferate when the chemotherapeutic agent clears away the other cells that compete for space and nutrients. Cells with a preexisting resistance to the drugs are selected for and produce more daughter cells with the same resistant characteristics, requiring the use of more than one chemotherapeutic agent.

Scientists estimate that cancer cells become resistant at a rate of approximately one cell per million. Because tumors contain about 1 billion cells, the average tumor will have close to 1000 resistant cells. Therefore, treating a cancer patient with a combination of chemotherapeutic agents aimed at different mechanisms increases the chances of destroying all the cancerous cells in a tumor.

Unfortunately, normal cells that divide rapidly are also affected by chemotherapy treatments. Hair follicles, cells that produce red and white blood cells, and cells that line the intestines and stomach are often damaged or destroyed. The effects of chemotherapy therefore include temporary hair loss, anemia (dizziness and fatigue due to decreased numbers of red blood cells), and lowered protection from infection due to decreases in the number of white blood cells. In addition, damage to the cells of the stomach and intestines can lead to nausea, vomiting, and diarrhea.

Several hours after each chemotherapy treatment, Nicole became nauseated; she often had diarrhea and vomited for a day or so after her treatments. Midway through her chemotherapy treatments, Nicole lost most of her hair.

Radiation Therapy. Cancer patients often undergo radiation treatments as well as chemotherapy. **Radiation therapy** uses high-energy particles to injure or destroy cells by damaging their DNA, making it impossible for these cells to continue to grow and divide. Radiation is applied directly to the tumor when possible. A typical course of radiation involves a series of 10 to 20 treatments performed after the surgical removal of the tumor, although sometimes radiation is used before surgery to decrease the size of the tumor. Radiation therapy is typically used only when cancers are located close to the surface of the body because it is difficult to focus a beam of radiation on internal organs such as an ovary. Therefore, Nicole's physician recommended chemotherapy only.

Stop & Stretch One risk of radiation therapy is an increased likelihood of secondary cancer emerging 5 to 15 years later. Why might this treatment increase cancer risk?

Nicole's treatments consisted of many different chemotherapeutic agents, spread over many months. The treatments took place at the local hospital on Wednesdays and Fridays. She usually had a friend drive her to the hospital

Figure 6.15 Chemotherapy. Many chemotherapeutic agents, such as Nicole's Taxol, are administered through an intravenous (IV) needle.

very early in the morning and return later in the day to pick her up. The drugs were administered through an intravenous (IV) needle into a vein in her arm (Figure 6.15). During the hour or so that she was undergoing chemotherapy, Nicole usually studied for her classes. She did not mind the actual chemotherapy treatments that much. The hospital personnel were kind to her, and she got some studying done. It was the aftermath of these treatments that she hated. Most days during her chemotherapy regimen, Nicole was so exhausted that she did not get out of bed until late morning, and on the day after her treatments, she often slept until late afternoon. Then she would get up and try to get some work done or make some phone calls before going back to bed early in the evening. After 6 weeks of chemotherapy, Nicole's CA125 levels started to drop. After another 2 months of chemotherapy, her CA125 levels were back down to their normal, precancerous level. If Nicole has normal CA125 levels for 5 years, she will be considered to be in **remission,** or no longer suffering negative impacts from cancer. After 10 years of normal CA125 levels, she will be considered cured of her cancer. Because Nicole's cancer responded to chemotherapy, she was spared from having to undergo any other, more experimental treatments.

Even though her treatments seemed to be going well, Nicole had other worries. She worried that her remaining ovary would not recover from the surgery and chemotherapy, which meant that she would never be able to have children. Nicole had always assumed that she would have children someday, and although she did not currently have a strong desire to have a child, she wondered if her feelings would change. Even though she was not planning to marry anytime soon, she also wondered how her future husband would feel if she were not able to become pregnant.

In addition to her concerns about being able to become pregnant, Nicole also became worried that she might pass on mutated, cancer-causing genes to her children. For Nicole, or anyone, to pass on genes to his or her children, reproductive cells must be produced by another type of nuclear division called meiosis.

6.6 Meiosis

Recall that mitosis is the division of a somatic cell's nucleus that ultimately results in two daughter nuclei that are exact copies of their parent. **Meiosis** is a form of cell division that *reduces* the number of chromosomes to produce specialized cells called **gametes** with only one-half the number of chromosomes of the parent cell. Gametes further differ from somatic cells in that gametes are produced only within the **gonads,** or sex organs.

In humans, and in most animals, the male gonads are the testes, and the female gonads are the ovaries. In animals, the male gametes are the sperm cells, while the gametes produced by the female are the egg cells. Because human somatic cells have 46 chromosomes and meiosis reduces the chromosome number by one-half, the gametes produced during meiosis contain 23 chromosomes each. (When an egg cell and a sperm cell combine their 23 chromosomes at fertilization, the developing embryo will then have the required 46 chromosomes.)

The placement of chromosomes into gametes is not random. For example, meiosis in humans does not simply place any 23 of the 46 human chromosomes into a gamete. Instead, meiosis apportions chromosomes in a very specific manner. Chromosomes in somatic cells occur in pairs. For example, the 46 chromosomes in human somatic cells are actually 23 different pairs of chromosomes. Meiosis produces gamete cells that contain one chromosome of every pair.

Autosomes (22 pairs)

Sex chromosomes (1 pair)

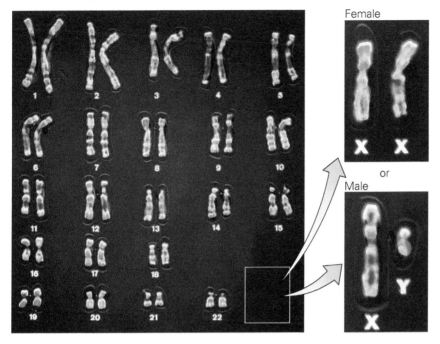

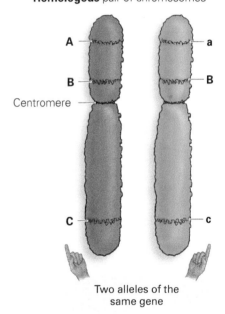

Figure 6.16 Karyotype. The pairs of chromosomes in this karyotype are arranged in order of decreasing size and numbered from 1 to 22. The X and Y sex chromosomes are the 23rd pair. The sex chromosomes from a female and a male are shown in the insets.
Visualize This: How do the X and Y chromosomes differ in terms of structure?

Homologous pair of chromosomes

Two alleles of the
same gene

It is possible to visualize chromosome pairs by preparing a **karyotype,** a highly magnified photograph of the chromosomes arranged in pairs. A human karyotype is usually prepared from chromosomes that have been removed from the nuclei of white blood cells, which have been treated with chemicals to stop mitosis at metaphase (**Figure 6.16**). Because these chromosomes are at metaphase of mitosis, they are composed of replicated sister chromatids and are shaped like the letter X. The 46 human chromosomes can be arranged into 22 pairs of nonsex chromosomes, or **autosomes,** and one pair of **sex chromosomes** (the X and Y chromosomes) to make a total of 23 pairs. Human males have an X and a Y chromosome, while females have two X chromosomes.

Each chromosome is paired with a mate that is the same size and shape and has its centromere in the same position. These pairs of chromosomes are called **homologous pairs.** Each member of a homologous pair of chromosomes carries the same genes along its length, although not necessarily the same versions of those genes. Different versions of the same gene are called **alleles** of a gene (**Figure 6.17**). In your cells, one member of each pair was inherited from your mother and the other from your father.

There are normal and mutant alleles of the *BRCA*2 gene. Note that there is a difference between the same type of information in the sense that both alleles of this gene code for a cell-cycle control protein, but they happen to code for different versions of the same protein. Alleles are alternate forms of a gene in the same way that chocolate and vanilla are alternate forms of ice cream.

When a chromosome is replicated, during the S phase of the cell cycle, the DNA is duplicated. Replication results in two copies, called *sister chromatids,* that are genetically identical. For this reason, we would find exactly the same

Figure 6.17 Homologous and non-homologous pairs of chromosomes. Homologous pairs of chromosomes have the same genes (shown here as A, B, and C) but may have different alleles. The dominant allele is represented by an uppercase letter, while the recessive allele is shown with the same letter in lowercase. Note that the chromosomes of a homologous pair each have the same size, shape, and positioning of the centromere. One member of each pair is inherited from one's mother (and colored pink), while the other is inherited from one's father (and colored blue).

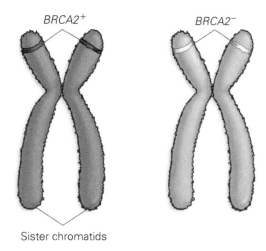

Figure 6.18 Duplicated chromosomes. This homologous pair of chromosomes has been duplicated. Note that the normal version of the *BRCA2* gene (symbolized *BRCA2+*) is present on both sister chromatids of the chromosome on the left. Its homologue on the right carries the mutant version of this allele (symbolized *BRCA2−*).

information on the sister chromatids that comprise a replicated chromosome (Figure 6.18).

Meiosis separates the members of a homologous pair from each other. Once meiosis is completed, there is one copy of each chromosome (1–23) in every gamete. When only one member of each homologous pair is present in a cell, we say that the cell is **haploid (n)**—both egg cells and sperm cells are haploid. All somatic cells in humans contain homologous pairs of chromosomes and are therefore diploid. For a diploid cell in a person's testes or ovary to become a haploid gamete, it must go through meiosis. After the sperm and egg fuse, the fertilized cell, or **zygote,** will contain two sets of chromosomes and is said to be **diploid (2n)** (Figure 6.19). Like mitosis, meiosis is preceded by an interphase stage that includes G_1, S, and G_2. Interphase is followed by two phases of meiosis, called meiosis I and meiosis II, in which divisions of the nucleus take place (Figure 6.20). Meiosis I separates the members of a homologous pair from each other. Meiosis II separates the chromatids from each other. Both meiotic divisions are followed by cytokinesis, during which the cytoplasm is divided between the resulting daughter cells.

Interphase

The interphase that precedes meiosis consists of G_1, S, and G_2. This interphase of meiosis is similar in most respects to the interphase that precedes mitosis. The centrioles from which the microtubules will originate are present. The G phases are times of cell growth and preparation for

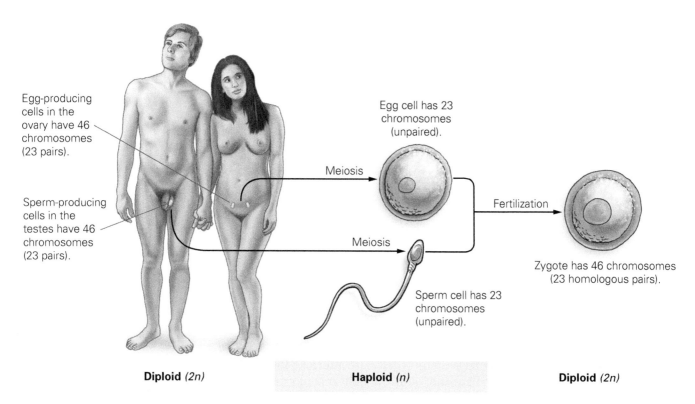

Figure 6.19 Gamete production. The diploid cells of the ovaries and testes undergo meiosis and produce haploid gametes. At fertilization, the diploid condition is restored.

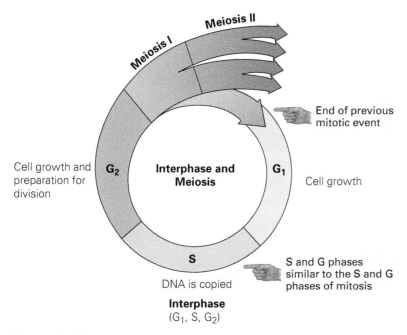

Figure 6.20 Interphase and meiosis. Interphase consists of G_1, S, and G_2 and is followed by 2 rounds of nuclear division, meiosis I and meiosis II. **Visualize This:** Does DNA duplication occur during the interphase preceding meiosis II?

division. The S phase is when DNA replication occurs. Once the cell's DNA has been replicated, it can enter meiosis I.

Meiosis I

The first meiotic division, meiosis I, consists of prophase I, metaphase I, anaphase I, and telophase I (**Figure 6.21** on the next page).

During prophase I of meiosis, the nuclear envelope starts to break down, and the microtubules begin to assemble. The previously replicated chromosomes condense so that they can be moved around the cell without becoming entangled. The condensed chromosomes can be seen under a microscope. At this time, the homologous pairs of chromosomes exchange genetic information in a process called *crossing over*, which will be explained in a moment.

At metaphase I, the homologous pairs line up at the cell's equator, or middle of the cell. Microtubules bind to the metaphase chromosomes near the centromere. Homologous pairs are arranged arbitrarily regarding which member faces which pole. This process is called *random alignment*. At the end of this section, you will find detailed descriptions of crossing over and random alignment along with their impact on genetic diversity.

At anaphase I, the homologous pairs are separated from each other by the shortening of the microtubules, and at telophase I, nuclear envelopes re-form around the chromosomes. DNA is then partitioned into each of the two daughter cells by cytokinesis. Because each daughter cell contains only one copy of each member of a homologous pair, at this point the cells are haploid. Now both of these daughter cells are ready to undergo meiosis II.

Stop & Stretch How is meiosis I similar to mitosis? How is it different?

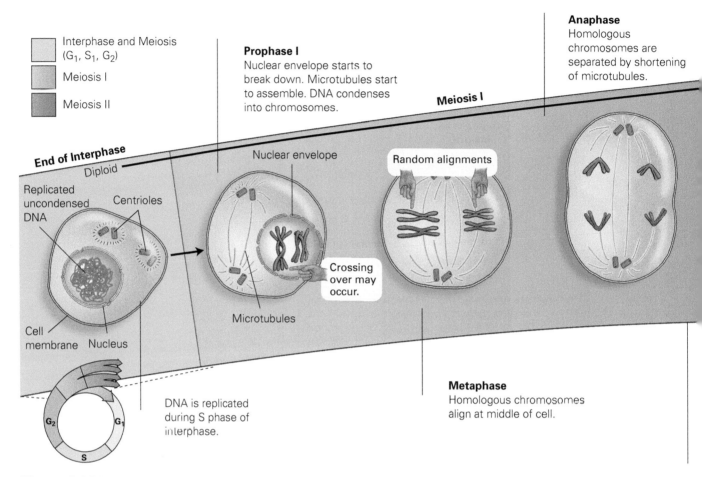

Figure 6.21 Sexual cell division. This diagram illustrates interphase, meiosis I, meiosis II, and cytokinesis in an animal cell.

Meiosis II

Meiosis II consists of prophase II, metaphase II, anaphase II, and telophase II. This second meiotic division is virtually identical to mitosis and serves to separate the sister chromatids of the replicated chromosome from each other.

At prophase II of meiosis, the cell is readying for another round of division, and the microtubules are lengthening again. At metaphase II, the chromosomes align in single file across the equator in much the same way that they do during mitosis—not as pairs, as was the case with metaphase I. At anaphase II, the sister chromatids separate from each other and move to opposite poles of the cell. At telophase II, the separated chromosomes each become enclosed in their own nucleus. In this fashion, half of a person's genes are physically placed into each gamete; thus, children carry one-half of each parent's genes.

Each parent can produce millions of different types of gametes due to two events that occur during meiosis I—crossing over and random alignment. Both of these processes greatly increase the number of different kinds of gametes that an individual can produce and therefore increase the variation in individuals that can be produced when gametes combine.

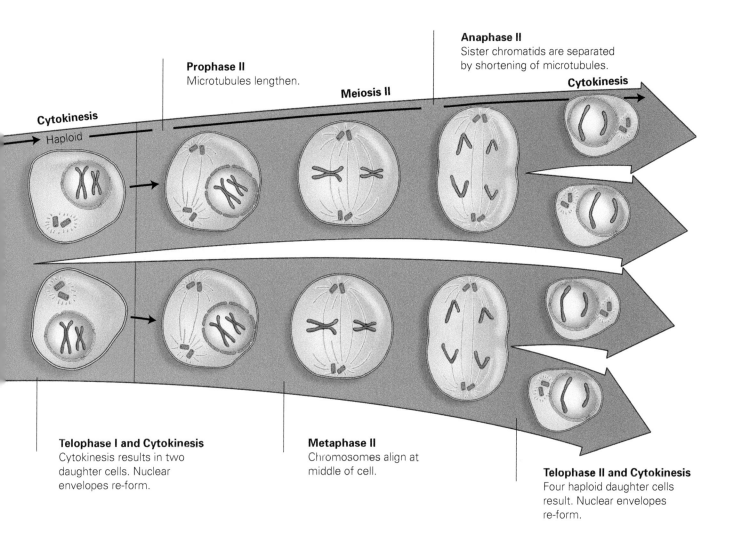

Anaphase II
Sister chromatids are separated by shortening of microtubules.

Prophase II
Microtubules lengthen.

Meiosis II

Cytokinesis

Cytokinesis
Haploid

Telophase I and Cytokinesis
Cytokinesis results in two daughter cells. Nuclear envelopes re-form.

Metaphase II
Chromosomes align at middle of cell.

Telophase II and Cytokinesis
Four haploid daughter cells result. Nuclear envelopes re-form.

Crossing Over and Random Alignment

Crossing over occurs during prophase I of meiosis I. It involves the exchange of portions of chromosomes from one member of a homologous pair to the other member. Crossing over is believed to occur several times on each homologous pair during each occurrence of meiosis.

To illustrate crossing over, consider an example using genes involved in the production of flower color and pollen shape in sweet pea plants. These two genes are on the same chromosome and are called **linked genes.** Linked genes move together on the same chromosome to a gamete, and they may or may not undergo crossing over.

If a pea plant has red flowers and long pollen grains, the chromosomes may appear as shown in **Figure 6.22** (on the next page). It is possible for this plant to produce four different types of gametes with respect to these two genes. Two types of gametes would result if no crossing over occurred between these genes—the gamete containing the red flower and long pollen chromosome and the gamete containing the white flower and short pollen chromosome. Two additional types of gametes could be produced if crossing over did occur—one type containing the red flower and short pollen grain chromosome and the other type containing the reciprocal white flower and long pollen grain chromosome. Therefore, crossing over increases genetic diversity by increasing the number of distinct combinations of genes that may be present in a gamete.

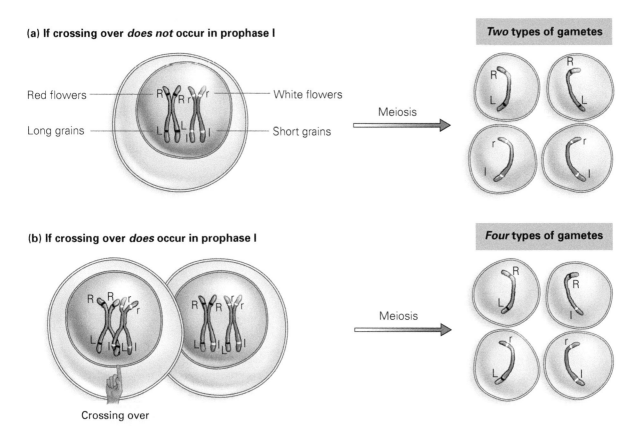

(a) If crossing over *does not* occur in prophase I

Red flowers — — White flowers

Long grains — — Short grains

Meiosis

Two types of gametes

(b) If crossing over *does* occur in prophase I

Crossing over

Meiosis

Four types of gametes

Figure 6.22 Crossing over. If a flower with the above arrangement of alleles undergoes meiosis, it can produce (a) two different types of gametes for these two genes if crossing over does not occur or (b) four different types of gametes for these two genes if crossing over occurs at L.

Random alignment of homologous pairs also increases the number of genetically distinct types of gametes that can be produced. Using Nicole's chromosomes as an example (Figure 6.23), let us assume that she did in fact inherit mutant versions of both the *BRCA2* and *HER2* genes and that these genes are located on different chromosomes. The arrangement of homologous pairs of chromosomes at metaphase I determines which chromosomes will end up together in a gamete. If we consider only these two homologous pairs of chromosomes, then two different alignments are possible, and four different gametes can be produced. For example, when Nicole produces egg cells, the two chromosomes that she inherited from her dad could move together to the gamete, leaving the two chromosomes she inherited from her mom to move to the other gamete. It is equally probable that Nicole could undergo meiosis in which one chromosome from each parent would align randomly together, resulting in two more types of gametes being produced. (In Chapters 7 and 8, you will see how random alignment and crossing over affect the inheritance of genes in greater detail.)

Mistakes in Meiosis

Sometimes mistakes occur during meiosis that result in the production of offspring with too many or too few chromosomes. Too many or too few chromosomes can result when there is a failure of the homologues (or sister chromatids) to separate during meiosis. This failure of chromosomes to separate is termed **nondisjunction** (Figure 6.24). The presence of an extra chromosome is termed **trisomy.** The absence of one chromosome of a homologous pair is termed **monosomy.** Nondisjunction can occur on autosomes or sex chromosomes.

Because the X and Y sex chromosomes do not carry the same genes and are not the same size and shape, they are not considered to be a homologous pair.

(a) One possible metaphase I alignment

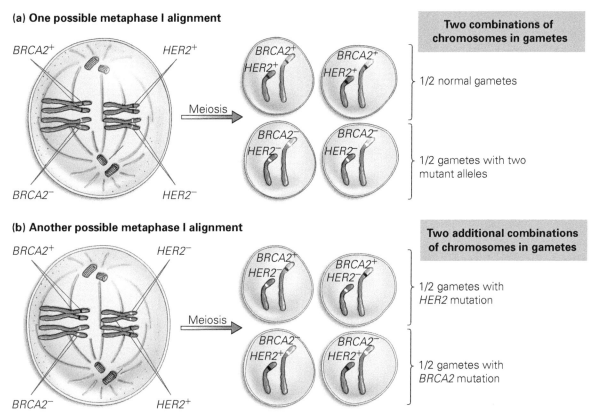

Two combinations of chromosomes in gametes

BRCA2⁺
HER2⁺

BRCA2⁺
HER2⁺

} 1/2 normal gametes

BRCA2⁻
HER2⁻

BRCA2⁻
HER2⁻

} 1/2 gametes with two mutant alleles

(b) Another possible metaphase I alignment

Two additional combinations of chromosomes in gametes

BRCA2⁺
HER2⁻

BRCA2⁺
HER2⁻

} 1/2 gametes with *HER2* mutation

BRCA2⁻
HER2⁺

BRCA2⁻
HER2⁺

} 1/2 gametes with *BRCA2* mutation

Figure 6.23 Random alignment. Two possible alignments, (a) and (b), can occur when there are two homologous pairs of chromosomes. These different alignments can lead to novel combinations of genes in the gametes.
Visualize This: How many different alignments are possible with three homologous pairs of chromosomes?

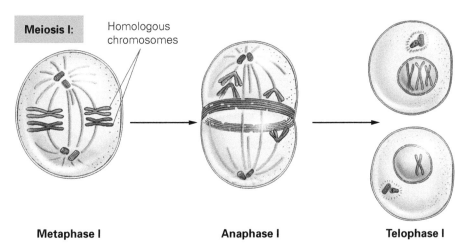

Meiosis I: Homologous chromosomes

Metaphase I   Anaphase I   Telophase I

Figure 6.24 Nondisjunction. Nondisjunction during meiosis I produces gametes with too many and too few chromosomes.

There is, however, a region at the tip of each chromosome that is similar enough so that they can pair up during meiosis. Typically, early embryos with too many or too few chromosomes will die because they have too much or too little genetic information. However, in some situations, such as when the extra or missing chromosome is very small (such as in chromosomes 13, 18, and 21) or contains very little genetic information (such as the Y chromosome or an X chromosome that will later be inactivated), the embryo can survive. **Table 6.2** (on the next page) lists some chromosomal anomalies in humans that are compatible with life.

TABLE 6.2

Autosomal and sex-linked chromosomal anomalies.

Conditions Caused by Nondisjunction of Autosomes	Comments
Trisomy 21—Down syndrome	Affected individuals are mentally retarded and have abnormal skeletal development and heart defects. The frequency of Down syndrome increases with parental age. Down syndrome among the children of parents over 40 occurs in 6 per 1000 births, which is six times higher than the rate found among the children of couples under 35 years old.
Trisomy 13—Patau syndrome	Affected individuals are mentally retarded and deaf and have a cleft lip and palate. Approximately 1 in 5000 newborns is affected.
Trisomy 18—Edwards syndrome	Affected individuals have malformed organs, ears, mouth, and nose, leading to an elfin appearance. These babies usually die within 6 months of birth. Approximately 1 in 6000 newborns is affected.

Conditions Caused by Nondisjunction of Sex Chromosomes	Comments
XO—Turner Syndrome	Females with one X chromosome can be sterile if their ovaries fail to develop. Webbing of the neck, shorter stature, and hearing impairment are also common. Approximately 1 in 5000 female newborns is born with only one X chromosome. Turner syndrome is the only human monosomy that is viable.
Trisomy X: Meta female	Females with three X chromosomes tend to develop normally. Approximately 1 in 1000 females is born with an extra X chromosome.
XXY—Kleinfelter syndrome	Males with the XXY genotype are less fertile than XY males; have small testes, sparse body hair, some breast enlargement; and may have mental retardation. Testosterone injections can reverse some of the anatomical abnormalities in the approximately 1 in 1000 males with this condition.
XYY condition	Males with two Y chromosomes tend to be taller than average but have an otherwise normal male phenotype. Approximately 1 in 1000 newborn males has an extra Y chromosome.

From the previous discussions, you have learned that cells undergo mitosis for growth and repair and meiosis to produce gametes. **Figure 6.25** compares the significant features of mitosis and meiosis.

It is now possible to revisit the question of whether Nicole will pass on cancer-causing genes to any children she may have. Because Nicole developed cancer at such a young age, it seems likely that she may have inherited at least one mutant cell-cycle control gene; thus, she may or may not pass that gene on. If Nicole has both a normal and a mutant version of a cell-cycle control gene, then she will be able to make gametes with and without the mutant allele. Therefore, she could pass on the mutant allele if a gamete containing that allele is involved in fertilization. We have also seen that it takes many "hits" or mutations for a cancer to develop. Therefore, even if Nicole does pass

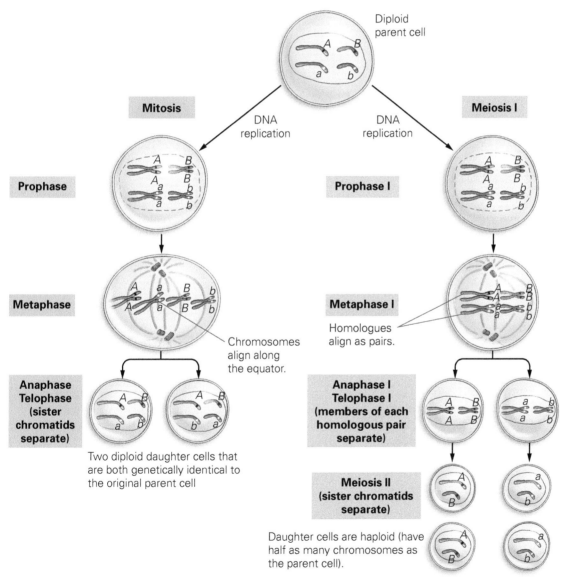

Figure 6.25 Comparing mitosis and meiosis. Mitosis is a type of cell division that occurs in somatic cells and gives rise to daughter cells that are exact genetic copies of the parent cell. Meiosis occurs in cells that will give rise to gametes and decreases the chromosome number by one-half. To do this and still ensure that each gamete receives one member of each homologous pair, the two members of each homologous pair align across the equator at the metaphase I of meiosis and are separated from each other during anaphase I.

on one or a few mutant versions of cell-cycle control genes to a child, environmental conditions will dictate whether enough other mutations will accumulate to allow a cancer to develop.

Mutations caused by environmental exposures are not passed from parents to children unless the mutation happens to occur in a cell of the gonads that will be used to produce a gamete. Nicole's cancer occurred in the ovary, the site of meiosis, but not all cells in the ovary undergo meiosis. Nicole's cancer originated in the outer covering of the ovary, a tissue that does not undergo meiosis. The cells involved in ovulation are located inside the ovary. A skin cancer that develops from exposure to ultraviolet light will not be passed on, and the same is true for most of the mutations that Nicole obtained from environmental exposures. Therefore, for any children that Nicole (or any of us) might have, it is the combined effects of inherited mutant alleles and any mutations induced by environmental exposures that will determine whether cancers will develop.

SAVVY READER

Alternative Cancer Treatments

One website claims that 80% of cancer therapies are blocked by the patient's emotions and that these blocks can be removed by a technique called tapping. Just buy the CD, follow the tapping instructions, and you will feel better in no time. Another website claims that sharks do not get cancer and that taking shark-cartilage supplements can help treat cancer. In a study of the effectiveness of shark cartilage, the study authors report that 15 of 29 patients diagnosed with terminal cancer were still alive 1 year after beginning to take the supplements, which is "a remarkable result by any measure," according to the study authors.

Websites for alternative cancer treatment centers offer unproven treatments to patients that cost tens of thousands of dollars and are not covered by insurance. These treatments are often supervised by actual medical doctors. These sites are filled with testimonials, but no data, about the effectiveness of such treatments.

1. Is there any guarantee that something written on a website is true?

2. The website that suggested that emotions can block cancer therapies presented no evidence to back up this claim. What would you do if you wanted to know whether that claim had any merit?

3. The study on shark cartilage was published in a non-peer reviewed journal. Does this fact add or subtract from the credibility of this article?

4. What other information do you need to determine whether the one-year survival rate of the shark cartilage study has any real meaning?

5. Would the fact that the author of the shark cartilage study owns a company that sells the product make you more or less skeptical of his findings? What about the fact that sharks actually do get cancer?

6. While most medical doctors have dedicated their lives to helping people in need, a few will promote products and services simply to make money. Should you always believe the word of someone with an advanced degree?

7. Why do you think that so many people fall prey to dubious cancer cures?

Chapter Review
Learning Outcomes

LO1 Describe the cellular basis of cancer (Section 6.1).

- Unregulated cell division can lead to the formation of a tumor (p. 113).

LO2 Compare and contrast benign and malignant tumors (Section 6.1).

- Benign or noncancerous tumors stay in one place and do not prevent surrounding organs from functioning. Malignant tumors are those that are invasive or those that metastasize to surrounding tissues, starting new cancers (p. 113).

LO3 List several risk factors for cancer development (Section 6.1).

- Risk factors for cancer include smoking, a poor diet, lack of exercise, obesity, alcohol use, and aging (pp. 114–117).

LO4 List the normal functions of cell division (Section 6.2).

- Cell division is a process required for growth and development (p. 118).

LO5 Describe the structure and function of chromosomes (Section 6.2).

- Chromosomes are composed of DNA wrapped around proteins. Chromosomes carry genes (p. 118).

LO6 Outline the process of DNA replication (Section 6.2).

- During DNA replication or synthesis, one strand of the double-stranded DNA molecule is used as a template for the synthesis of a new daughter strand of DNA. The newly synthesized DNA strand is complementary to the parent strand. The enzyme DNA polymerase ties together the nucleotides on the daughter strand. (pp. 119–120)

LO7 Describe the events that occur during interphase of the cell cycle (Section 6.3).

- Interphase consists of two gap phases of the cell cycle (G_1 and G_2), during which the cell grows and prepares to enter mitosis or meiosis, and the S (synthesis) phase, during which time the DNA replicates. The S phase of interphase occurs between G_1 and G_2 (pp. 120–121).

LO8 Diagram two chromosomes as they proceed through mitosis of the cell cycle (Section 6.3).

- During mitosis, the sister chromatids are separated from each other into daughter cells. During prophase, the replicated DNA condenses into linear chromosomes. At metaphase, these replicated chromosomes align across the middle of the cell. At anaphase, the sister chromatids separate from each other and align at opposite poles of the cells. At telophase, a nuclear envelope re-forms around the chromosomes lying at each pole (pp. 121–123).

LO9 Describe the process of cytokinesis in animal and plant cells (Section 6.3).

- Cytokinesis is the last phase of the cell cycle. During cytokinesis, the cytoplasm is divided into two portions, one for each daughter cell (pp. 123–124).

LO10 Describe how the cell cycle is regulated and how dysregulation can lead to tumor formation (Section 6.4).

- When cell division is working properly, it is a tightly controlled process. Normal cells divide only when conditions are favorable. Proteins survey the cell and its environment at checkpoints as the cell moves through G_1, G_2, and metaphase, and can halt cell division if conditions are not favorable. Mistakes in regulating the cell cycle arise when genes that control the cell cycle are mutated. Proto-oncogenes regulate the cell cycle. Oncogenes are mutated versions of these genes. Tumor suppressors are normal genes that can encode proteins to stop cell division if conditions are not favorable and can repair damage to the DNA. They serve as backups in case the proto-oncogenes undergo mutation (pp. 126–127).

LO11 Explain how genes and environment both impact cancer risk (Section 6.4).

- Mutated genes can be inherited, or mutations can be caused by exposure to carcinogens (pp. 128–129).

LO12 Discuss the various methods of cancer detection and treatment (Section 6.5).

- A biopsy is a common method for detecting cancer. It involves removing some cells or tissues suspected of being cancerous and analyzing them. Typical cancer treatments include chemotherapy, which involves injecting chemicals that kill rapidly dividing cells, and radiation, which involves killing tumor cells by exposing them to high-energy particles (pp. 129–132).

LO13 **Explain what types of cells undergo meiosis, the end result of this process, and how meiosis increases genetic diversity (Section 6.6).**

- Meiosis is a type of sexual cell division, occurring in cells, that gives rise to gametes. Gametes contain half as many chromosomes as somatic cells do. The reduction of chromosome number that occurs during meiosis begins with diploid cells and ends with haploid cells (pp. 132–134).
- Meiosis is preceded by an interphase stage in which the DNA is replicated. During meiosis I, the members of a homologous pair of chromosomes are separated from each other. During meiosis II, the sister chromatids are separated from each other (pp. 134–137).

LO14 **Diagram four chromosomes from a diploid organism undergoing meiosis (Section 6.6).**

- Homologues align in pairs during meiosis I and as individual chromosomes at metaphase II (pp. 135–137).

LO15 **Explain the significance of crossing over and random alignment in terms of genetic diversity (Section 6.6).**

- Homologous pairs of chromosomes exchange genetic information during crossing over at prophase I of meiosis, thereby increasing the number of genetically distinct gametes that an individual can produce. The alignment of members of a homologous pair at metaphase I is random with regard to which member of a pair faces which pole. This random alignment of homologous chromosomes increases the number of different kinds of gametes an individual can produce (pp. 137–138).

Roots to Remember

The following roots of words come mainly from Latin and Greek and will help you decipher terms:

cyto- and -cyte relate to cells. Chapter term: cytoplasm

-kinesis means motion. Chapter term: cytokinesis

meio- means to make smaller. Chapter term: meiosis

mito- means a thread. Chapter term: mitosis

onco- means cancer. Chapter term: oncogene

proto- means before. Chapter term: proto-oncogene

soma- and -some mean body. Chapter terms: somatic and chromosome

telo- means end or completion. Chapter term: telophase

Learning the Basics

1. **LO8 LO14** List the ways in which mitosis and meiosis differ.

2. **LO12** What property of cancer cells do chemotherapeutic agents attempt to exploit?

3. **LO8** A cell that begins mitosis with 46 chromosomes produces daughter cells with _____.

    A. 13 chromosomes; **B.** 23 chromosomes;
    C. 26 chromosomes; **D.** 46 chromosomes

4. **LO5** The centromere is a region at which _____.

    A. sister chromatids are attached to each other;
    B. metaphase chromosomes align; **C.** the tips of chromosomes are found; **D.** the nucleus is located

5. **LO4 LO8** Mitosis _____.

    A. occurs in cells that give rise to gametes; **B.** produces haploid cells from diploid cells; **C.** produces daughter cells that are exact genetic copies of the parent cell; **D.** consists of two separate divisions, mitosis I and mitosis II

6. **LO4 LO8** At metaphase of mitosis, _____.

    A. the chromosomes are condensed and found at the poles; **B.** the chromosomes are composed of one sister chromatid; **C.** cytokinesis begins; **D.** the chromosomes are composed of two sister chromatids and are lined up along the equator of the cell

7. **LO5** Sister chromatids _____.

    A. are two different chromosomes attached to each other; **B.** are exact copies of one chromosome that are attached to each other; **C.** arise from the centrioles; **D.** are broken down by mitosis; **E.** are chromosomes that carry different genes

8. **LO6** DNA polymerase _____.

    A. attaches sister chromatids at the centromere;
    B. synthesizes daughter DNA molecules from fats and phospholipids; **C.** is the enzyme that facilitates DNA synthesis; **D.** causes cancer cells to stop dividing

9. **LO13** After telophase I of meiosis, each daughter cell is _____.

    A. diploid, and the chromosomes are composed of one doublestranded DNA molecule; **B.** diploid, and the chromosomes are composed of two sister chromatids; **C.** haploid, and the chromosomes are composed of one double-stranded DNA molecule; **D.** haploid, and the chromosomes are composed of two sister chromatids

10. **LO15** List two things that happen during meiosis that increase genetic diversity.

11. **LO10** Define the terms *proto-oncogene* and *oncogene*.

12. **LO7 LO9** In what ways is the cell cycle similar in plant and animal cells, and in what ways does it differ?

13. **LO14** State whether the chromosomes depicted in parts (a)–(d) of **Figure 6.26** are haploid or diploid.

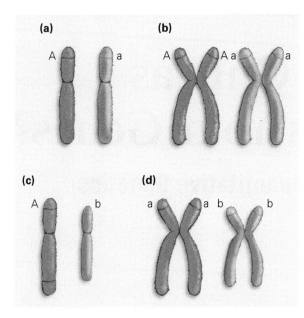

(a)

(b)

(c)

(d)

Figure 6.26 Haploid or diploid chromosomes?

Analyzing and Applying the Basics

1. **LO3 LO11** Would a skin cell mutation that your father obtained from using tanning beds make you more likely to get cancer? Why or why not?

2. **LO2** Will all tumors progress to cancers?

3. **LO12** Why are some cancers treated with radiation therapy while others are treated with chemotherapy?

Connecting the Science

1. Should members of society be forced to pay the medical bills of smokers when the cancer risk from smoking is so evident and publicized? Explain your reasoning.

2. Would you want to be tested for the presence of cell-cycle control mutations? How would knowing whether you had some mutated proto-oncogenes be of benefit or harm?

Answers to **Stop & Stretch, Visualize This, Savvy Reader,** and **Chapter Review** questions can be found in the **Answers** section at the back of the book.

Are You Only as Smart as Your Genes?

Mendelian and Quantitative Genetics

Can a woman create the
perfect child if she
chooses the right sperm?

The Fairfax Cryobank is a nondescript brick building located in a quiet, tree-lined suburb of Washington, DC. Stored inside this unremarkable edifice are the hopes and dreams of thousands of women and their partners. The Fairfax Cryobank is a sperm bank; inside its many freezers are vials containing sperm collected from hundreds of men. Women can order these sperm for a procedure called artificial insemination, which may allow them to conceive a child despite the lack of a fertile male partner.

If a woman chooses a donor with the right genes, her child may look like her partner.

Women who purchase sperm from the Fairfax Cryobank can choose from hundreds of potential donors. The donors are categorized into three classes, and their sperm is priced accordingly. Most women who choose artificial insemination want detailed information about the donor before they purchase a sample; while all Fairfax Cryobank donors submit to comprehensive physical exams and disease testing and provide a detailed family health history, not all provide childhood pictures, audio CDs of their voices, or personal essays. Sperm samples from men who did not provide this additional information are sold at a discount because most women seek a donor who seems compatible in interests and aptitudes.

If she chooses a donor with the right genes, will her child be a genius?

However, in addition to the information-rich donors, there is also a set of premium sperm donors referred to by Fairfax Cryobank as its "Doctorate" category. These men either are in the process of earning or have completed a doctoral degree in medicine, law, or academia. A sperm sample from this donor category is 30% more expensive than sperm from the standard donor, and because several samples are typically needed, ensuring pregnancy from one of these donors is likely to cost hundreds of additional dollars.

Or is a child's intelligence more influenced by his environment?

147

Why would some women be willing to pay significantly more for sperm from a donor who has an advanced degree? Because academic achievement is associated with intelligence. These women want intelligent children, and they are willing to pay more to provide their offspring with "extra-smart" genes. But are these women putting their money in the right place? Is intelligence about genes, or is it a function of the environmental conditions in which a baby is raised? In other words, is who we are a result of our "nature" or our "nurture"? As you read this chapter, you will see that the answer to this question is not a simple one—our characteristics come from both our biological inheritance and the environment in which we developed.

7.1 The Inheritance of Traits

Most of us recognize similarities between our birth parents and ourselves. Family members also display resemblances—for instance, all the children of a single set of parents may have dimples. However, it is usually quite easy to tell siblings apart. Each child of a set of parents is unique, and none of us is simply the "average" of our parents' traits. We are each more of a combination—one child may be similar to her mother in eye color and face shape, another similar to mom in height and hair color.

To understand how your parents' traits were passed to you and your siblings, you need to understand the human life cycle. A **life cycle** is a description of the growth and reproduction of an individual (**Figure 7.1**).

A human baby is typically produced from the fusion of a single sperm cell produced by the male parent and a single egg cell produced by the female parent. Egg and sperm (called *gametes*) fuse at **fertilization**, and the resulting cell, called a zygote, duplicates all of the genetic information it contains and undergoes mitosis to produce two identical daughter cells. Each of these daughter cells

Figure 7.1 Instructions inside. Both parents contribute genetic information to their offspring via their sperm and egg. The single cell that results from the fusion of one sperm with one egg contains all of the instructions necessary to produce an adult human.

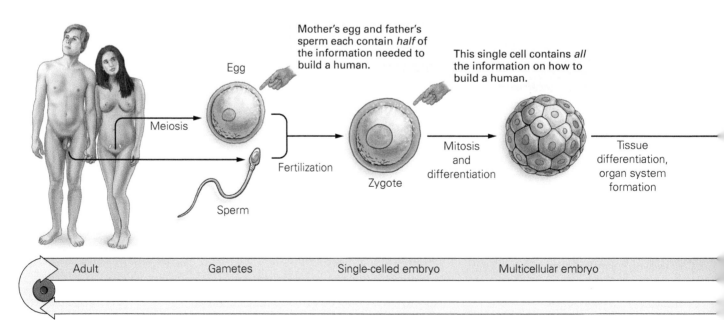

Mother's egg and father's sperm each contain *half* of the information needed to build a human.

This single cell contains *all* the information on how to build a human.

Egg

Meiosis

Fertilization

Sperm

Zygote

Mitosis and differentiation

Tissue differentiation, organ system formation

Adult   Gametes   Single-celled embryo   Multicellular embryo divides dozens of times in the same way. The cells in this resulting mass then differentiate into specialized cell types, which continue to divide and organize to produce the various structures of a developing human, called an embryo. Continued division of this single cell and its progeny leads to the production of a full-term infant and eventually an adult.

We are made up of trillions of individual cells, all of them the descendants of that first product of fertilization and nearly all containing exactly the same information originally found in the zygote. All of our traits are influenced by the information contained in that tiny cell.

Genes and Chromosomes

Each normal sperm and egg contains information about "how to build an organism." A large portion of that information is in the form of genes, segments of DNA that generally code for proteins.

Imagine genes as roughly equivalent to the words used in an instruction manual. These words are contained on chromosomes, which are roughly analogous to pages in the manual.

Prokaryotes such as bacteria typically contain a single, circular chromosome that floats freely inside the cell and is passed in its entirety to each offspring. In contrast, eukaryotes carry their genes on more than one linear chromosome. The number of chromosomes in eukaryotes can vary greatly, from 2 in the jumper ant (*Myrmecia pilosula*) to an incredible 1260 in the stalked adder's tongue, a species of fern (*Ophioglossum reticulatum*, **Figure 7.2**). Human cells contain 46 chromosomes, most of which carry thousands of genes. Thus, each cell has 46 pages of instructions, with each page containing thousands of words.

The instruction manual inside a cell is different from the instruction manual that comes with, for instance, a kit for building a model car. You would read the manual for building the car beginning at page 1 and follow an orderly set of steps to produce the final product. The human instruction manual is much more complicated—the pages and words to be read are different for different types of cell and may even change according to the situation. The "final product" of any given cell depends on the words used and the order in which the words are read from this common instruction manual.

For instance, eye cells and pancreas cells in mammals both carry instructions for the protein rhodopsin, which helps detect light, but rhodopsin is

(a) Stalked adder's tongue (a fern)

(b) Single fern cell

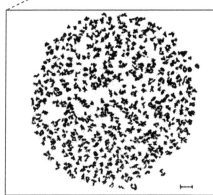

Figure 7.2 Variation in chromosome number. The amount of genetic information in an organism does not correlate to its complexity, as can be seen by examination of the 1260 chromosomes contained in a single cell of the stalked adder's tongue fern.

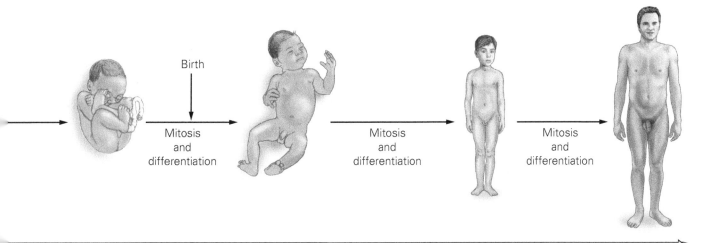

Birth

Mitosis and differentiation

Mitosis and differentiation

Mitosis and differentiation

Fetus   Baby   Child   Adult

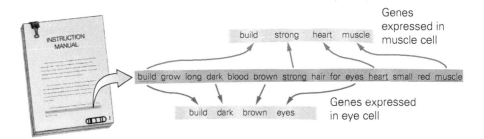

Figure 7.3 **Genes as words in an instruction manual.** Different words from the manual are used in different parts of the body, and identical words may be used in distinctive combinations in different cells.
Visualize This: Write several more instructions that can be extracted from this group of 14 words.

produced only in eye cells, not in pancreas cells. Rhodopsin requires assistance from another protein, called transducin, to translate the light that strikes it into the actions of the eye cell. Transducin is also produced in pancreas cells, but there it is combined with a third protein where it may help coordinate release of the hormone insulin.

Thus, a protein may serve two or more different functions depending on its context. Because genes, like words, can be used in many combinations, the instruction manual for building a living organism is very flexible (Figure 7.3).

Producing Diversity in Offspring

During reproduction, genes from both parents are copied and transmitted to the next generation. The copying and transmittal of genes from one generation to the next introduces variation among genes. It is this **genetic variation** that most interests individuals seeking sperm donors; the genes in the sperm they select will provide information about the traits of their offspring.

Gene Mutation Creates Genetic Diversity. Recall the process of DNA replication. Like a retyped instruction manual page, copies of chromosomes are rewritten rather than "photocopied." As a result, there is a chance of a typographical error, or mutation, every time a cell divides. Mutations in genes lead to different versions, or **alleles**, of the gene. The various types of mutation are described in Figure 7.4. As you can see from this figure, many mutations result in nonsensical instructions, that is, dysfunctional alleles, and thus are often harmful to the individual who possesses them. Dysfunctional alleles tend to be "lost" over time because individuals with them do not function as well as those without the mutation. In short, individuals with dysfunctional mutations may not survive or may reproduce at very low rates.

However, some mutations are neutral in effect, or even beneficial in certain situations, while some harmful mutations can be hidden if the individuals who carry them also carry a functional allele. These mutations tend to persist over generations. Because mutations occur at random and are not expected to occur in the same genes in different individuals, each of us should have a unique set of alleles reflecting the mutations passed along to us by our unique set of ancestors.

By contributing to differences among families over many generations, mutation creates genetic variation in a population in the form of new alleles. When a novel characteristic increases an individual's chance of survival and reproduction, the mutation contributes to a population's adaptation to its environment (Chapter 11). Genetic misspellings are thus the engine that drives evolution itself.

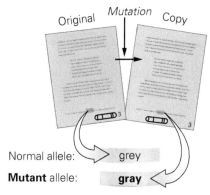

(a) **The mutant allele has the same meaning** (mutant allele function the same as the original allele).

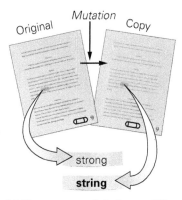

(b) **The mutant allele has a different meaning** (mutant allele functions differently than the original allele).

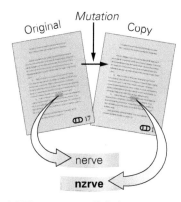

(c) **The mutant allele has no meaning** (mutant allele is no longer functional).

Figure 7.4 The formation of different alleles. Different alleles for a gene may form as a result of copying errors. In this analogy, misspellings (mutations) do not change the meaning of the word (allele), but some may result in altered meanings (different allele function) or have no meaning at all (no allele function).

> **Stop & Stretch** Not all of the DNA in a cell codes for proteins—some DNA functions as binding sites for gene promoters (proteins that "turn on" genes), other segments may provide structural support for chromosomes, and other parts may be meaningless. What do you think the effect of "misspellings" is on these nongene segments of DNA?

Segregation and Independent Assortment Create Gamete Diversity.

Both parents contribute genetic instructions to each child, but they do not contribute their entire manual. If they did, the genetic instructions carried in human cells would double every generation, making for a pretty crowded cell. Instead, the process of meiosis reduces the number of chromosomes carried in gametes by one-half (Chapter 6).

Although they are only transmitting half of their genetic information in a gamete, each parent actually gives a complete copy of the instruction manual to each child (**Figure 7.5**). This can occur because, in effect, our body cells each contain two copies of the manual—that is, each has two versions of each page, with each version containing essentially the same words. In other words, the 46 chromosomes each cell contains are actually 23 pairs of chromosomes, with each member of a pair containing essentially the same genes. Each set of two equivalent chromosomes is referred to as a **homologous pair.** The members of a homologous pair are equivalent, but not identical, because even though both have the same genes, each contains a unique set of alleles inherited from one or the other parent.

The process of meiosis separates homologous pairs of chromosomes and also places chromosomes independently into each gamete. These two events explain why siblings are not identical (with the exception of identical twins). For the most part, it is because parents do not give all of their offspring exactly the same set of alleles.

When homologous pairs are separated during meiosis, the alleles carried on the members of the pair are separated as well. The separation of pairs of alleles during the production of gametes is called **segregation.** Thus, a parent with two different alleles of a gene will produce gametes with a 50% probability of containing one version of the allele and a 50% probability of containing the other version.

The segregation of chromosomes during meiosis leads to **independent**

Figure 7.5 Equivalent information from parents. Each parent provides a complete set of instructions to each offspring.

Egg		Sperm		Zygote
	+		=	

The 23 pages of each instruction manual are roughly equivalent to the 23 chromosomes in each egg and sperm.

The zygote has 46 pages, equivalent to 46 chromosomes.

Parent cells have 2 copies of each chromosome — that is, 2 full sets of instruction manual pages, 1 from each parent.

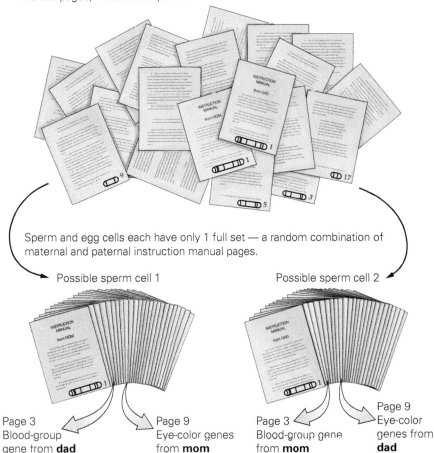

Sperm and egg cells each have only 1 full set — a random combination of maternal and paternal instruction manual pages.

Possible sperm cell 1

Possible sperm cell 2

Page 3
Blood-group
gene from **dad**

Page 9
Eye-color genes
from **mom**

Page 3
Blood-group gene
from **mom**

Page 9
Eye-color
genes from
dad

Figure 7.6 Each egg and sperm is unique. Because each sperm is produced independently, the set of pages in each sperm will be a unique combination of the pages that the man inherited from his mom and dad.
Visualize This: This man could produce four distinctly different kinds of sperm cell when considering these two genes and four alleles. List the other two possibilities not pictured here.

assortment, in which the alleles for each gene are inherited (mostly) independently of each other. Independent assortment arises from the *random alignment* of chromosomes during meiosis, which is the uncoordinated "lining up" of chromosome pairs before the first division of meiosis. As a result of random alignment, each homologous chromosome pair is segregated into daughter cells independently of all the other pairs during the production of gametes. Therefore, genes that are on different chromosomes are inherited independently of each other.

Due to independent assortment, the instruction manual contained in a single sperm cell is made up of a unique combination of pages from the manuals a man received from each of his parents. In fact, almost every sperm he makes will contain a unique subset of chromosomes—and thus a unique subset of his alleles. **Figure 7.6** illustrates this. In the figure, you can see that independent assortment causes an allele for an eye-color gene to end up in a sperm cell independently from an allele for the blood-group gene.

The independent assortment of segregated chromosomes into daughter cells is repeated every time a sperm is produced, and thus the set of alleles that a child receives from a father is different for all of his offspring. The sperm that contributed half of your genetic information might have carried an eye-color allele from your father's mom and a blood-group allele from his dad, while the sperm that produced your sister might have contained both the allele for eye color and the allele for blood group from your paternal grandmother. As a result of independent assortment, only about 50% of an individual's alleles are identical to those found in another offspring of the same parents—that is, for each gene, you have a 50% chance of being like your sister or brother.

Stop & Stretch In Figure 7.6, you can see that a man can produce four different combinations of alleles in his sperm when considering two genes on two different chromosomes. How many different allele combinations within gametes can be produced when considering three genes on three different chromosomes? What is the relationship between chromosome number (n) and the number of possible combinations that can be produced via independent assortment?

Random Fertilization Results in a Large Variety of Potential Offspring. As a result of the independent assortment of 23 pairs of chromosomes, each individual human can make at least 8 million different types of either egg or sperm. Consider now that each of your parents was able to produce such an enormous diversity of gametes. Further, any sperm produced by your father had an equal chance (in theory) of fertilizing any egg produced by your mother.

In other words, gametes combine without regard to the alleles they carry, a process known as **random fertilization.** Hence, the odds of your receiving your particular combination of chromosomes are 1 in 8 million times 1 in 8 million—or 1 in 64 trillion. Remarkably, your parents together could have made more than 64 trillion genetically different children, and you are only one of the possibilities.

Mutation creates new alleles, and independent assortment and random fertilization result in unique combinations of alleles in every generation. These processes help to produce the diversity of human beings.

A Special Case—Identical Twins. Although the process of sexual reproduction can produce two siblings who are very different from each other, an event that may occur after fertilization can result in the birth of two children who share 100% of their genes.

Identical twins are referred to as **monozygotic twins** because they develop from one zygote—the product of a single egg and sperm. Recall that after fertilization the zygote grows and divides, producing an embryo made up of many daughter cells containing the same genetic information. Monozygotic twinning occurs when cells in an embryo separate from each other. If this happens early in development, each cell or clump of cells can develop into a complete individual, yielding twins who carry identical genetic information (**Figure 7.7a**).

In contrast to identical twins, nonidentical twins (also called fraternal twins) occur when two separate eggs fuse with different sperm. These twins are called **dizygotic,** and although they develop simultaneously, they are genetically no more similar than siblings born at different times (**Figure 7.7b**). In humans, about 1 in every 80 pregnancies produces dizygotic twins, while only approximately 1 of every 285 pregnancies results in identical twins.

Identical twins provide a unique opportunity to study the relative effects of our genes and environment in determining who we are. Because identical twins carry the same genetic information, researchers are able to study how important genes are in determining health, tastes, intelligence, and personality.

We will examine the results of some twin studies later in the chapter, as we continue to explore the question of predicting the heredity of the complex genetic traits possessed by a particular sperm donor. Our review of the relationship among parents, offspring, and genetic material has now prepared us to examine the inheritance of traits controlled by a single gene.

7.2 Mendelian Genetics: When the Role of Genes Is Clear

A few human genetic traits have easily identifiable patterns of inheritance. These traits are said to be "Mendelian" because Gregor Johann Mendel (**Figure 7.8**) was the first person to accurately describe their inheritance.

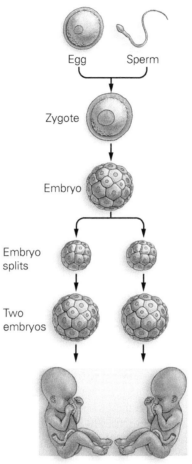

(a) Monozygotic (identical) twins

Egg   Sperm

Zygote

Embryo

Embryo splits

Two embryos

100% genetically identical

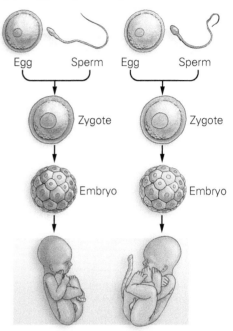

(b) Dizygotic (fraternal) twins

Egg   Sperm   Egg   Sperm

Zygote   Zygote

Embryo   Embryo

50% identical (no more similar than siblings born at different times)

Figure 7.7 The formation of twins. (a) Monozygotic twins form from one fertilization event and thus are genetically identical. (b) Dizygotic twins form from two independent fertilizations, resulting in two embryos who are as genetically similar as any other siblings.

Figure 7.8 Gregor Mendel. The father of the science of genetics.

① A pea flower normally self-pollinates.

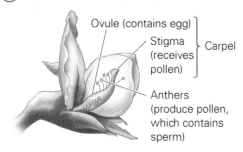

Ovule (contains egg)
Stigma (receives pollen) } Carpel
Anthers (produce pollen, which contains sperm)

② Pollen containing structures can be removed to prevent self-fertilization.

Tweezers

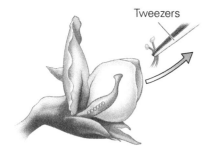

③ Pollen from another flower is dabbed on to stigma.

Paint brush

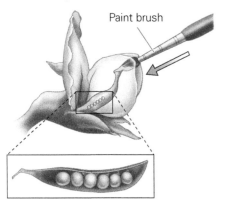

The resulting seeds will contain information on flower color, seed shape and color, and plant height from both parents.

Figure 7.9 Peas and genes. Pea plants were an ideal study organism for Mendel because their reproduction is easy to control, they complete their life cycle in a matter of weeks, and a single plant can produce thousands of offspring.

Mendel was born in Austria in 1822. Because his family was poor and could not afford private schooling, he entered a monastery to obtain an education. After completing his monastic studies, Mendel attended the University of Vienna. There he studied math and botany in addition to other sciences. Mendel attempted to become an accredited teacher but was unable to pass the examinations. After leaving the university, he returned to the monastery and began his experimental studies of inheritance in garden peas.

Mendel studied close to 30,000 pea plants over a 10-year period. His careful experiments consisted of controlled matings between plants with different traits. Mendel was able to control the types of mating that occurred by hand-pollinating the peas' flowers—that is, by taking pollen, which produces sperm, from the anthers of one pea plant and applying it to the stigma of the carpel (the egg-containing structure) of another pea plant. By growing the seeds that resulted from these controlled matings, he could evaluate the role of each parent in producing the traits of the offspring (**Figure 7.9**).

Although Mendel did not understand the chemical nature of genes, he was able to determine how traits were inherited by carefully analyzing the appearance of parent pea plants and their offspring. His patient, scientifically sound experiments demonstrated that both parents contribute equal amounts of genetic information to their offspring.

Mendel published the results of his studies in 1865, but his scientific contemporaries did not fully appreciate the significance of his work. Mendel eventually gave up his genetic studies and focused his attention on running the monastery until his death in 1884. His work was independently rediscovered by three scientists in 1900; only then did its significance to the new science of genetics become apparent.

The pattern of inheritance Mendel described occurs primarily in traits that are the result of a single gene with a few distinct alleles. **Table 7.1** lists some of the traits Mendel examined in peas; we will examine the principles he discovered, such as dominance and recessiveness, by looking at human disease genes that are of interest to prospective parents.

Genotype and Phenotype

We call the genetic composition of an individual his **genotype** and his physical traits his **phenotype.** The genotype is a description of the alleles for a particular gene carried on each member of a homologous pair of chromosomes (see Chapter 6, Figure 6.17). An individual who carries two different alleles for a gene has a **heterozygous** genotype. An individual who carries two copies of the same allele has a **homozygous** genotype.

The effect of an individual's genotype on her phenotype depends on the nature of the alleles she carries. Some alleles are **recessive,** meaning that their effects can be seen only if a copy of a dominant allele (described below) is not also present. For example, in pea plants the allele that codes for wrinkled seeds is recessive to the allele for round seeds. Wrinkled seeds will only appear when seeds carry only the wrinkled allele and no copies of the round allele.

A typical recessive allele is one that codes for a nonfunctional protein. Homozygotes having two copies of such an allele produce no functional protein. In contrast, heterozygotes carrying one copy of the functional allele have normal phenotypes because the normal protein is still produced. The functional allele in the case of the pea plant with round seeds prevents water from accumulating in the seed. When seeds containing one or two copies of this allele dry, they look much the same as when they first matured. However, wrinkled seeds result from two recessive, nonfunctional alleles (the

TABLE 7.1

Pea traits studied by Mendel.		
Character Studied	**Dominant Trait**	**Recessive Trait**
Seed shape	Round	Wrinkled
Seed color	Yellow	Green
Flower color	Purple	White
Stem length	Tall	Dwarf

absence of a functional allele). When the functional allele is missing from a seed, water flows into the seed, inflating it and causing its coat to increase in size. When this seed dries, it deflates and wrinkles, much like the surface of a balloon becomes wrinkly after it is blown up once and then deflated.

Dominant alleles are so named because their effects are seen even when a recessive allele is present. In the wrinkled seed example, the dominant allele is the one that produces a functional protein. The dominant allele does not always code for the "normal" condition of an organism, however. Sometimes mutations can create abnormal dominant alleles that essentially mask the effects of the recessive, normal allele. For example, "American Albino" horses—known for their snow-white coats, pink skin, and dark eyes—result from an allele that stops a horse's hair-color genes from being expressed during the horse's development. Because the allele prevents normal coat color development, it has its effect even if the animal carries only one copy—in other words, albinism in horses is dominant to the normal coat color (**Figure 7.10**). Dominant conditions in humans include cheek dimples and, surprisingly, the production of six fingers and toes.

Mendel worked with traits in peas that expressed only simple dominance and recessive relationships. However, for some genes, more than one dominant allele may be produced, and for others, a dominant allele may have different effects in a heterozygote than a homozygote. These situations are referred to as *codominance* and *incomplete dominance*, respectively (and are discussed in Chapter 8). For now, like Mendel, we will focus on simple dominant and recessive traits only.

Figure 7.10 A dominant allele. American Albino horses occur when an individual carries an allele that prevents normal coat color development. Because the product of this allele actively interferes with a biochemical pathway, horses that have only one copy of the allele are albino. (Photo by Linda Gordon.)

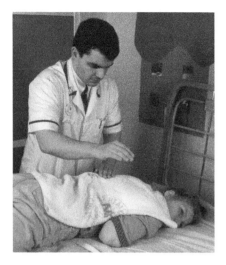

Figure 7.11 Treatment for cystic fibrosis. Percussive therapy consists of pounding the back of a patient with CF to loosen mucus inside the lungs. The mucus is then coughed up, reducing opportunities for bacterial infections to take hold.

Genetic Diseases in Humans

Most alleles in humans do not cause disease or dysfunction; they are simply alternative versions of genes. The diversity of alleles in the human population contributes to diversity among us in our appearance, physiology, and behaviors.

While women who use sperm banks are likely interested in a wide variety of traits in donors, sperm banks are primarily concerned with providing sperm that do not carry a risk of common genetic diseases. Some genetic diseases are produced by recessive alleles, while others are the result of dominant alleles.

Cystic Fibrosis Is a Recessive Condition.

Individuals with cystic fibrosis (CF) cannot transport chloride ions into and out of cells lining the lungs, intestines, and other organs. As a result of this dysfunction, the balance between sodium and chloride in the cell is disrupted, and the cell produces a thick, sticky mucus layer instead of the thin, slick mucus produced by cells with the normal allele. Affected individuals suffer from progressive deterioration of their lungs and have difficulties absorbing nutrients across the lining of their intestines. Most children born with CF suffer from recurrent lung infections and have dramatically shortened life spans (Figure 7.11).

Cystic fibrosis results from the production of a mutant chloride ion transporter protein that cannot embed in a cell membrane. The CF allele is recessive because individuals who carry only one copy of the normal allele can still produce the functional chloride transporter protein. The disease affects only homozygous individuals with two mutant alleles and thus no functional proteins.

Cystic fibrosis is among the most common genetic diseases in European populations; nearly 1 in 2500 individuals in these populations is affected with the disease, and 1 in 25 is heterozygous for the allele. Heterozygotes for a recessive disease are called **carriers** because even though they are unaffected, these individuals can pass the trait to the next generation. Sperm banks can test donor sperm for several recessive disorders, including CF; any men who are carriers of CF are excluded from most donor programs, including Fairfax Cryobank. Other relatively well-known recessive conditions in humans include albinism—absence of pigment in skin and hair—and Tay-Sachs disease, a fatal condition that causes the relentless deterioration of mental and physical abilities of affected infants as a result of the accumulation of wastes in the brain.

Huntington's Disease Is Caused by a Dominant Allele.

Early symptoms of Huntington's disease include restlessness, irritability, and difficulty in walking, thinking, and remembering. These symptoms typically begin to manifest in middle age. Huntington's disease is progressive and incurable—the nervous, mental, and muscular symptoms gradually become worse and eventually result in the death of the affected individual. Huntington's disease is an example of a fatal genetic condition caused by a dominant allele.

The Huntington's allele causes production of a protein that forms clumps inside the nuclei of cells. Nerve cells in areas of the brain that control movement are especially likely to contain these protein clumps, and these cells gradually die off over the course of the disease (Figure 7.12). Because this dysfunctional allele produces a mutant protein that damages cells, the presence of the normal allele cannot compensate or correct for this mutant version. An individual needs only one copy of the Huntington's allele to be affected by the disease; that is, even heterozygotes exhibit the symptoms of Huntington's.

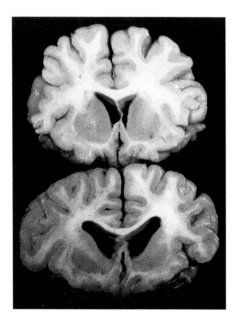

Figure 7.12 Huntington's and the brain. Cells containing mutant proteins die, shrinking and deforming the brain. This image compares the brain of a Huntington's sufferer (top) with an unaffected individual (bottom).

Stop & Stretch Typically, dominant mutations that result in death cannot be passed on from one generation to the next. What characteristic of Huntington's disease allows this allele to persist in the human population?

Only since the mid-1980s has genetic testing allowed people with a family history of Huntington's disease to learn whether they are affected before they show signs of the disease. Although most sperm banks do not test for the presence of the Huntington's allele because it is a rare condition, the detailed family medical histories required of sperm bank donors enable Fairfax Cryobank to exclude men with a family history of Huntington's disease from their donor list.

Using Punnett Squares to Predict Offspring Genotypes

Traits such as cystic fibrosis and Huntington's disease are the result of a mutation in a single gene, and the inheritance of these conditions and of other single-gene traits is relatively easy to understand. We can predict the likelihood of inheritance of small numbers of these single-gene traits by using a tool developed by Reginald Punnett, a British geneticist. A **Punnett square** is a table that lists the different kinds of sperm or eggs parents can produce relative to the gene or genes in question and then predicts the possible outcomes of a **cross**, or mating, between these parents (**Figure 7.13**).

Using a Punnett Square with a Single Gene. Imagine a woman and a sperm donor who are both carriers of the CF allele. Different alleles for a gene are symbolized with letters or number codes that refer to a trait that the gene affects. For instance, the CF gene is symbolized *CFTR* for *cystic fibrosis transmembrane regulator*. The dysfunctional *CFTR* allele is called *CFTR-ΔF508*, so both carriers would have the genotype *CFTR/CFTR-ΔF508*. However, to make this easier to follow, we will use a simpler key: the letters *F* and *f*, representing the dominant functional allele and recessive nonfunctional allele, respectively. A carrier for CF has the genotype *Ff*. A genetic cross between two carriers could then be symbolized as follows:

$$Ff \times Ff$$

We know that the female in this cross can produce eggs that carry either the *F* or *f* allele since the process of meiosis will segregate the two alleles from

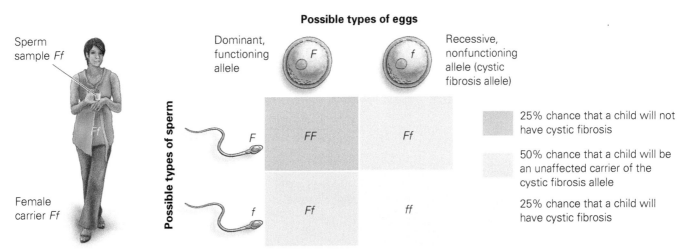

Figure 7.13 A cross between heterozygote individuals. What are the risks associated with sperm donated by a man who carries one copy of the cystic fibrosis allele? This Punnett square illustrates the likelihood that a woman who carries the cystic fibrosis allele would have a child with cystic fibrosis if the sperm donor were also a carrier.

each other. We place these two egg types across the horizontal axis of what will become a Punnett square. The male can also make two types of sperm, containing either the *F* or the *f* allele. We place these along the vertical axis. Thus, the letters on the horizontal and vertical axes represent all the possible types of eggs and sperm that the mother and father can produce by meiosis if we consider only the gene that codes for the chloride transport protein.

Inside the Punnett square are all the genotypes that can be produced from a cross between these two heterozygous individuals. The content of each box is determined by combining the alleles from the egg column and the sperm row.

Note that for a cross involving a single gene with two different alleles, there are three possible offspring types. The chance of this cross producing a child affected with CF is one in four, or 25%, because the *ff* combination of alleles occurs once out of the four possible outcomes. You can see why sperm banks would exclude carriers of CF from their donor rolls: The risk of an affected child born to a woman who might not know she is a carrier is too high. The *FF* genotype is also represented once out of four times, meaning that the probability of a homozygous unaffected child is also 25%. The probability of producing a child who is a **carrier** of cystic fibrosis is one in two, or 50%, since two of the possible outcomes inside the Punnett square are unaffected heterozygotes—one produced by an *F* sperm and an *f* egg and the other produced by an *f* sperm and an *F* egg.

When parents know which alleles they carry for a single-gene trait, they can easily determine the probability that a child they produce will have the disease phenotype (**Figure 7.14**). You should note that this probability is generated independently for each child. In other words, each offspring of two carriers has a 25% chance of being affected.

Punnett Squares for Crosses with More Genes. **Dihybrid crosses** are genetic crosses involving two traits. Let's go back to Mendel's peas as an example. Seed color and seed shape are each determined by a single gene, and each is carried on different chromosomes. The two seed-color gene alleles Mendel studied are designated here as *Y*, which is dominant and codes for yellow color, and *y*, the recessive allele, which results in green seeds when homozygous. The two seed-shape alleles Mendel studied are designated as *R*, the dominant allele, which codes for a smooth, round shape, and *r*, which is recessive and codes for a wrinkled shape.

Because the genes for seed color and seed shape are on different chromosomes, they are placed in eggs and sperm independently of each other. In other words, a pea plant that is heterozygous for both genes (genotype *YyRr*) can make four different types of eggs: one carrying dominant alleles for both genes (*YR*), one carrying recessive alleles for both genes (*yr*), one carrying the dominant allele for seed color and the recessive allele for seed shape (*Yr*), and one carrying the recessive allele for color and the dominant allele for shape (*yR*).

As with the Punnett square discussed above, the analysis of a dihybrid cross places all possible sperm genotypes on one axis of the square and all possible egg genotypes on the other other axis. Thus, a Punnett square for a cross between two individuals who are heterozygous for both seed-color and seed-shape genes would have four columns representing the four possible egg genotypes and four rows representing the four possible sperm genotypes; resulting in 16 boxes within the square describing four different possible phenotypes (**Figure 7.15**).

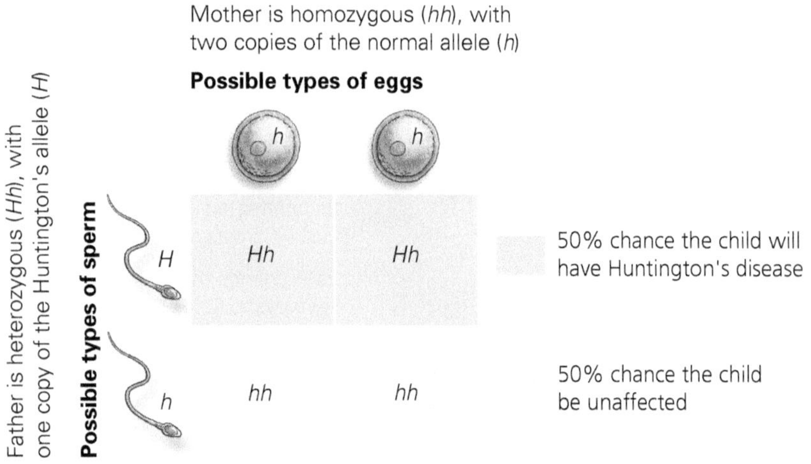

Figure 7.14 A cross between a heterozygote individual and a homozygote individual. This Punnett square illustrates the outcome of a cross between a man who carries a single copy of the dominant Huntington's disease allele and an unaffected woman.
Visualize This: If one child produced by these parents carries the Huntington's allele, what is the likelihood that a second child produced by this couple will have Huntington's disease?

Mother is homozygous (*hh*), with two copies of the normal allele (*h*)

Possible types of eggs

Father is heterozygous (*Hh*), with one copy of the Huntington's allele (*H*)

Possible types of sperm

	h	*h*
H	*Hh*	*Hh*
h	*hh*	*hh*

50% chance the child will have Huntington's disease

50% chance the child be unaffected

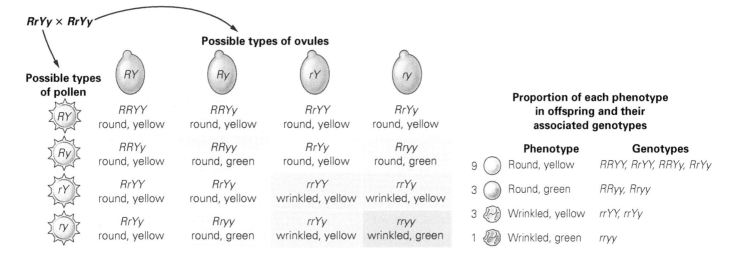

RrYy × RrYy

Possible types of ovules

Possible types of pollen

	RY	Ry	rY	ry
RY	*RRYY* round, yellow	*RRYy* round, yellow	*RrYY* round, yellow	*RrYy* round, yellow
Ry	*RRYy* round, yellow	*RRyy* round, green	*RrYy* round, yellow	*Rryy* round, green
rY	*RrYY* round, yellow	*RrYy* round, yellow	*rrYY* wrinkled, yellow	*rrYy* wrinkled, yellow
ry	*RrYy* round, yellow	*Rryy* round, green	*rrYy* wrinkled, yellow	*rryy* wrinkled, green

Proportion of each phenotype in offspring and their associated genotypes

	Phenotype	Genotypes
9	Round, yellow	*RRYY, RrYY, RRYy, RrYy*
3	Round, green	*RRyy, Rryy*
3	Wrinkled, yellow	*rrYY, rrYy*
1	Wrinkled, green	*rryy*

Figure 7.15 A dihybrid cross. Punnett squares can be used to predict the outcome of a cross involving two different genes. This cross involves two pea plants that are both heterozygous for the seed-color and seed-shape genes.
Visualize This: Curly hair (*CC*) is dominant over wavy (*Cc*) or straight hair (*cc*), and darkly pigmented eyes (*DD* or *Dd*) are dominant over blue eyes (*dd*). (Eye color is actually determined by several different genes, but we'll imagine it is controlled by only one in this example.) What fraction of the offspring of a mother and father who are heterozygous for both of these traits will have curly hair and dark eyes?

The phenotypes produced by a dihybrid cross result in a 9:3:3:1 **phenotypic ratio.** In this case, 9/16 include those genotypes produced by both dominant alleles (Y_R_ where the dashes indicate that the second allele for each gene could be either dominant or recessive); 3/16 include those produced by dominant alleles of one gene only (Y_rr); 3/16 include those produced by dominant alleles of the other gene (yyR_); and 1/16 have the phenotype produced by possessing recessive alleles only (yyrr).

Stop & Stretch Punnett squares can be produced for any number of genes, each of which has two alleles. Imagine a cross between two heterozygous tall, yellow, and round-seeded pea plants (TtYyRr). How many different types of gametes can each plant make? How many different cells would the resulting Punnett square table contain?

As you might imagine, as the number of genes in a Punnett square analysis increases, the number of boxes in the square increases, as does the number of possible genotypes. With two genes, each with two alleles, the number of unique gametes produced by a heterozygote is four, the number of boxes in the Punnett square is 16, and the number of unique genotypes of offspring that can be produced is nine. With three genes, each with two alleles, the Punnett square has 64 boxes and 22 different possible genotypes. With four genes, the square has 256 boxes, and with five genes, there are over 1000 boxes! Predicting the outcome of a cross becomes significantly more difficult as the number of genes we are following increases.

As scientists identify more genes and alleles, the amount of information about the genes of sperm donors—or any potential parent—will also increase. Identifying and testing for particular genes in potential parents will allow us to predict the *likelihood* of numerous genotypes in their offspring. Unfortunately, this increase in genetic testing is not necessarily equaled by an increase in our understanding of how more complex traits develop, as we shall see in the next section.

7.3 Quantitative Genetics: When Genes and Environment Interact

The single-gene traits discussed in the previous section have a distinct off-or-on character; individuals have either one phenotype (for example, the pea seed is round) or the other (the pea seed is wrinkled). Traits like this are known as *qualitative* traits. However, many of the traits that interest women who are choosing a sperm donor, such as height, weight, eye color, musical ability, susceptibility to cancer, and intelligence, do not have this off-or-on character. These traits are called **quantitative traits** and show **continuous variation;** that is, we can see a large range of phenotypes in a population—for instance, from very short people to very tall people. Wide variation in quantitative traits leads to the great diversity we see in the human population.

The distribution of phenotypes of a quantitative trait in a population can be displayed on a graph. These data often take the form of a curve called a *normal distribution.* **Figure 7.16a** illustrates the normal distribution of heights in a college class. Each individual is standing behind the label (at the bottom) that indicates their height. The curved line drawn across the photo summarizes these data—note the similarity of this curve to the outline of bell, leading to its common name, a bell curve.

A bell curve contains two important pieces of information. The first is the highest point on the curve, which generally corresponds to the average, or **mean,** value for data. The mean is calculated by adding all of the values for a trait in a population and dividing by the number of individuals in that population. The second is in the width of the bell itself, which illustrates the variability of a population. The variability is described with a mathematical measure called **variance,** which is essentially the average distance any one individual in the population is from the mean. If a low variance for a trait indicates a small amount of variability in the population, a high variance indicates a large amount of variability (**Figure 7.16b**).

(a) Normal distribution of student height in one college class

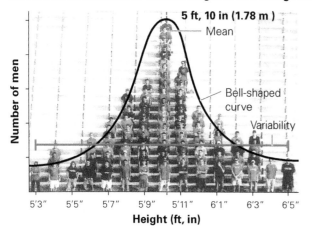

(b) Variance describes the variability around the mean.

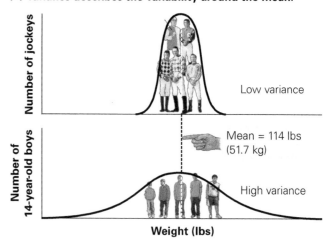

Figure 7.16 A quantitative trait. (a) This photo of men arranged by height illustrates a normal distribution. The highest point of the bell curve is also the mean height of 5 feet, 10 inches. (b) Fourteen-year-old boys and professional jockeys have the same average weight—approximately 114 pounds. However, to be a jockey, you must be within about 4 pounds of this average. Thus, the variance among jockeys in weight is much smaller than the variance among 14-year-olds.

Visualize This: Examine Figure 7.16a closely: Does an average height of 5 feet, 10 inches in this particular population imply that most men were this height? Were most men in this population close to the mean, or was there a wide range of heights?

An Overview: Why Traits Are Quantitative

A range of phenotypes may exist for a trait because numerous genotypes are found among the individuals in the population. This happens when a trait is influenced by more than one gene; traits influenced by many genes are called **polygenic traits.**

As we saw above, when a single gene with two alleles determines a trait, only three possible genotypes are present: *FF, Ff,* and *ff,* for example. But when more than one gene, each with more than one allele, influences a trait, many genotypes are possible. For example, eye color in humans is a polygenic trait influenced by at least three genes, each with more than one allele. These genes help produce and distribute the pigment melanin, a brown color, to the iris. When different alleles for the genes for eye pigment production and distribution interact, a range of eye colors, from dark brown (lots of melanin produced) to pale blue (very little melanin produced), is found in humans. The continuous variation in eye color among people is a result of several genes, each with several alleles, influencing the phenotype.

Continuous variation also may occur in a quantitative trait due to the influence of environmental factors. In this case, each genotype is capable of producing a range of phenotypes depending on outside influences. For a clear example of the effect of the environment on phenotype, see Figure 7.17. These identical twins share 100% of their genes but are quite different in appearance. Their difference is entirely due to variations in their environment—one twin smoked cigarettes and had high levels of sun exposure, whereas the other did not smoke and spent less time in the sun.

Most traits that show continuous variation are influenced by both genes and the environment. Skin color in humans is an example of this type of trait. The shade of an individual's skin is dependent on the amount of melanin present near the skin's surface. As with eye color, a number of genes have an effect on skin-color phenotype—those that influence melanin production and those that affect melanin distribution. However, the environment, particularly the amount of exposure to the sun during a season or lifetime, also influences the skin color of individuals (Figure 7.18). Melanin production increases, and any melanin that is present darkens in sun-exposed skin. In fact, after many years of intensive sun exposure, skin may become permanently darker.

Stop & Stretch The average height of American men has stayed the same over the past 50 years, while average heights in other countries (for example, Denmark) have increased. Height is a classic quantitative trait. Provide both a genetic and an environmental hypothesis for why the height of the American population is not increasing along with other countries.

Women choosing Doctorate category sperm donors from Fairfax Cryobank are presumably interested in having smart, successful children, but intelligence has both a genetic and an environmental component. With an important role for both influences, how can we predict if the child of a father with a doctorate will also be capable of earning a doctorate?

To determine the role of genes in determining quantitative traits, scientists must calculate the **heritability** of the trait. To estimate heritability for quantitative traits in most populations, researchers use correlations between individuals with varying degrees of genetic similarity. A correlation determines how accurately one can predict the measure of a trait in an

Figure 7.17 The effect of the environment on phenotype. These identical twins have exactly the same genotype, but they are quite different in appearance due to environmental factors. The twin on the right was a life-long smoker, while the one on the left never smoked.

(a) Genes

(b) Environment

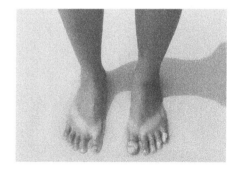

Figure 7.18 Skin color is influenced by genes and environment. (a) The difference in skin color between these two women is due primarily to variations in several alleles that control skin pigment production. (b) The difference in color between the sun-protected and sun-exposed portions of the individual in this picture is entirely due to environmental effects.

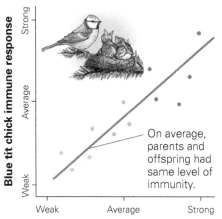

Points represent parent-offspring pairs with matching immunity levels.
• Weak • Average • Strong

On average, parents and offspring had same level of immunity.

Blue tit parent immune response

Figure 7.19 Using correlation to calculate heritability. The close correlation of immune system response between parents and offspring in the blue tit, a European bird, indicates that immune response is highly heritable.

Visualize This: How would a graph that showed a low correlation between parents and offspring differ from this graph?

individual when its measure in a related individual is known. For example, Figure 7.19 shows a correlation between parent birds and their offspring in the strength of their response to tetanus vaccine. An individual that responded strongly produced a large number of antitetanus proteins, called antibodies, while ones that responded weakly produced a lower number of antibodies.

As you can see from the graph, parents with weak responses tended to have offspring with weak responses, and parents with strong responses had offspring with strong responses. This strong correlation indicates that the ability to respond to tetanus is highly heritable—most of the difference between birds in their immune system response results from genetic differences.

Heritability in humans is typically measured by examining correlations between groups. These studies calculate how similar or different parents are to their children, or siblings are to each other, in the value of a particular trait. When examined across an entire population, the strength of a correlation provides a measure of heritability (Figure 7.20). The uses and limitations of heritability are addressed more specifically in Section 7.4. *See the next section for A Closer Look at using correlations to calculate heritability of intelligence.*

For most traits, such as body size, people come in a wide variety of types.

How much of our differences is due to the environment, and how much is a result of different genes?

Correlations between relatives, (for instance, here between fathers and sons) can provide information about the importance of genes in determining variation among individuals.

Figure 7.20 Determining heritability in humans. Comparisons between parents and children can help us estimate the genetic component of a quantitative trait.

A Closer Look:
Using Correlations to Calculate the Heritability of Intelligence

Intelligence is often measured by performance on an IQ test. The French psychologist Alfred Binet developed the intelligence quotient (or IQ) in the early 1900s to identify schoolchildren who were in need of remedial help. Binet's IQ test was not based on any theory of intelligence and was not meant to comprehensively measure mental ability, but the tests remain a commonly used way to measure innate or "natural" intelligence. There is still significant controversy over the use of IQ tests in this way.

Even if IQ tests do not really measure general intelligence, IQ scores have been correlated with academic success—meaning that individuals at higher academic levels usually have higher IQs. So, even without knowing their IQ scores, we can reasonably expect that donors in the Doctorate category have higher IQs than do other available sperm donors. However, the question of whether the high IQ of a prospective sperm donor has a genetic basis still remains.

The average correlation between IQs of parents and their children—in other words, the estimated heritability of IQ by this method—is 0.42. However, parents and the children who live with them are typically raised in a similar social and economic environment. As a result, correlations of IQ between the two groups cannot distinguish the relative importance of genes from the importance of the environment. This is the problem found in most arguments about "nature versus nurture"—do children resemble their parents because they are "born that way" or because they are "raised that way"?

To avoid the problem of overlap in environment and genes between parents and children, researchers seek situations that remove one or the other overlap. These situations are called **natural experiments** because one factor is "naturally" controlled, even without researcher intervention. Human twins are one source of a natural experiment to test hypotheses about the heritability of quantitative traits in humans.

By comparing monozygotic twins, who share all of their alleles, to dizygotic twins, who share, on average, 50% of their alleles, researchers can begin to separate the effects of shared genes from the effects of shared environments. Because twins raised in the same family have similar childhood experiences, one would expect that the only real difference between monozygotic and dizygotic twins is their genetic similarity. The average heritability of IQ calculated from a number of twin studies of this type is about 0.52. According to these studies, 52% of the variability in IQ among humans is due to differences in genotypes. Surprisingly, this value is even higher than the 42% calculated from the correlation between parents and children.

However, monozygotic twins and dizygotic twins likely *do* differ in more than just genotype. In particular, identical twins are treated more similarly than nonidentical twins. This occurs both because they look very similar and because other people may presume that they are identical in all other respects. If monozygotic twins are *expected* to be more alike than dizygotic twins, their IQ scores may be similar because they are encouraged to have the same experiences and to achieve at the same level.

There is one natural experiment that can address this problem, however. By comparing identical and nonidentical twins raised apart, the problem of differential treatment of the two *types* of twins is minimized because no one would know that the individual members of a pair have a twin (**Figure 7.21**). If variation in genes does not explain much of the variation among peoples' IQ scores, then identical twins raised apart should be no more similar than any two unrelated people of the same age and sex.

Unfortunately for scientists (but perhaps fortunately for children), the frequency of early twin separation is extremely low. Researchers have estimated the heritability of IQ at a remarkable 0.72 in a small sample of twins raised apart. This study and the other correlations appear to support the hypothesis that differences in our genes explain the majority of the variation in IQ among people. **Table 7.2** summarizes the estimates of IQ heritability but previews the cautions discussed in the next section of this chapter.

Figure 7.21 Twins separated at birth. Tamara Rabi (left) and Adriana Scott were reunited at age 20 when a mutual friend noticed their remarkable resemblance. Tamara was raised on the Upper West Side of Manhattan and Adriana in suburban Long Island, but they both have similar outgoing personalities, love to dance, and even prefer the same brand of shampoo.

(continued on the next page)

(A Closer Look continued)

TABLE 7.2

To what extent is IQ heritable? A summary of various estimates of IQ heritability, their shortcomings, and the problems with using them to understand the role of genes in determining an individual's potential intelligence.

Method of Measurement	Estimated Percentage of Genetic Influence	Warnings When Interpreting This Result	Warnings That Apply to All Measurements of Heritability
Correlation between parents' IQ and children's IQ in a population	42%	Because parents and children are similar in genes and environment, a correlation cannot be used to indicate the relative importance of genes and environment in determining IQ.	• Heritability values are specific to the populations for which they were measured. • High heritability for a trait does not mean that it is not heavily influenced by environmental conditions; we cannot predict how the trait will respond to a change in the environment. • Heritability is a measure of a population, not an individual.
Natural experiment comparing IQ in pairs of identical twins versus nonidentical twins	52%	Identical twins are treated as more alike than nonidentical twins. Therefore, their environment is different from that of nonidentical twins—the heritability value could be an overestimate.	
Natural experiment comparing IQ of identical twins raised apart versus nonidentical twins raised apart	72%	Small sample size may skew results.	

7.4 Genes, Environment, and the Individual

Perhaps we can now determine the importance of a sperm donor father who has earned a doctorate to his child's intellectual development. We know that a sperm donor will definitely influence some of his child's traits—eye and skin color and perhaps even susceptibility to certain diseases. In addition, according to the studies discussed above, the donor will probably pass on some intellectual traits to the child. In fact, with a heritability of IQ at above 50%, it appears that genes are primary in determining an individual's intelligence. Perhaps it is a good idea to pay a premium price for Doctorate category sperm after all.

However, we need to be very careful when applying the results of twin studies to questions about the individual sperm donors. To understand why, we will take a closer look at the practical significance of heritability.

The Use and Misuse of Heritability

A calculated heritability value is unique to the population in which it was measured and to the environment of that population. We should be very cautious when using heritability to measure the *general* importance of genes to the development of a trait. The following sections illustrate why.

Differences Between Groups May Be Entirely Environmental. A "thought experiment" can help illustrate this point. Body weight in laboratory mice has a strong genetic component, with a calculated heritability of about 0.90. In a population of mice in which weight is variable, bigger mice have bigger offspring, and smaller mice have smaller offspring.

Imagine that we randomly divide a population of variable mice into two groups—one group is fed a rich diet, and the other group is fed a poor diet. Otherwise, the mice are treated identically. As you might predict, regardless of their genetic predispositions, the well-fed mice become fat, while the poorly fed mice become thin. Consider the outcome if we were to keep the mice in these same conditions and allowed the two groups to reproduce. Not surprisingly, the second generation of well-fed mice is likely to be much heavier than the second generation of poorly fed mice. Now imagine that another researcher came along and examined these two populations of mice without knowing their diets. Knowing that body weight is highly heritable, the researcher might logically conclude that the groups are genetically different. However, we know this is not the case—both are grandchildren of the same original source population. It is the environment of the two populations that differs (Figure 7.22).

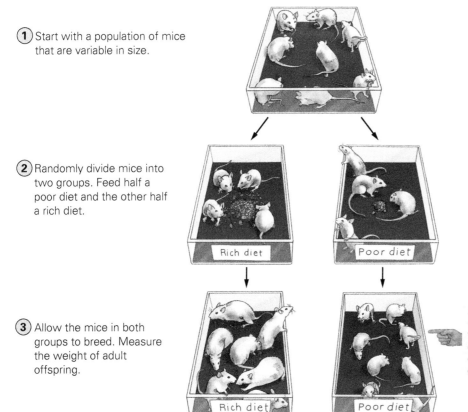

① Start with a population of mice that are variable in size.

② Randomly divide mice into two groups. Feed half a poor diet and the other half a rich diet.

Rich diet

Poor diet

③ Allow the mice in both groups to breed. Measure the weight of adult offspring.

Rich diet

Poor diet

Figure 7.22 The environment can have powerful effects on highly heritable traits. If genetically similar populations of mice are raised in radically diverse environments, then differences between the populations are entirely due to environment.

Average weight of the mice in the rich-diet environment is twice the average weight of the population in the poor-diet environment. However, there is no genetic difference between the two groups.

Now extend the same thought experiment to human groups. Imagine that we have two groups of humans, and we have determined that IQ had high heritability. In this case, people in one group were affluent, and their average IQ was higher. The other group was impoverished, and their average IQ was lower. What conclusions could you draw about the genetic differences between these two populations? The answer to the question above is none— as with the laboratory mice, these differences could be entirely due to environment. The high heritability of IQ cannot tell us if two human groups in differing social environments vary in IQ because of variations in genes or because of differences in environment.

A Highly Heritable Trait Can Still Respond to Environmental Change. A high heritability for IQ might seem to imply that IQ is not strongly influenced by environmental conditions. However, intelligence in other animals can be demonstrated to be both highly heritable and strongly influenced by the environment.

Rats can be bred for maze-running ability, and researchers have produced rats that are "maze bright" and rats that are "maze dull." Maze-running ability is highly heritable in the laboratory environment; that is, bright rats have bright offspring, and dull rats have dull offspring. The results of an experiment that measured the number of mistakes made by maze-bright and maze-dull rats raised in different environments are presented in **Table 7.3**.

In the typical lab environment, bright rats were much better at maze running than dull rats. But in both a very boring or restricted environment and a very enriched environment, the two groups of rats did about the same. In fact, no rats excelled in a restricted environment, and all rats did better at maze running in enriched environments, with the duller rats improving most dramatically.

What this example demonstrates is that we cannot predict the response of a trait to a change in the environment, even when that trait is highly heritable. Thus, even if IQ has a strong genetic component, environmental factors affecting IQ can have big effects on an individual's intelligence.

TABLE 7.3

A highly heritable trait is not identical in all environments. This table describes the average number of mistakes made by rats of two different genotypes in three different environments.

| | | Number of Mistakes in ... | | |
Phenotype		Normal Environment	Restricted Environment	Enriched Environment
Maze-bright rats		115	170	112
Maze-dull rats		165	170	122
Explanation of Results		Maze-dull rats made more mistakes than maze-bright rats when running a maze.	Both groups made the same number of mistakes when running a maze.	Both groups made fewer mistakes when running the maze. The maze-dull rats improved the most.

Stop & Stretch Some commentators have argued that given IQ's high heritability, policies that increase financial resources to failing schools will ultimately fail to increase achievement because such a predominantly genetic trait will not respond well to environmental change. Use your understanding of the proper application of heritability to refute this argument.

Heritability Does Not Tell Us Why Two Individuals Differ. High heritability of a trait is often presumed to mean that the difference between two individuals is mostly due to differences in their genes. However, even if genes explain 90% of the population variability in a particular environment, the reason one individual differs from another may be entirely a function of environment (as an example of this, look back at the identical twins in Figure 7.18).

Currently, there is no way to determine if a particular child is a poor student because of genes, a poor environment, or a combination of both factors. There is also no way to predict whether a child produced from the sperm of a man with a doctorate will be an accomplished scholar. All we can say is that given our current understanding of the heritability of IQ and the current social environment, the alleles in Doctorate category sperm *may* increase the probability of having a child with a high IQ.

How Do Genes Matter?

We know that genes can have a strong influence on eye color, risk of genetic diseases such as CF, and even the structure of the brain. But what really determines who we are—nature or nurture?

Even with single-gene traits, the outcome of a cross between a woman and a sperm donor is not a certainty; it is only a probability. Couple this with traits being influenced by more than one gene, and independent assortment greatly increases the offspring types possible from a single mating (recall how the number of cells in a Punnett square increased exponentially when more genes were added). Knowing the phenotype of potential parents gives you relatively little information about the phenotype of their children. So, even if genes have a strong effect on traits, we cannot "program" the traits of children by selecting the traits of their parents (**Figure 7.23**).

In truth, we are really asking ourselves the wrong question when we wonder if nature or nurture has a more powerful influence on who we are. Both nature and nurture play an important role. Our cells carry instructions for all the essential characteristics of humanity, but the process of developing from embryo to adult takes place in a physical and social environment that influences how these genes are expressed. Scientists are still a long way from answering questions about how all of these complex, interacting circumstances result in who we are.

What is the lesson for women and couples who are searching for a sperm donor from Fairfax Cryobank? Donors in the Doctorate category may indeed have higher IQs than donors in the cryobank's other categories, but there is no real way to predict if a particular child of one of these donors will be smarter than average. According to the current data on the heritability of IQ, sperm from high-IQ donors may increase the odds of having an offspring with a high IQ, but only if parents provide them with a stimulating, healthy, and challenging environment in which to mature. This, of course, would be good for children with any alleles.

Figure 7.23 Genes are not destiny. Even when the traits of both parents are known, the children they produce may be very different in interests and aptitudes.

SAVVY READER

Who Picks Your Friends?

Researchers [have] found genetic links among groups of friends that suggest people tend to form friendships with others who share at least some genetic markers with them.

"People's friends may not only have similar traits, but actually resemble each other on a genotypic level," write researcher James H. Fowler, of the division of medical genetics at the University of California, San Diego and colleagues, in the *Proceedings of the National Academy of Sciences.*

In two separate groups, one comprised of adolescents and the other adults, researchers found correlations in two of six genetic markers tested in those who forged friendships.

The study showed those who carried the DRD2 genetic marker, which has been previously associated with alcoholism, tended to form friendships with others who shared the same marker. Similarly, those who lacked this gene formed friendships with others who also did not have the gene.

Researchers say this finding supports the notion that people are attracted to others with similar traits—the "birds of feather flock together" theory.

In contrast, a second genetic correlation provided evidence to support the belief that "opposites attract." The results showed that people who had a

gene associated with an open personality, CYAP26, tended to have friends who did not share this gene.

Researchers say together these results could have important implications in explaining the role genes play in shaping human behavior.

"For example, a person with a genotype that makes her susceptible to alcoholism may be directly influenced to drink," write the researchers. "However, she may also be indirectly influenced to drink because she chooses friends with the same genotype."

1. Decode this article. What do the researchers' results seem to indicate about the role of heritable influence versus environmental influence in determining our choices of friends?

2. How might these results inform studies of alcoholism prevention?

3. Put this study in context with what you learned about heritability in this chapter. Are your friendships determined by your genes, your environment, or both?

Source: Web MD Health News: "Genes May Link Friends, Not Just Family" by Jennifer Warner. January 18, 2011. (http://www.webmd.com/news/20110118/genes-may-link-friends-not-just-family: Accessed Jan 25, 2011)

Chapter Review

Learning Outcomes

LO1 Describe the relationship between genes, chromosomes, and alleles (Section 7.1).

*   Children resemble their parents in part because they inherit their parents' genes, segments of DNA that contain information about how to make proteins (pp. 148–149).

Go to the Study Area at www.masteringbiology.com for practice quizzes, myeBook, BioFlix™ 3-D animations, MP3Tutor sessions, videos, current events, and more.

*   Chromosomes contain genes. Different versions of a gene are called alleles (p. 150).
*   Mutations in genes generate a variety of alleles. Each allele typically results in a slightly different protein product (p. 151).

LO2 Explain why cells containing identical genetic information can look different from each other (Section 7.1).

- Although nearly all cells in an individual contain the exact same genetic information, the genes that are "read" in those cells differ. Even when the same gene is expressed in two different cells, their products may have very different effects (p. 150).

LO3 Define *segregation* and *independent assortment* and explain how these processes contribute to genetic diversity (Section 7.1).

- Segregation separates alleles during the production of eggs and sperm, meaning that a parent contributes only half of its genetic information to an offspring (p. 151).
- In independent assortment, individual chromosomes are sorted into eggs and sperm independent of each other, meaning that the subset of information passed on by parents is unique for each gamete (pp. 151–152).

LO4 Distinguish between homozygous and heterozygous genotypes and describe how recessive and dominant alleles produce particular phenotypes when expressed in these genotypes (Section 7.2).

- An individual heterozygous for a particular gene carries two different alleles for the gene, while one who is homozygous carries two identical alleles (p. 154).
- A dominant allele is expressed even when the genotype is heterozygous, while a recessive allele is only expressed when the individual carries no copies of the dominant allele—that is, when it is homozygous recessive (pp. 154–155).

LO5 Demonstrate how to use a Punnett square to predict the likelihood of a particular offspring genotype and phenotype from a cross of two individuals with known genotype (Section 7.2).

- A Punnett square helps us determine the probability that two parents of known genotype will produce a child with a particular genotype (pp. 157–159).

LO6 Define *quantitative trait* and describe the genetic and environmental factors that cause this pattern of inheritance (Section 7.3).

- Many traits—such as height, IQ, and musical ability—show quantitative variation, which results in a range of values for the trait within a given population (p. 160).
- Quantitative variation in a trait may be generated because the trait is influenced by several genes, because the trait can be influenced by environmental factors, or due to a combination of both factors (p. 161).

LO7 Describe how heritability is calculated and what it tells us about the genetic component of quantitative traits (Section 7.4).

- The role of genes in determining the phenotype for a quantitative trait is estimated by calculating the heritability of the trait (pp. 161–162).
- Heritability is calculated by examining the correlation between parents and offspring or by comparing pairs of monozygotic twins to pairs of dizygotic twins (pp. 163–164).

LO8 Explain why a high heritability still does not always mean that a given trait is determined mostly by the genes an individual carries (Section 7.4).

- Calculated heritability values are unique to a particular population in a particular environment. The environment may cause large differences among individuals, even if a trait has high heritability (pp. 165–167).
- Our current understanding of the relationship between genes and complex traits does not allow us to predict the phenotype of a particular offspring from the phenotype of its parents (p. 167).

Roots to Remember

The following roots of words come mainly from Latin and Greek and will help you decipher terms:

di-	means two. Chapter term: dizygotic
hetero-	means the other, another, or different. Chapter term: heterozygous
homo-	means the same. Chapter term: homozygous
mono-	means one. Chapter term: monozygotic
pheno-	comes from a verb meaning "to show." Chapter term: phenotype
poly-	means many. Chapter term: polygenic
-zygous	derives from *zygote*, the "yoked" cell resulting from the union of an egg and sperm. Chapter terms: monozygous, heterozygote

Learning the Basics

1. **LO2** What is the relationship between genotype and phenotype?

2. **LO6** What factors cause quantitative variation in a trait within a population?

3. **LO1** Which of the following statements correctly describes the relationship between genes and chromosomes?

    A. Genes are chromosomes; B. Chromosomes contain many genes; C. Genes are made up of hundreds or thousands of chromosomes; D. Genes are assorted independently during meiosis, but chromosomes are not; **E.** More than one of the above is correct.

4. LO1 An allele is a _____.

A. version of a gene; **B.** dysfunctional gene; **C.** protein; **D.** spare copy of a gene; **E.** phenotype

5. LO3 Sperm or eggs in humans always _____.

A. each have 2 copies of every gene; **B.** each have 1 copy of every gene; **C.** each contain either all recessive alleles or all dominant alleles; **D.** are genetically identical to all other sperm or eggs produced by that person; **E.** each contain all of the genetic information from their producer

6. LO2 Scientists have recently developed a process by which a skin cell from a human can be triggered to develop into a human heart muscle cell. This is possible because _____.

A. most cells in the human body contain the genetic instructions for making all types of human cells; **B.** a skin cell is produced when all genes in the cell are expressed; turning off some genes in the cell results in a heart cell; **C.** scientists can add new genes to old cells to make them take different forms; **D.** a skin cell expresses only recessive alleles, so it can be triggered to produce dominant heart cell alleles; **E.** it is easy to mutate the genes in skin cells to produce the alleles required for other cell types

7. LO3 What is the physical basis for the independent assortment of alleles into offspring?

A. There are chromosome divisions during gamete production; **B.** Homologous chromosome pairs are separated during gamete production; **C.** Sperm and eggs are produced by different sexes; **D.** Each gene codes for more than one protein; **E.** The instruction manual for producing a human is incomplete.

8. LO4 Among heritable diseases, which genotype can be present in an individual without causing a disease phenotype in that individual?

A. heterozygous for a dominant disease; **B.** homozygous for a dominant disease; **C.** heterozygous for recessive disease; **D.** homozygous for a recessive disease; **E.** all of the above

9. LO6 A quantitative trait _____.

A. may be one that is strongly influenced by the environment; **B.** varies continuously in a population; **C.** may be influenced by many genes; **D.** has more than a few values in a population; **E.** all of the above

10. LO7 LO8 When a trait is highly heritable, _____.

A. it is influenced by genes; **B.** it is not influenced by the environment; **C.** the variance of the trait in a population can be explained primarily by variance in genotypes; **D.** A and C are correct; **E.** A, B, and C are correct

Genetics Problems

1. LO1 A single gene in pea plants has a strong influence on plant height. The gene has two alleles: tall (*T*), which is dominant, and short (*t*), which is recessive. What are the genotypes and phenotypes of the offspring of a cross between a *TT* and a *tt* plant?

2. LO5 What are the genotypes and phenotypes of the offspring of *Tt* × *Tt*?

3. LO4 The "*P*" gene controls flower color in pea plants. A plant with either the *PP* or *Pp* genotype has purple flowers, while a plant with the *pp* genotype has white flowers. What is the relationship between *P* and *p*?

4. LO4 LO5 Albinism occurs when individuals carry 2 recessive alleles (*aa*) that interfere with the production of melanin, the pigment that colors hair, skin, and eyes. If an albino child is born to two individuals with normal pigment, what is the genotype of each parent?

5. LO5 Pfeiffer syndrome is a dominant genetic disease that occurs when certain bones in the skull fuse too early in the development of a child, leading to distorted head and face shape. If a man heterozygous for the allele that causes Pfeiffer syndrome marries a woman who is homozygous for the nonmutant allele, what is the chance that their first child will have this syndrome?

6. LO4 LO5 A cross between a pea plant that produces yellow peas and a pea plant that produces green peas results in 100% yellow pea offspring.

A. Which allele is dominant in this situation? **B.** What are the likely genotypes of the yellow pea and green pea plants in the initial cross?

7. LO5 A cross between a pea plant that produces round (*R*), yellow (*Y*) peas and a pea plant that produces wrinkled (*r*), green (*y*) peas results in 50% yellow, round pea offspring and 50% green, wrinkled pea offspring. What are the genotypes of the plants in the initial cross?

8. LO5 A woman who is a carrier for the cystic fibrosis allele marries a man who is also a carrier.

A. What percentage of the woman's eggs will carry the cystic fibrosis allele? **B.** What percentage of the man's sperm will carry the cystic fibrosis allele? **C.** The probability that this couple will have a child who carries two copies of the cystic fibrosis allele is equal to the percentage of eggs that carry the allele times the percentage of sperm that carry the allele. What is this probability? **D.** Is this the same result you would generate when doing a Punnett square of this cross?

9. LO5 LO6 The allele *BRCA2* was identified in families with unusually high rates of breast and ovarian cancer. Up to 80% of women with one copy of the *BRCA2* allele develop one of these cancers in their lifetime.

A. Is *BRCA2* a dominant or a recessive allele? **B.** How is *BRCA2* different from the typical pattern of Mendelian inheritance?

Analyzing and Applying the Basics

1. **LO5 LO6** Two parents both have brown eyes, but they have two children with brown eyes and two with blue eyes. How is it possible that two people with the same eye color can have children with different eye color? If eye color in this family is determined by differences in genotype for a single gene with two alleles, what percentage of the children are expected to have blue eyes? If the ratio of brown to blue eyes in this family does not conform to expectations, why does this result not refute Mendelian genetics?

2. **LO7 LO8** Does a high value of heritability for a trait indicate that the average value of the trait in a population will not change if the environment changes? Explain your answer.

3. **LO8** The heritability of IQ has been estimated at about 72%. If John's IQ is 120 and Jerry's IQ is 90, does John have stronger "intelligence" genes than Jerry does? Explain your answer.

Connecting the Science

1. If scientists find a gene that is associated with a particular "undesirable" personality trait (for instance, a tendency toward aggressive outbursts), will it mean greater or lesser tolerance toward people with that trait? Will it lead to proposals that those affected by the "disorder" should undergo treatment to be "cured" and that measures should be taken to prevent the birth of other individuals who are also afflicted?

2. The higher price for Doctorate sperm at the Fairfax Cryobank seems to imply that these donors are rare and highly desirable. If you were a woman who was looking for a sperm donor, would you focus your selection process on the Doctorate donors? What might you miss by focusing only on these donors?

3. Down syndrome is caused by a mistake during meiosis and results in physical characteristics such as a short stature and distinct facial features as well as cognitive impairment (also known as mental retardation). Does the fact that Down syndrome is a genetic condition that results in low IQ mean that we should put fewer resources into education for people with Down syndrome? How does your answer to this question relate to questions about how we should treat individuals with other genetic conditions?

Answers to **Stop & Stretch, Visualize This, Savvy Reader,** and **Chapter Review** questions can be found in the **Answers** section at the back of the book.

CHAPTER 8

DNA Detective

Complex Patterns of Inheritance and DNA Fingerprinting

The Romanov family ruled Russia until their overthrow, exile, and 1918 execution.

On the night of July 16, 1918, the tsar of Russia, Nicholas II, his wife, Alexandra Romanov, their five children, and four family servants were executed in a small room in the basement of the house to which they had been exiled. These murders ended three centuries of rule by the Romanov family over the Russian Empire.

In February 1917, in the wake of protests throughout Russia, Nicholas II had relinquished his power by abdicating for both himself and on behalf of his only son, Alexis, then 13 years old. The tsar hoped that these abdications would protect his son, the heir to the throne, as well as the rest of the family from harm.

The political climate in Russia at that time was explosive. During the summer of 1914, Russia and other European countries became embroiled in World War I. This war proved to be a disaster for the imperial government. Russia faced severe food shortages, and the poverty of the common people contrasted starkly with the luxurious lives of their leaders. The Russian people felt deep resentment toward the tsar's family. This sentiment sparked the first Russian Revolution in February 1917. Following Nicholas's abdication, the imperial family was kept under guard at one of their palaces outside St. Petersburg.

In November 1917, the Bolshevik Revolution brought the communist regime, led by Vladimir Lenin, to power. Ridding the country of the last vestige of Romanov rule became a priority for Lenin and his political party. Lenin believed that doing so would solidify his regime as well as garner support among people who felt that the exiled Romanovs and their opulent lifestyle represented all that was wrong with Imperial Russia.

The fall of the communist Soviet Union prompted the desire for a proper burial of the Romanov family.

Photo by Dr. Sergey Nikitin. People believed that bones found in a grave in Ekaterinburg were those of the slain Romanovs.

Forensic evidence, like that used on popular crime shows, helped confirm that the bones buried in the shallow grave belonged to the Romanovs.

LEARNING OUTCOMES

LO1 Differentiate incomplete dominance from codominance.

LO2 Explain how polygenic and pleiotropic traits differ.

LO3 Outline the inheritance of the multiple-allelic ABO blood system.

LO4 Describe the mechanism of sex determination in humans.

LO5 Explain how the inheritance of sex-linked genes differs from the inheritance of autosomal genes.

LO6 Describe how X-inactivation changes patterns of inheritance.

LO7 Explain how scientists use pedigrees.

LO8 Explain the significance of DNA fingerprinting and how the process works.

Fearing any attempt by pro-Romanov forces to save the family, Lenin ordered them to the town of Ekaterinburg in Siberia.

Shortly after midnight on July 16, the family was awakened and asked to dress. Nicholas, Alexandra, and their children—Olga, Tatiana, Maria, Anastasia, and Alexis—along with the family physician, cook, maid, and valet, were escorted to a room in the basement of the house in which they had been kept. Believing they were to be moved, the family waited. A soldier entered the room and read a short statement indicating they were to be killed. Armed men stormed into the room, and after a hail of bullets, the royal family and their entourage lay dead.

After the murders, the men loaded the bodies of the Romanovs and their servants into a truck and drove to a remote, wooded area in Ekaterinburg. Historical accounts differ regarding whether the bodies were dumped down a mineshaft, later to be removed, or were immediately buried. There is also some disagreement regarding the burial of two of the people who were executed. Some reports indicated that all eleven people were buried together, and two of them either were badly decomposed by acid placed on the ground of the burial site or were burned to ash. Other reports indicated that two members of the family were buried separately. Some people even believe that two victims escaped the execution. In any case, the bodies of at least nine people were buried in a shallow grave, where they lay undisturbed until 1991.

The bodies were not all that remained buried. For decades, details of the family's murder were hidden in the Communist Party archives in Moscow. However, after the dissolution of the Soviet Union, postcommunist leaders allowed the bones to be exhumed so that they could be given a proper burial. This exhumation took on intense political meaning because the people of Russia hoped to do more than just give the family a proper burial. The event took on the symbolic significance of laying to rest the brutality of the communist regime that took power after the murders of the Romanov family.

Because all that remained of the bodies when they were exhumed was a pile of bones, it was difficult to know if these were the remains of the royal family. A great deal of circumstantial evidence provided by forensic science pointed to that conclusion.

Forensic science is a branch of science that helps answer questions of interest to the legal system. Many of the techniques typically used by forensic scientists were not possible to use in the case of the Romanovs due to the decay of the bodies and the crime scene. For instance, no fingerprint, toxicology, footprint, or ballistic (firearm) evidence remained at the scene. Evidence that did remain at the scene included dental and bone structure. Dental evidence showed that five of the bodies had gold, porcelain, and platinum dental work, which had been available only to aristocrats.

The bones seemed to indicate that they belonged to six adults and three children. Investigators electronically superimposed the photographs of the skulls on archived photographs of the family. They compared the skeletons' measurements with clothing known to have belonged to the family. These and other data were consistent with the hypothesis that the bodies could be those of the tsar, the tsarina, three of their five children, and the four servants.

Karyotype analysis (Chapter 6) requires dividing cells, which also did not exist at the crime scene. Therefore, scientists tried to determine the sex of the buried individuals based on pelvic bone structure. Females have evolved to have wider pelvic openings to accommodate the passage of a child through the birth canal. Russian scientists thought that all three of the children's skeletons and two of the adults' skeletons were probably female (and four of the adult skeletons were male). However, the pelvises had decayed, so it was impossible to be certain.

By using the forensic evidence available to them, Russian scientists had shown only that these skeletons might be the Romanovs. They had not yet shown with any degree of certainty that these bodies *did* belong to the slain royals. The new Russian leaders did not want to make a mistake when symbolically burying a former regime. Unassailable proof was necessary because so much was at stake politically.

Before more sophisticated techniques became available to scientists, solving the puzzle of the Ekaterinburg bones required that scientists be able to show relatedness between the tsar and tsarina's skeletons and their children's skeletons.

8.1 Extensions of Mendelian Genetics

Patterns of inheritance are fairly simple to predict when genes are inherited in a straightforward manner. Predictions about traits that are controlled by one or two genes with dominant and recessive alleles can be made by using Punnett squares (Chapter 7). Patterns of inheritance that are a little more complex are said to be *extensions* of Mendelian genetics.

For some genes, two identical copies of a dominant allele are required for expression of the full effect of a phenotype. In this case, the phenotype of the heterozygote is intermediate between both homozygotes—a situation called **incomplete dominance.** For example, the alleles that determine flower color in snapdragons are incompletely dominant: One homozygote produces red flowers, the other, presumably carrying two nonfunctional copies of a color gene, produces white flowers; the heterozygote, carrying one "red" and one "white" allele, produces pink flowers (**Figure 8.1** on the next page).

Alternatively, the phenotype of a heterozygote in which neither allele is dominant to the other may be a combination of both fully expressed traits. This situation, by which two different alleles of a gene are both expressed in an individual, is known as **codominance.** In cattle, for example, the allele

Flower color in snapdragons

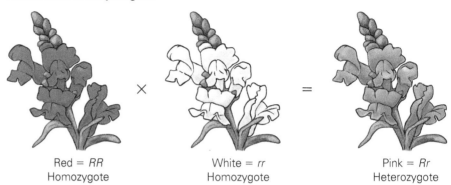

Red = *RR*
Homozygote

White = *rr*
Homozygote

Pink = *Rr*
Heterozygote

Figure 8.1 Incomplete dominance. Snapdragons show incomplete dominance in the inheritance of flower color. The heterozygous flower has a phenotype that is in between that of the two homozygotes.

Coat color in cattle

Red = *R¹R¹*

White = *R²R²*

Roan = *R¹R²*
(patchy red and white coat)

Figure 8.2 Codominance. Roan coat color in cattle is an example of codominance. Both alleles are equally expressed in the heterozygote, so the conventional uppercase and lowercase nomenclature for alleles no longer applies.

that codes for red hair color and the allele that codes for white hair color are both expressed in a heterozygote. These individuals have patchy coats that consist of an approximately equal mixture of white hairs and red hairs (**Figure 8.2**).

Because all that was left of the Romanovs was a pile of bones, the scientists could study only a few genetic traits to show the relatedness of the adult skeletons to two of the four children's skeletons. Genetic traits that were obvious, such as bone size and structure, are controlled by many genes and affected by environmental components like nutrition and physical activity level. Traits that are affected by the interactions between many are called **polygenic traits.** This made using bone size and structure to predict which of the adults' skeletons were related to the children's skeletons a matter of guesswork. The scientists had to use more sophisticated analyses.

A technique that scientists use to help determine relatedness of people is blood typing, which involves determining if certain carbohydrates are located on the surface of red blood cells. These surface markers are part of the **ABO blood system.** The ABO blood system displays two extensions of Mendelism—codominance and **multiple allelism,** which occurs when there are more than two alleles of a gene in the population. In fact, three distinct alleles of one blood-group gene code for the enzymes that synthesize the sugars found on the surface of red blood cells. Two of the three alleles display codominance to each other, and one allele is recessive to the other two.

Figure 8.3 summarizes the possible genotypes and phenotypes for the ABO blood system. The three alleles of this blood-type gene are I^A, I^B, and i. A given individual will carry only two alleles, one on each of his or her homologous pairs of chromosomes, even though three alleles are being passed on in the entire population. In other words, one person may carry the I^A and I^B alleles, and another might carry the I^A and i alleles. There are three different alleles, but each individual can carry only two alleles.

The symbols used to represent these alternate forms of the blood-type gene tell us something about their effects. The lowercase i allele is recessive to both the I^A and I^B alleles. Therefore, a person with the genotype I^Ai has type A blood, and a person with the genotype I^Bi has type B blood. A person with both recessive alleles, genotype ii, has type O blood. The uppercase I^A and I^B alleles display codominance in that neither masks the expression of the other. Both of these alleles are expressed. Thus, a person with the genotype I^AI^B has type AB blood.

Stop & Stretch   Imagine a new allele I^C that is codominant with I^A and I^B and dominant to i. List all of the possible blood-group genotypes and phenotypes in a population containing all four alleles.

Another molecule on the surface of red blood cells is called the **Rh factor.** Someone who is positive (+) for this trait has the Rh factor on his or her red blood cells, while someone who is negative (–) does not. This trait, unlike the ABO blood system, is inherited in a straightforward two-allele, completely dominant manner with Rh^+ dominant to Rh^-. Persons who are Rh-positive can have the genotype Rh^+Rh^+ or Rh^+Rh^-. An Rh-negative individual has the genotype Rh^-Rh^-.

Blood typing is often used to help establish whether a given set of parents could have produced a particular child. For example, a child with type AB blood and parents who are type A and type B could be related, but a child with type O blood could not have a parent with type AB blood. Likewise, if a child has blood type B and the known mother has blood type AB, then the father of that child could have type AB, A, B, or O blood, which does not help to establish parentage.

If a child has a blood type consistent with alleles that he or she may have inherited from a man who might be his or her father, this finding does not mean that the man is the father. Instead, it is only an indication that the man could be. In fact, many other men would also have that blood type (**Table 8.1**). Therefore, blood-type analysis can be used only to eliminate people from consideration. Blood typing cannot be used to positively identify someone as the parent of a particular child.

Clinicians must take ABO blood groups into account when performing blood transfusions. Persons receiving transfusions from incompatible blood

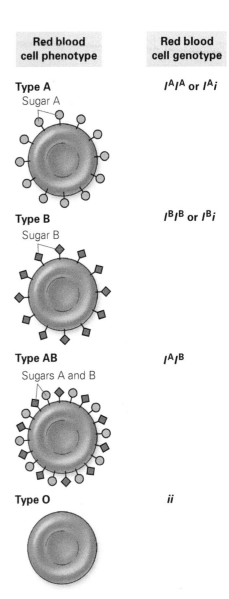

Red blood cell phenotype	Red blood cell genotype
Type A Sugar A	I^AI^A or I^Ai
Type B Sugar B	I^BI^B or I^Bi
Type AB Sugars A and B	I^AI^B
Type O	ii

Figure 8.3 ABO blood system. Red blood cell phenotypes and corresponding genotypes. Alleles I^A and I^B are codominant, and both are dominant to i.

TABLE 8.1

Frequency of blood types in U.S. population. Percentages of the population with a given blood type are listed from most to least common. The negative and positive superscripts refer to the absence or presence of the Rh factor.							
Blood Type	**O⁺**	**A⁺**	**B⁺**	**O⁻**	**A⁻**	**AB⁺**	**B⁻**	**AB⁻**
Frequency in U.S. Population (%)	40	32	11	7	5	3	1.5	0.5

TABLE 8.2

Blood transfusion compatibilities.

Recipient	Recipient Can Receive	Recipient Cannot Receive
Type O	Type O	Type A
		Type B
		Type AB
Type A	Type O	Type B
	Type A	Type AB
Type B	Type O	Type A
	Type B	Type AB
Type AB	Type O	None
	Type A	
	Type B	
	Type AB	

groups will mount an immune response against those sugars that they do not carry on their own red blood cells. The presence of these foreign red blood cell sugars causes a severe reaction in which the donated, incompatible red blood cells form clumps. This clumping can block blood vessels and kill the recipient. **Table 8.2** shows the types of blood transfusions individuals of various blood types can receive.

Blood-typing analysis can sometimes provide scientists with some information about potential relatedness of victims. However, it was not an option in the case of the Romanovs. The very old remains contained no blood, so blood typing was not possible.

The Romanov family was known to carry a trait that demonstrates another extension of Mendelism: pleiotropy. **Pleiotropy** is the ability of a single gene to cause multiple effects on an individual's phenotype. Alexis was the last of several Romanovs to have **hemophilia,** a pleiotropic blood-clotting disorder. A person with the most common form of hemophilia cannot produce a protein called clotting factor VIII. When this protein is absent, blood does not form clots to stop bleeding from a cut or internal blood vessel damage. Affected individuals bleed excessively, even from small cuts.

Due to the direct effects of excessive bleeding, hemophilia can lead to excessive bruising, pain and swelling in the joints, vision loss from bleeding into the eye, and anemia, resulting in fatigue. In addition, neurological problems may occur if bleeding or blood loss occurs in the brain.

Historical records indicate that Alexis, heir to the throne, was so ill with hemophilia that his father actually had to carry him to the basement room where he was executed.

8.2 Sex Determination and Sex Linkage

It appears that Alexis inherited the hemophilia allele from his mother. We can deduce this pattern of inheritance because we now know that the clotting factor gene is inherited in a sex-specific manner. The clotting factor VIII gene (the gene that, when mutated, causes hemophilia) is located in the X chromosome. Of the 23 pairs of chromosomes present in the cells of human males, 22 pairs are **autosomes,** or nonsex chromosomes, and one pair, X and Y, are the **sex chromosomes.** Males have 22 pairs of autosomes and one X and one Y sex chromosome. Females also have 22 pairs of autosomes, but their sex chromosomes are comprised of two X chromosomes.

Chromosomes and Sex Determination

The X and Y chromosomes are involved in producing the sex of an individual through a process called **sex determination.** When men produce sperm and the chromosome number is divided in half through meiosis, their sperm cells contain one member of each autosome and either an X or a Y chromosome. Females produce gametes with 22 unpaired autosomes and one of their two X chromosomes. Therefore, human egg cells normally contain one copy of an X chromosome, but sperm cells can contain either an X or a Y chromosome.

The sperm cell determines the sex of the offspring resulting from a particular fertilization. If an X-bearing sperm unites with an egg cell, the resulting offspring will be female (XX). If a sperm bearing a Y chromosome unites with an egg cell, the resulting offspring will be male (XY). **Figure 8.4** summarizes the process of sex determination in humans, and **Table 8.3** (on the next page) outlines some mechanisms of sex determination in nonhumans.

Sex Linkage

Genes located on the X or Y chromosome are called **sex-linked genes** because biological sex is inherited along with, or "linked to," the X or Y chromosome. Sex-linked genes found on the X chromosomes are said to be X linked, while those on the Y chromosome are Y linked. The X chromosome is much larger than the Y chromosome, which carries very little genetic information (**Figure 8.5**).

X-Linked Genes. X-linked genes are located on the X chromosome. The fact that males have only one X chromosome leads to some peculiarities in inheritance of sex-linked genes. Males always inherit their X chromosome from their mother because they must inherit the Y chromosome from their father to be male. Thus, males will inherit X-linked genes only from their mothers. Males are more likely to suffer from diseases caused by recessive alleles on the X chromosome because they have only one copy of any X-linked gene. Females are less likely to suffer from these diseases because they carry two copies of the X chromosome and thus have a greater likelihood of carrying at least one functional version of each X-linked gene.

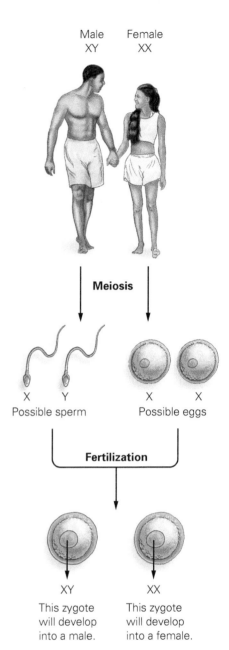

Figure 8.4 Sex determination in humans. In humans, sex is determined by the male because males produce sperm that carry either an X or a Y chromosome (in addition to 22 autosomes). The egg cell always carries an X chromosome along with the 22 unpaired autosomes. When an X-bearing sperm fertilizes the egg cell, a female (XX) results. When a Y-bearing sperm fertilizes an egg cell, an XY male results. **Visualize This:** What is the probability that a couple will have a boy? What is the probability that a couple with four boys will have a girl for their fifth child?

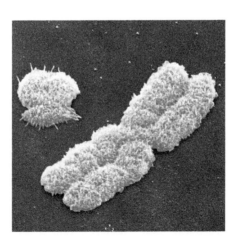

Figure 8.5 The X and Y chromosomes. The Y chromosome (left) is smaller than the X chromosome and carries fewer genes.

TABLE 8.3

Sex determination in nonhuman organisms.

Type of Organism		Mechanism of Sex Determination
Vertebrates (fish, amphibians, reptiles, birds, and mammals)		The male may have two of the same chromosomes and the female two different chromosomes. In these cases, the female determines the sex of the offspring.
Egg-laying reptiles		In many egg-laying species, two organisms with the same suite of sex chromosomes could become different sexes. Sex depends on which genes are activated during embryonic development. For example, the sex of some reptiles is determined by the incubation temperature of the egg.
Wasps, ants, bees		Sex is determined by the presence or absence of fertilization. In bees, males (drones) develop from unfertilized eggs. Females (workers or queens) develop from fertilized eggs.
Bony fishes		Some species of bony fishes change their sex after maturation. All individuals will become females unless they are deflected from that pathway by social signals such as dominance interactions.
Caenorhabditis elegans		The nematode *C. elegans* can either be male or have both male and female reproductive organs. Such individuals are called hermaphrodites.

A **carrier** of a recessively inherited trait has one copy of the recessive allele and one copy of the normal allele and will not exhibit symptoms of the disease. Only females can be carriers of X-linked recessive traits because males with a copy of the recessive allele will have the trait. Both males and females can be carriers of non-sex-linked, autosomal traits.

Even though female carriers of an X-linked recessive trait will not display the recessive trait, they can pass the trait on to their offspring. For this reason, most women carrying the hemophilia allele will not even realize that they are a carrier until their son becomes ill. Figure 8.6a illustrates that a cross between a male who does not have hemophilia and a female carrier can produce unaffected females, carrier females, unaffected males, and affected males. Figure 8.6b illustrates that no male children produced by a cross between an affected male and an unaffected

(a) Unaffected male × Carrier female

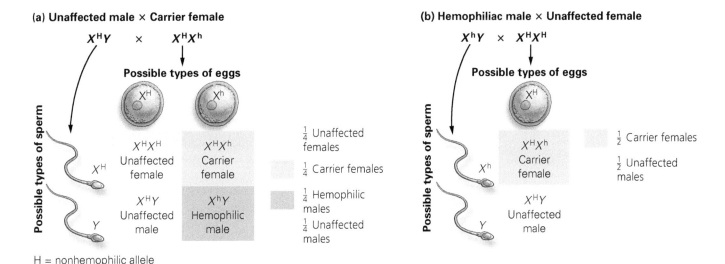

X^HY × X^HX^h

Possible types of eggs

Possible types of sperm

X^H

$X^H X^H$ Unaffected female

$X^H X^h$ Carrier female

$X^H Y$ Unaffected male

$X^h Y$ Hemophilic male

$\frac{1}{4}$ Unaffected females

$\frac{1}{4}$ Carrier females

$\frac{1}{4}$ Hemophilic males

$\frac{1}{4}$ Unaffected males

H = nonhemophilic allele
h = hemophilic allele

(b) Hemophiliac male × Unaffected female

X^hY × X^HX^H

Possible types of eggs

X^H

$X^H X^h$ Carrier female

$X^H Y$ Unaffected male

$\frac{1}{2}$ Carrier females

$\frac{1}{2}$ Unaffected males

Figure 8.6 Genetic crosses involving the X-linked hemophilia trait. Cross (a) shows the possible outcomes and associated probabilities of a mating between a nonhemophilic male and a female carrier of hemophilia. Cross (b) shows the possible outcomes and associated probabilities of a cross between a hemophilic male and an unaffected female. **Visualize This:** What genetic cross would result in the highest frequency of affected males?

female would have hemophilia. All daughters produced by this cross would be carriers of the trait. In the United States, there are over 20,000 hemophiliacs, nearly all of whom are male.

Stop & Stretch What must be true of the parents of a female who has an X-linked recessive disease?

Red-green colorblindness is another X-linked trait. This trait affects approximately 4% of males. Red blindness is the inability to see red as a distinct color. Green blindness is the inability to see green as a distinct color. When normal (in this case, the dominant alleles are normal), these genes code for the production of proteins called opsins that help absorb different wavelengths of light. A lack of opsins causes insensitivity to light of red and green wavelengths.

Duchenne muscular dystrophy is a progressive, fatal X-linked disease of muscle wasting that affects approximately 1 in 3500 males. The onset of muscle wasting occurs between 1 and 12 years of age, and by age 12, affected boys are often confined to a wheelchair. The affected gene is one that normally codes for the dystrophin protein. When at least one allele is normal, dystrophin stabilizes cell membranes during muscle contraction. It is thought that the absence of normal dystrophin protein causes muscle cells to break down and muscle tissue to die.

Most of the protein products of over 100 genes on the X chromosome have nothing at all to do with the production of biological sex differences. Accordingly, females and males should require equal doses of the products of X-linked genes. How can we account for the fact that females, with their two X chromosomes, could receive two doses of X-linked genes, while males receive only one? The answer comes from a phenomenon called **X inactivation** that occurs in all of the cells of a developing female embryo. This inactivation guarantees that all females actually receive only one dose of the proteins produced by genes on the X chromosomes. Inactivation of the genes on one of the two X chromosomes takes place in the embryo at about the time that the embryo implants in the uterus. One chromosome is inactivated when a string of RNA is wrapped around it (**Figure 8.7**).

Active X chromosome

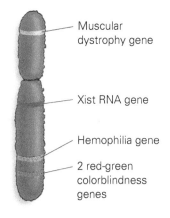

Muscular dystrophy gene

Xist RNA gene

Hemophilia gene

2 red-green colorblindness genes

Inactive X chromosome

Xist RNA

Figure 8.7 X inactivation. One of a female's two X chromosomes is inactivated when strands of RNA wrap around the chromosome.

This inactivation is random with respect to the parental source—either of the two X chromosomes can be inactivated in a given cell. Inactivation is also irreversible and, as such, is inherited during cell replication. In other words, once a particular X chromosome is inactivated in a cell, all descendants of that cell continue inactivating the same chromosome.

Cats with tortoiseshell coat coloring illustrate the effects of X inactivation. The coats of tortoiseshell cats are a mixture of black and orange patches. The genes for fur color are located on the X chromosomes. If a cat with orange fur mates with a cat with black fur, a female kitten could have one X chromosome with the gene for orange fur, and one X chromosome with the gene for black fur (Figure 8.8a). Early in development, when the embryo consists of about 16 cells, one of the two X chromosomes is randomly inactivated in each cell. Thus, some cells will be expressing the orange fur-color gene, and others will be expressing the black fur-color gene.

The pattern of inactivation (the X chromosome that the kitten inherited from its mother or the X chromosome that the kitten inherited from its father) is passed on to the daughter cells of the 16-celled embryo, resulting in the patches of orange and black fur color seen in tortoiseshell cats (Figure 8.8b). Because this pattern of coat coloration requires the expression of both alleles of

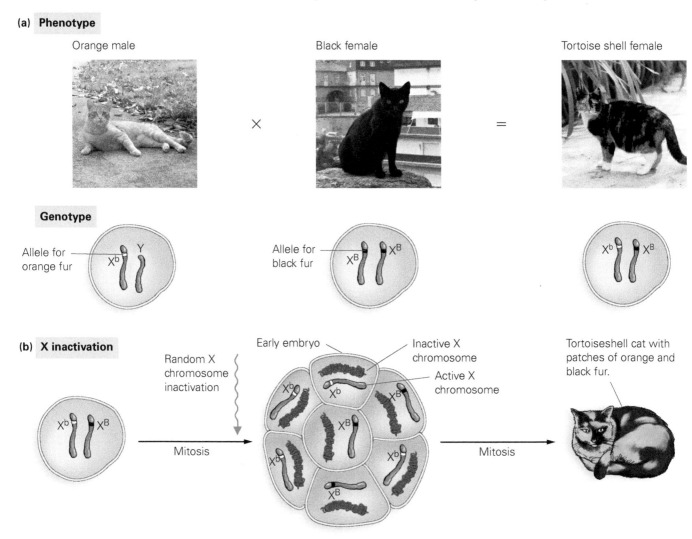

(a) Phenotype

Orange male   Black female   Tortoise shell female

×   =

Genotype

Allele for orange fur — X^b Y

Allele for black fur — X^B X^B

X^b X^B

(b) X inactivation

X^b X^B

Random X chromosome inactivation

Mitosis

Early embryo   Inactive X chromosome

Active X chromosome

X^b X^b X^B

X^B

X^b X^B X^b

X^B

Mitosis

Tortoiseshell cat with patches of orange and black fur.

Figure 8.8 X inactivation and patchy gene expression. (a) An orange male cat and a black female cat can produce female offspring with tortoiseshell coats. (b) Random inactivation of one X chromosome early in development leads to patches of coat color.

the color gene in different patches of cells, a cat must have two different alleles of this X-linked gene. Therefore, these cats are almost always female. On rare occasions, male cats can have this pattern of coloring. This can happen only if a male cat has two X chromosomes and one Y chromosome—a situation that does occur, though infrequently, via nondisjunction (Chapter 6)—and results in sterility.

Y-Linked Genes. **Y-linked genes** are located on the Y chromosome and are passed from fathers to sons. Although this distinctive pattern of inheritance should make Y-linked genes easy to identify, very few genes have been localized to the Y chromosomes. One gene known to be located exclusively on the Y chromosome is called the *SRY* gene (for sex-determining region of the Y chromosome). The expression of this gene triggers a series of events leading to development of the testes and some of the specialized cells required for male sexual characteristics. Genes other than *SRY*, on chromosomes other than the Y, code for proteins that are unique to males but are not expressed unless testes develop.

Stop & Stretch Sometimes chromosomal sex does not match biological sex. For instance, males can sometimes have two X chromosomes and no Y chromosome. Most of these males carry a copy of the SRY gene on one X chromosome. What event during meiosis can explain this phenomenon? Some genes on the Y chromosome code for sperm production. What does this imply about XX males?

In the case of the bones thought to belong to the Romanovs, because karyotype (review Figure 6.18) and pelvic bone analyses were difficult to perform on the decayed bones, scientists analyzed DNA from the bones for sequences known to be present only on the Y chromosome. When DNA that was isolated from the children's remains was analyzed, it became clear that the children's bones all belonged to girls. If these bones did belong to the Romanovs, one of the two missing children was Alexis, the Romanovs' only son.

8.3 Pedigrees

Another line of evidence was provided by the extensive family trees of the Romanovs and their relatives. Because the hemophilia gene is X linked, Alexis Romanov inherited the disease from his mother, who must have been a carrier of the disease. We can trace the lineage of this disease through the Romanov family by using a chart called a pedigree. A **pedigree** is a family tree that follows the inheritance of a genetic trait for many generations of relatives. Pedigrees are often used in studying human genetics because it is impossible and unethical to set up controlled matings between humans the way one can with fruit flies or plants. Pedigrees allow scientists to study inheritance by analyzing matings that have already occurred. **Figure 8.9** identifies some of the symbols used in pedigrees, and **Figure 8.10** (on the next page) shows how scientists can use pedigrees to determine whether a trait is inherited as autosomal dominant or recessive or as sex-linked recessive.

Information is available about the Romanovs' ancestors because they were royalty and because scientists interested in hemophilia had kept very good records of the inheritance of that trait. Hemophilia was common among European royal families but rare among the rest of the population. This was because members of the royal families intermarried to preserve the royal bloodlines. The tsarina must have been a carrier of the hemophilia allele because her son had

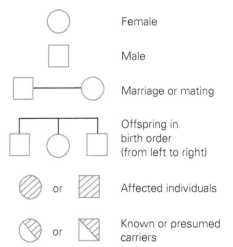

Pedigree analysis symbols

○ Female

□ Male

□—○ Marriage or mating

Offspring in birth order (from left to right)

⊘ or ▨ Affected individuals

⊘ or ▧ Known or presumed carriers

Figure 8.9 Pedigree analysis. Symbols used in pedigrees. **Visualize This:** Draw a pedigree showing a female who has a boy with her first husband and a girl with her second husband.

(a) Dominant trait: Polydactyly

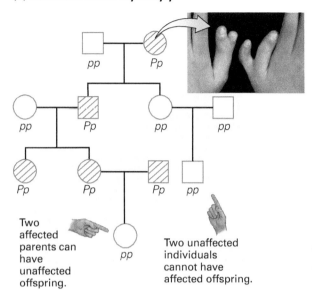

Two affected parents can have unaffected offspring.

Two unaffected individuals cannot have affected offspring.

(b) Recessive trait: Attached earlobes

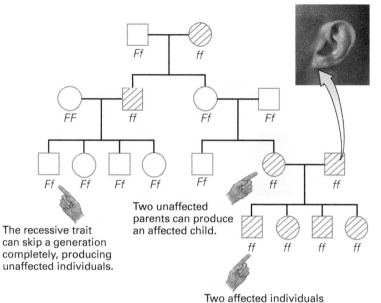

The recessive trait can skip a generation completely, producing unaffected individuals.

Two unaffected parents can produce an affected child.

Two affected individuals have affected offspring.

(c) Sex-linked trait: Muscular dystrophy

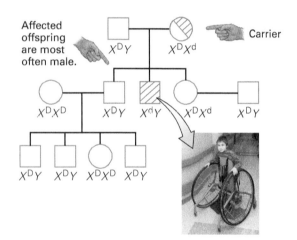

Affected offspring are most often male.

Carrier

Figure 8.10 Pedigrees showing different modes of inheritance. (a) Polydactyly is a dominantly inherited trait. People with this condition have extra fingers or toes. (b) Having attached earlobes is a recessively inherited trait. (c) Muscular dystrophy is inherited as an X-linked recessive trait.

the trait. Her mother, Alice, must also have been a carrier because the tsarina's brother Fred had the disease, as did two of her sister Irene's sons, Waldemar and Henry. The tsarina's grandmother, Queen Victoria, seems to have been the first carrier of this allele in the family, because there is no evidence of this disease before her eighth child, Leopold, was affected.

Queen Victoria's mother, Princess Victoria, most likely incurred a mutation to the clotting factor VIII gene while the cells of her ovaries were undergoing DNA synthesis to produce egg cells. The egg cell that carried the mutant clotting factor VIII gene was passed from Princess Victoria to her daughter, who came to be crowned Queen Victoria. When the inherited mutant cell divided by mitosis to produce Queen Victoria's body (somatic) cells, the mutant clotting factor VIII gene was passed on to each of those cells. When the cells in her ovaries underwent meiosis to produce gametes, the Queen passed the mutant version on to three of her nine children (**Figure 8.11**). The extensive pedigree available to the scientists working on the Romanov case, in concert with a powerful technique called DNA fingerprinting, would provide the key data in solving the mystery of the buried bones.

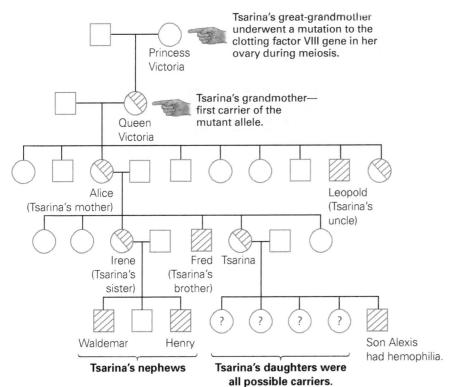

Tsarina's great-grandmother underwent a mutation to the clotting factor VIII gene in her ovary during meiosis.

Tsarina's grandmother—first carrier of the mutant allele.

Tsarina's nephews

Tsarina's daughters were all possible carriers.

Figure 8.11 Origin and inheritance of the hemophilia allele. This abbreviated pedigree shows the origin of the hemophilia allele and its inheritance among the tsarina's family. It appears that the tsarina's great-grandmother underwent a mutation that she passed on to her daughter, who then passed it on to three of her nine children—one of whom was the tsarina's mother, Alice.

8.4 DNA Fingerprinting

Limits on the power of conventional forensic techniques such as blood typing and karyotyping to identify the bones found in the Ekaterinburg grave necessitated the use of more sophisticated techniques. To do so, scientists took advantage of the fact that any two individuals who are not identical twins have small differences in the sequences of nucleotides that comprise their DNA. To test the hypothesis that the bones buried in the Ekaterinburg grave belonged to the Romanov family, the scientists had to answer the following questions:

1. Which of the bones from the pile are actually different bones from the same individuals?
2. Which of the adult bones could have been from the Romanovs, and which bones could have belonged to their servants?
3. Are these bones actually from the Romanovs, not some other related set of individuals?

An Overview: DNA Fingerprinting

All of these questions were answered using **DNA fingerprinting.** This technique allows unambiguous identification of people in the same manner that traditional fingerprinting has been used in the past. To begin this process, it is necessary to isolate the DNA to be fingerprinted. Scientists can isolate DNA from blood, semen, vaginal fluids, a hair root, skin, and even (as was the case in Ekaterinburg) degraded skeletal remains.

Because often there is not much DNA present, it must first be copied to generate larger quantities for analysis. Since each person has a unique sequence of DNA, particular regions of DNA, which vary in size, are selected for copying. When the copied DNA fragments are separated and stained, a unique pattern is produced. *See the next section for* ***A Closer Look*** *at DNA Fingerprinting.*

A Closer Look:
DNA Fingerprinting

When very small amounts of DNA are available, as is often the case, scientists can make many copies of the DNA by first performing a DNA-amplifying reaction.

Polymerase Chain Reaction (PCR)

The **polymerase chain reaction (PCR)** is used to amplify, or produce many copies of, a particular region of DNA (**Figure 8.12**). To perform PCR, scientists place four essential components into a tube: (1) the **template,** which is the double-stranded DNA to be copied; (2) the **nucleotides** adenine (A), cytosine (C), guanine (G), and thymine (T), which are the individual building-block subunits of DNA; (3) short single-stranded pieces of DNA called **primers,** which are complementary to the ends of the region of DNA to be copied; and (4) an enzyme called *Taq* polymerase.

***Taq* polymerase** is a DNA polymerase that uses one strand of DNA as a template for the synthesis of a daughter strand that carries complementary nucleotides (A:T base pairs are complementary, as are G:C base pairs) (Chapter 6). This enzyme was given the first part of its name (*Taq*) because it was first isolated from *Thermus aquaticus*, a bacterium that lives in hydrothermal vents and can withstand very high temperatures. The second part of the enzyme's name (polymerase) describes its synthesizing activity—it acts as a DNA polymerase.

The main difference between human DNA polymerase and *Taq* polymerase is that the *Taq* polymerase is resistant to extremely high temperatures, temperatures at which human DNA polymerase would be inactivated. The heat-resistant qualities of *Taq* polymerase thus allow PCR reactions to be run at very high temperatures. High temperatures are

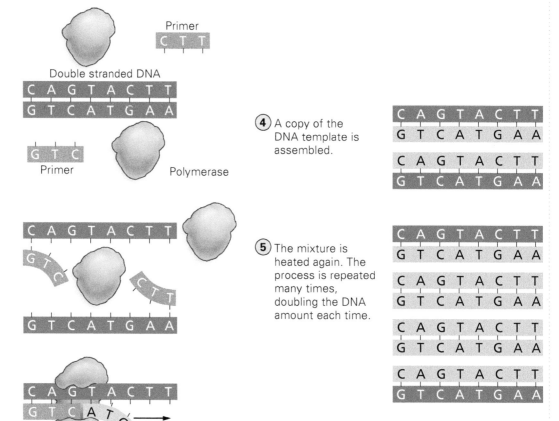

1 PCR is used to amplify, or make copies of, a specific region of DNA. During a PCR reaction, primers (short stretches of single-strand DNA), free nucleotides, and template DNA are mixed with heat-tolerant *Taq* polymerase.

2 The DNA is heated to separate, or denature, the two strands. As the mixture cools, the primers bond to the region of the DNA template they complement.

3 The *Taq* polymerase uses the primers to initiate synthesis.

4 A copy of the DNA template is assembled.

5 The mixture is heated again. The process is repeated many times, doubling the DNA amount each time.

Figure 8.12 The polymerase chain reaction (PCR). PCR is used to amplify, or make copies of, DNA. Each round of PCR doubles the number of DNA molecules. This type of exponential growth can yield millions of copies of DNA.

necessary because the DNA molecule being amplified must first be **denatured,** or split up the middle of the double helix, to produce single strands. After heating, the DNA solution is allowed to cool, and the primers bond to the region of the DNA template they complement. *Taq* polymerase then uses the primers to initiate DNA synthesis, producing double-stranded DNA molecules. This cycle of heating and cooling the tube is repeated many times, with each round of PCR doubling the amount of double-stranded DNA present in the tube.

Once scientists have produced enough DNA by PCR, they can analyze the differences between individuals. Because each individual has distinct nucleotide sequences, copying different people's DNA with the same enzyme can produce fragments of different sizes.

Analysis of Variable Number Tandem Repeats (VNTRs)

During PCR, the copied regions of DNA have different lengths due to the presence of DNA sequences that vary in number, called variable number tandem repeats (VNTRs) (**Figure 8.13**). These are nucleotide sequences that all of us carry, but in different numbers. For example, one person may have four copies of the following sequence (CGATCGA) on one chromosome

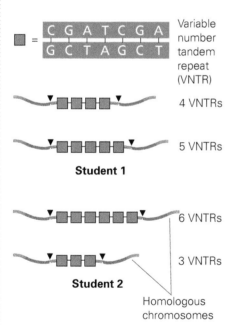

Figure 8.13 Variable number tandem repeats (VNTRs). Student 1 has four repeat sequences comprising the VNTRs on one of his chromosomes and five on the other. Student 2 has six copies of the same repeat sequence on one of her chromosomes and three on the other member of the homologous pair. The repeat sequence is represented as a box. PCR amplifies the region of DNA containing the VNTRs, generating fragments of different sizes. PCR primer sites around the VNTRs are shown as arrowheads.

and five copies of the same sequence on the other, homologous, chromosome. Another person may have six repeating copies of that sequence on one chromosome and three copies on the other member of the homologous pair. Within a population, there are variable numbers of these known tandem repeat sequences. When PCR is used to copy the region of DNA containing a VNTR, those segments of DNA that carry more repeats will be longer than those that carry fewer repeats. Because it is impossible to determine the size of DNA fragments by simply looking at the DNA in the tube, techniques that allow for the separation and visualization of the DNA are required.

Stop & Stretch   VNTRs do not code for proteins. If two individuals have different numbers of VNTRs around a particular gene, does that mean that they carry different alleles for the gene? Explain your answer.

Gel Electrophoresis

The amplified fragments of DNA generated by PCR can be separated from each other by allowing the fragments to migrate through a solid support called an **agarose gel,** which resembles a thin slab of gelatin. When an electric current is applied, the gel impedes the progress of the larger DNA fragments more than it does the smaller ones, facilitating the size-based separation. The size-based separation of molecules when an electric current is applied to a gel is a technique called **gel electrophoresis.** Segments of DNA with more repeats would be longer than those with fewer repeats. Longer DNA segments will not migrate as far in the gel as shorter DNA fragments. Thus, agarose gel electrophoresis separates the DNA fragments on the basis of their size (**Figure 8.14**).

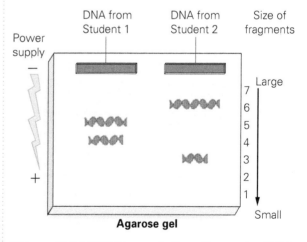

Figure 8.14 DNA fingerprinting. Fragments of DNA that were amplified by PCR can be separated by gel electrophoresis. The DNA itself is not visible to the unaided eye and must be stained to produce the DNA fingerprint.

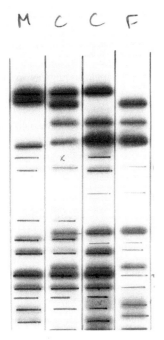

Figure 8.15 DNA fingerprint. This photograph of a DNA fingerprint shows a mother (M), a father (F), and their two children (C). Note that every band present in a child must also be present in one of the parents.

DNA Fingerprinting Evidence Helped Solve the Romanov Mystery

In 1992, a team of Russian and English scientists used the DNA fingerprinting technique to determine which of the bones discovered at Ekaterinburg belonged to the same skeleton. They used a slightly different DNA fingerprinting technique in that they isolated DNA from mitochondria located in the bone cells to confirm that the pile of decomposed bones in the Ekaterinburg grave belonged to nine different individuals.

Once scientists had established that the bones from nine different people were buried in the grave, they tried to determine which bones might belong to the adult Romanovs and which belonged to the servants. For the answer to these questions, scientists took advantage of the fact that Romanov family members would have more DNA sequences in common with each other than they would with the servants.

In fact, each DNA region that a child carries is inherited from one of his or her parents. Therefore, each band produced in a DNA fingerprint of a child must be present in the DNA fingerprint of one of that child's parents (Figure 8.15). By comparing DNA fingerprints made from the smaller skeletons, scientists were able to determine which of the six adult skeletons could have been the tsar and tsarina. Figure 8.16 shows a hypothetical DNA fingerprint that illustrates how the banding patterns produced can be used to determine which of the bones belonged to the parents of the smaller skeletons and which bones may have belonged to the unrelated servants. Notice that Figure 8.16 has more than two bands in each gel lane. This is because the PCR reaction was done in a way that amplified many different VNTR sites at the same time.

Stop & Stretch Why doesn't a child possess all of the same fragments found in one parent?

Figure 8.16 Hypothetical fingerprint of adult- and child-sized skeletons. A hypothetical DNA fingerprint of DNA from the bone cells of individuals found in the Ekaterinburg grave is shown. From the results of this fingerprint, it is evident that children 1, 2, and 3 are the offspring of adults 1 and 3. Note that each band from each child has a corresponding band in either adult 1 or adult 3. The remaining DNA from adults does not match any of the children, so these adults are not the parents of any of these children.

DNA evidence was further used to help put to rest claims made by many pretenders to the throne. People from all over the world had alleged that they were either a Romanov who had escaped execution or a descendant of an escapee.

The most compelling of these claims was made by a young woman who was rescued from a canal in Berlin, Germany, two years after the murders. This young woman suffered from amnesia and was cared for in a mental hospital, where the staff named her Anna Anderson (**Figure 8.17**). She later came to believe that she was Anastasia Romanov, a claim she made until her death in 1984. The 1956 Hollywood film *Anastasia,* starring Ingrid Bergman, made Anna Anderson's claim seem plausible. A more recent animated version of the story of an escaped princess, also titled *Anastasia,* convinced many young viewers that Anna Anderson was indeed the Romanov heiress.

Because the sex-typing analysis showed only that one daughter was missing from the grave, but not which daughter, scientists again looked to the fingerprinting data. DNA fingerprinting had been done in the early 1990s on intestinal tissue removed during a surgery performed before Anna Anderson's death. The analysis showed that Anna was not related to anyone buried in the Ekaterinburg grave. She could not be Anastasia.

Thus far in our narrative, scientists have answered two of the questions posed. They have determined (1) that nine different individuals were buried in the Ekaterinburg grave and (2) that two of the adult skeletons were the parents of the three children. The last question is still unanswered. How did the scientists show that these were bones from the Romanov family, not just some other set of related individuals?

To answer this question, the scientists turned to living relatives of the Romanovs. DNA testing was performed on England's Prince Philip, who is a grandnephew of Tsarina Alexandra. In addition, Nicholas II's dead brother

(a) Anna Anderson

(b) Anastasia Romanov

Figure 8.17 Anna Anderson and Anastasia Romanov. Photos were not useful in determining whether Anna Anderson was Anastasia Romanov, as she claimed. DNA evidence was.

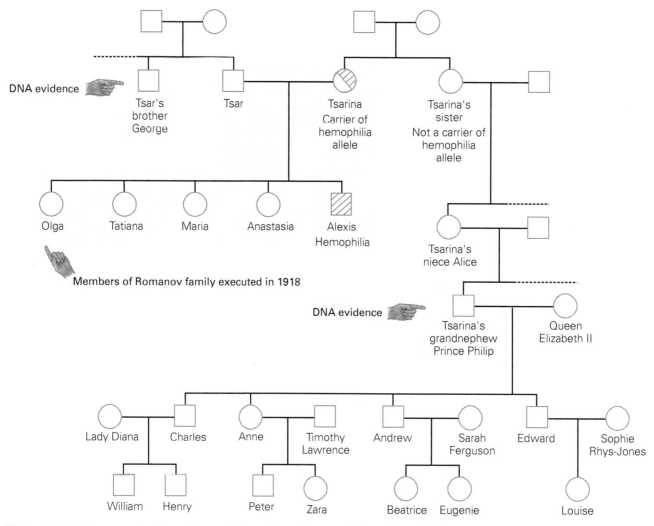

Figure 8.18 Romanov family pedigree. This pedigree shows only the pertinent family members. DNA from the tsar's brother George showed that he was related to the tsar. Note that Prince Philip is the tsarina's grandnephew. Prince Philip married Queen Elizabeth II. Together they had four children, Charles, Anne, Andrew, and Edward, the current British royal family. The tsarina's sister does not appear to have been a carrier of the hemophilia allele because none of her descendants has been affected by the disease.

George was exhumed, and his DNA was tested. **Figure 8.18** shows the Romanov family pedigree; you can see how these individuals are related to each other. The DNA testing performed on these individuals showed that George was genetically related to the adult male skeleton related to the children's skeletons, and that Prince Philip was genetically related to the adult female skeleton shown to be related to the children's skeletons. This evidence strongly supported the hypothesis that the adult skeletons were indeed those of the tsar and tsarina. The process of elimination suggests that the remaining four skeletons had to be the servants.

Table 8.4 summarizes how scientists used the scientific method to test the hypothesis that the remains were indeed those of the Romanovs.

In 1998, some 80 years after their execution, the Romanov family was finally laid to rest, and the people of postcommunist Russia symbolically laid to rest this part of their country's political history. Ten years after that, all cases of pretenders to the throne were dismissed when bone fragments unearthed in a forest near where the family was killed were found to belong to Alexis and his sister.

TABLE 8.4

The scientific method. A summary of tests and the conclusions that were drawn from them. **Hypothesis:** The bones found in the Ekaterinburg grave belonged to the Romanov family and their servants.	
Test	**Description of Results**
Analyze teeth	Expensive dental work was typically seen only in royalty.
Measure skeletons	The skeletons are those of six adults and three children.
Sex typing	The two children missing from the grave are a male and a female.
DNA fingerprinting	Children in the grave are related to two adults in the grave.
DNA fingerprinting	Claims to be one of the missing Romanov children or their descendants are disproved.
DNA fingerprinting	The two adults related to the children are related to known Romanov relatives.

Conclusion: When you look at each result individually, the evidence is less compelling than when you look at all the evidence together. As a whole, the evidence strongly supports the hypothesis that it was indeed the Romanovs who were buried in the Ekaterinburg grave.

SAVVY READER

Determining the Sex of Your Child

You are killing time in a pharmacy while your prescription is filled. You come across a product that claims to help you conceive a child of a particular sex. This product claims to have a success rate of 60%.

In a different aisle of the pharmacy, there is a magazine with a cover story claiming that more attractive parents have more girls. This article claims that Americans who are rated as attractive have close to 60% chance of having a daughter.

1. What are the odds of having a child of a particular gender by chance? Can we make any decisions about the validity of these assertions without data and information about the significance of any reported differences?

2. Can you think of any way your gametes could be directed to produce a male or female offspring?

3. A friend of yours believes that parents have children of the gender that the parents are best able to rear. Couples who have daughters have traits that are better for daughters to have and vice versa. Your friend believes this is consistent with the theory of natural selection. Based on your understanding of natural selection and gametogenesis, does this seem possible?

Chapter Review

Learning Outcomes

MasteringBIOLOGY

Go to the Study Area at www.masteringbiology.com for practice quizzes, myeBook, BioFlix™ 3-D animations, MP3Tutor sessions, videos, current events, and more.

LO1 **Differentiate incomplete dominance from codominance (Section 8.1).**

- Incomplete dominance is an extension of Mendelian genetics in which the phenotype of the progeny is intermediate to that of both parents (pp. 175–176).
- Codominance occurs when both alleles of a given gene are expressed (pp. 175–176).

LO2 **Explain how polygenic and pleiotropic traits differ (Section 8.1).**

- Polygenic inheritance occurs when many genes control one trait (p. 176).
- Pleiotropy occurs when a single gene leads to multiple effects (p. 178).

LO3 **Outline the inheritance of the multiple allelic ABO blood system (Section 8.1).**

- Genes that have more than two alleles segregating in a population are said to have multiple alleles (p. 176).
- The ABO blood system displays both multiple allelism (alleles I^A, I^B, and i) and codominance because both I^A and I^B are expressed in the heterozygote (pp. 176–178).

LO4 **Describe the mechanism of sex determination in humans (Section 8.2).**

- One mechanism of sex determination involves the suite of sex chromosomes present (p. 179).
- In humans, males have an X and a Y chromosome and can produce gametes containing either sex chromosome, and females have two X chromosomes and always produce gametes containing an X chromosome. When an X-bearing sperm fertilizes an egg cell, a female baby will result. When a Y-bearing sperm fertilizes an egg cell, a male baby will result (pp. 179–180).

LO5 **Explain the pattern of inheritance exhibited by sex-linked genes (Section 8.2).**

- Genes linked to the X and Y chromosomes show characteristic patterns of inheritance. Males need only one recessive X-linked allele to display the associated phenotype. Females can be carriers of an X-linked recessive allele and may pass an X-linked disease on to their sons (pp. 179–181).
- Y-linked genes are passed from fathers to sons (p. 183).

LO6 **Describe how X-inactivation changes patterns of inheritance (Section 8.2).**

- One of the two X chromosomes in females is inactivated, and the genes residing on it are not expressed. This inactivation is faithfully propagated to all daughter cells (pp. 181–183).

LO7 **Explain how scientists use pedigrees (Section 8.3).**

- Pedigrees are charts that scientists use to study the transmission of genetic traits among related individuals (pp. 183–185).

LO8 **Explain the significance of DNA fingerprinting and how the process works (Section 8.4).**

- DNA fingerprinting is used to show the relatedness of individuals based on similarities in their DNA sequences (p. 185).
- When small amounts of DNA are available, the polymerase chain reaction (PCR) can be used to make millions of copies of the DNA (p. 186).
- The lengths of repeated sequences are characteristic of a given individual (p. 187).
- DNA samples in an agarose gel subjected to an electric current will separate according to their size (p. 187).

Roots to Remember

The following roots of words come mainly from Latin and Greek and will help you decipher terms:

forensic	comes from a Latin word, *forum*, which was the place where legal disputes and lawsuits were heard.
hemo-	means blood. Chapter term: hemophilia
pleio- and poly-	both come from Greek words for many. Chapter terms: pleiotropy and polygenic

Learning the Basics

1. **LO3** What does it mean for a trait to have multiple alleles? Can an individual carry more than two alleles of one gene?

2. **LO4** How is sex determined in humans?

3. **LO8** Describe the technique of DNA fingerprinting.

4. LO3 If a man with blood type A and a woman with blood type B have a child with type O blood, what are the genotypes of each parent?

5. LO5 A man with type A⁺ blood whose father had type O⁻ blood and a woman with type AB⁻ blood could produce children with which phenotypes relative to these blood-type genes?

6. LO8 Which of the following is *not* part of the procedure used to make a DNA fingerprint?

A. DNA is amplified by PCR; **B.** DNA is placed in a gel and subjected to an electric current; **C.** The genes that encode fingerprint patterns are cloned into bacteria; **D.** DNA from blood, semen, vaginal fluids, or hair root cells can be used for analysis.

7. LO8 Which of the following statements is consistent with the DNA fingerprint shown in **Figure 8.19**?

A. B is the child of A and C; **B.** C is the child of A and B; **C.** D is the child of B and C; **D.** A is the child of B and C; **E.** A is the child of C and D.

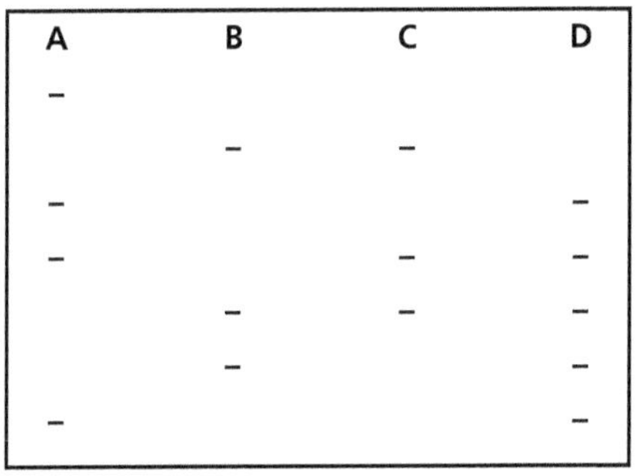

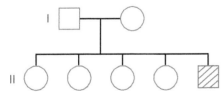

Figure 8.19 DNA fingerprint.

8. LO4 What is the probability that a family with two children will have one boy and one girl?

A. 100%; **B.** 75%; **C.** 50%; **D.** 25%

9. LO3 The pedigree in **Figure 8.20** illustrates the inheritance of hemophilia (sex-linked recessive trait) in the royal family. What is the genotype of individual II-5 (Alexis)?

A. $X^H X^H$; **B.** $X^H X^h$; **C.** $X^h X^h$; **D.** $X^H Y$; **E.** $X^h Y$

Figure 8.20 Pedigree.

10. LO5 A woman is a carrier of the X-linked recessive color-blindness gene. She mates with a man with normal color vision. Which of the following is true of their offspring?

A. All the males will be colorblind; **B.** All the females will be carriers; **C.** Half the females will be colorblind; **D.** Half the males will be colorblind.

11. LO2 Which of the following traits is most likely to be a polygenic trait?

A. a blood disease caused by one mutation; **B.** a complex trait like intelligence; **C.** a disease you get when exposed to radiation; **D.** having type AB blood

12. LO6 What causes tortoiseshell cats to have patches of color?

A. multiple alleles; **B.** environmental influences; **C.** pleiotropy; **D.** random X-inactivation

Analyzing and Applying the Basics

1. LO1 Compare and contrast codominance and incomplete dominance.

2. LO8 Draw a DNA fingerprint that might be generated by two sisters and their parents.

3. LO7 Draw a pedigree of a mating between first cousins and use the pedigree to explain why matings between relatives can lead to an increased likelihood of offspring with rare recessive diseases.

Connecting the Science

1. Science helped solve the riddle of who was buried in the Ekaterinburg grave, leading to a church burial for some of the Romanov family. In this manner, science played a role in helping the people of Russia come to terms with the brutal communist regime that followed the deaths of the royal family. Can you think of other examples for which science has been used to help answer a question with great social implications?

2. The Innocence Project is a nonprofit legal clinic at the Benjamin Cardozo School of Law in New York City that attempts to help prisoners whose claims of innocence can be verified by DNA testing because biological evidence from their cases still exists. Over 250 inmates have been exonerated since the project started in 1992. Thousands of inmates await the opportunity for testing. Does the fact that the Innocence Project has shown that the criminal justice system convicts innocent people change your opinion about the death penalty? Why or why not?

Answers to **Stop & Stretch, Visualize This, Savvy Reader,** and **Chapter Review** questions can be found in the **Answers** section at the back of the book.

Genetically Modified Organisms

Gene Expression, Mutation, and Cloning

Many people are concerned about genetic engineering.

Genetic engineering is the use of laboratory techniques to alter hereditary traits in an organism. Genetically modified foods, also called GM foods, have had one or more of their genes modified or a gene from another organism inserted alongside their normal complement of genes. Emotions run high in the debate over the altering of our food supply, with people on both sides of the issue making their cases dramatically.

Demonstrators dressed in biohazard suits toss genetically modified crackers, cereal, and pasta into a garbage can in front of a supermarket while shareholders inside vote on whether to remove such foods from the store shelves. Protesters hack down plots of genetically modified corn and coffee. Some parents refuse to send their kids to day-care centers that use milk from cows treated with growth hormone.

On the other side of the issue are proponents of genetically modifying foods who believe the technology holds tremendous promise. Advocates claim that genetically modified foods have the potential to help wipe out hunger. They see a future in which all children will be vaccinated through the use of edible vaccines, and common human diseases like heart disease will be prevented when people begin to eat genetically modified animals such as pigs that produce omega-3 fatty acids. Proponents even see environmental benefits because some genetically modified crops allow farmers to use far smaller amounts of chemicals that damage the environment.

Even more controversy surrounds the idea of genetically modifying humans. Scientists have already cloned many different animals; can humans be far behind?

Can you tell which of these foods have been genetically modified?

Is it acceptable to modify animals such as these pigs that can produce an essential nutrient for human consumption?

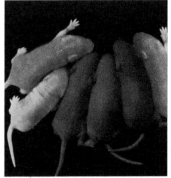

Or animals such as these luminescent mice?

195

How can the average citizen determine whether the benefits of genetic engineering outweigh the drawbacks? To answer these questions for yourself, you must learn about how genes typically produce their products and then how they can be modified to suit human desires.

9.1 Protein Synthesis and Gene Expression

To genetically modify foods, scientists can move a gene known to produce a certain protein from one organism to another. Alternatively, the scientists can change the amount of protein a gene produces. Regulating the amount of protein produced by a cell is also referred to as *regulating gene expression*.

One of the first examples of scientists controlling gene expression occurred in the early 1980s when genetic engineers began to produce **recombinant bovine growth hormone (rBGH)** in their laboratories. Recombinant (r) bovine growth hormone is a protein that has been made by genetically engineered bacteria. These bacterial cells have had their DNA manipulated so that it carries the instructions for, or encodes, a cow growth hormone that can be produced in the laboratory. Hormones are substances that are secreted from specialized glands and travel through the bloodstream to affect their target organs. Growth hormones act on many different organs to increase the overall size of the body. Bovine growth hormone that is produced in a laboratory can be injected into dairy cows to increase their milk production.

Production of growth hormone protein, or any protein, in the lab or in a cell, requires the use of the genetic information coded in the DNA.

From Gene to Protein

Protein synthesis involves using the instructions carried by a gene to build a particular protein. Genes do not build proteins directly; instead, they carry the instructions that dictate how a protein should be built. Understanding protein synthesis requires that we review a few basics about DNA, genes, and RNA. First, DNA is a polymer of nucleotides that make chemical bonds with each other based on their complementarity: adenine (A) to thymine (T) and cytosine (C) to guanine (G). Second, a gene is a sequence of DNA that encodes a protein. Proteins are large molecules composed of amino acids. Each protein has a unique function that is dictated by its particular structure. The structure of a protein is the result of the order of amino acids that constitute it because the chemical properties of amino acids cause a protein to fold in a particular manner. Before a protein can be built, the instructions carried by a gene are first copied. When the gene is copied, the copy is made up not of DNA (deoxyribonucleic acid) but of **RNA (ribonucleic acid).**

RNA, like DNA, is a polymer of nucleotides. A nucleotide is composed of a sugar, a phosphate group, and a nitrogen-containing base. Whereas the sugar in DNA is deoxyribose, the sugar in RNA is ribose. RNA has the nitrogenous base uracil (U) in place of thymine. RNA is usually single stranded, not double stranded like DNA (**Figure 9.1**).

When a cell requires a particular protein, a strand of RNA is produced using DNA as a guide or template. RNA nucleotides are able to make base pairs

(a) DNA

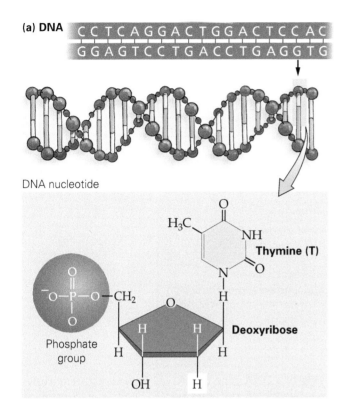

DNA nucleotide

Thymine (T)

Phosphate group

Deoxyribose

(b) RNA

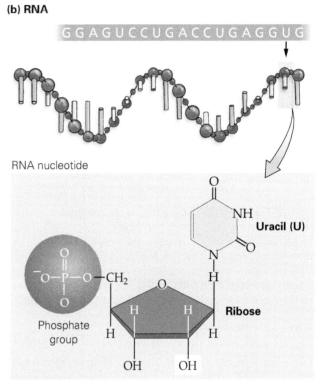

RNA nucleotide

Uracil (U)

Phosphate group

Ribose

with DNA nucleotides. C pairs with G, and A pairs with U. The RNA copy then serves as a blueprint that tells the cell which amino acids to join together to produce a protein. Thus, the flow of genetic information in a eukaryotic cell is from DNA to RNA to protein (**Figure 9.2**).

How does this flow of information actually take place in a cell? Going from gene to protein involves two steps. The first step, called **transcription,** involves producing the copy of the required gene. In the same way that a transcript of a speech is a written version of the oral presentation, transcription inside a cell produces a transcript of the original gene, with the RNA nucleotides substituted for DNA nucleotides. The second step, called **translation,** involves decoding the copied RNA sequence and producing the protein for which it codes. In the same way that a translator deciphers one language into another, translation in a cell involves moving from the language of nucleotides (DNA and RNA) to the language of amino acids and proteins.

Figure 9.1 DNA and RNA. (a) DNA is double stranded. Each DNA nucleotide is composed of the sugar deoxyribose, a phosphate group, and a nitrogen-containing base (A, G, C, or T). (b) RNA is single stranded. RNA nucleotides are composed of the sugar ribose, a phosphate group, and a nitrogen-containing base (A, G, C, or U). **Visualize This:** Point out the chemical difference between the sugar in DNA and the one in RNA and the difference between the nitrogenous bases thymine and uracil.

Figure 9.2 The flow of genetic information. Genetic information flows from a DNA to an RNA copy of the DNA gene, to the amino acids that are joined together to produce the protein coded for by the gene.

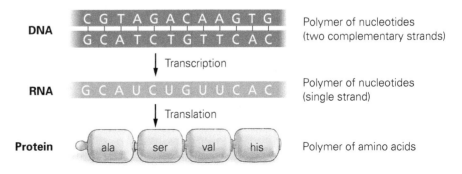

DNA — Polymer of nucleotides (two complementary strands)

Transcription

RNA — Polymer of nucleotides (single strand)

Translation

Protein — Polymer of amino acids

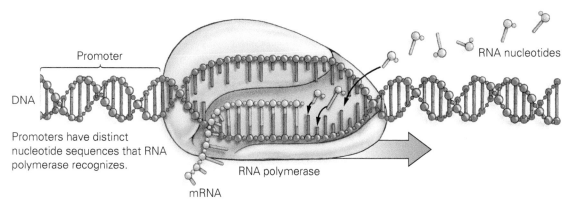

Promoter

DNA

Promoters have distinct nucleotide sequences that RNA polymerase recognizes.

RNA nucleotides

RNA polymerase

mRNA

Figure 9.3 Transcription. RNA polymerase ties together nucleotides within the growing RNA strand as they form hydrogen bonds with their complementary base on the DNA. When the RNA polymerase reaches the end of the gene, the mRNA transcript is released.
Visualize This: Propose a sequence for the DNA strand that is being used to produce the mRNA. Keep in mind that purines (A and G) are composed of two rings and thus are represented by longer pegs in this illustration. Once you have proposed a DNA sequence, determine the mRNA sequence that would be produced by transcription.

Transcription

Transcription is the copying of a DNA gene into RNA (Figure 9.3). The copy is synthesized by an enzyme called **RNA polymerase.** To begin transcription, the RNA polymerase binds to a nucleotide sequence at the beginning of every gene, called the **promoter.** Once the RNA polymerase has located the beginning of the gene by binding to the promoter, it then rides along the strand of the DNA helix that comprises the gene. As it is traveling along the gene, the RNA polymerase unzips the DNA double helix and ties together RNA nucleotides that are complementary to the DNA strand it is using as a template. This results in the production of a single-stranded RNA molecule that is complementary to the DNA sequence of the gene. This complementary RNA copy of the DNA gene is called **messenger RNA (mRNA)** because it carries the message of the gene that is to be expressed.

Translation

The second step in moving from gene to protein, translation, requires that the mRNA be used to produce the actual protein for which the gene encodes. For this process to occur, a cell needs mRNA, a supply of amino acids to join in the proper order, and some energy in the form of ATP. Translation also requires structures called ribosomes and transfer RNA molecules.

Ribosomes. **Ribosomes** are subcellular, globular structures (Figure 9.4) that are composed of another kind of RNA called **ribosomal RNA (rRNA),** which is wrapped around many different proteins. Each ribosome is composed of two subunits—one large and one small. When the large and small subunits of the ribosome come together, the mRNA can be threaded between them. In addition, the ribosome can bind to structures called **transfer RNA (tRNA)** that carry amino acids.

Transfer RNA (tRNA). Transfer RNA (Figure 9.5) is yet another type of RNA found in cells. An individual transfer RNA molecule carries one specific amino acid and interacts with mRNA to place the amino acid in the correct location of the growing polypeptide.

As mRNA moves through the ribosome, small sequences of nucleotides are sequentially exposed. These sequences of mRNA, called **codons,** are three nucleotides long and encode a particular amino acid. Transfer RNAs

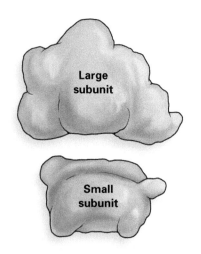

Large subunit

Small subunit

Figure 9.4 Ribosome. Ribosomes are composed of two subunits. Each subunit in turn is composed of rRNA and protein.

also have a set of three nucleotides, which will bind to the codon if the right sequence is present. These three nucleotides at the base of the tRNA are called the **anticodon** because they complement a codon on mRNA. The anticodon on a particular tRNA binds to the complementary mRNA codon. In this way, the codon calls for the incorporation of a specific amino acid. The ribosome moves along the mRNA sequentially exposing codons for tRNA binding.

When a tRNA anticodon binds to the mRNA codon, a peptide bond is formed. The ribosome adds the amino acid that the tRNA is carrying to the growing chain of amino acids that will eventually constitute the finished protein. The transfer RNA functions as a sort of cellular translator, fluent in both the language of nucleotides (its own language) and the language of amino acids (the target language).

To help you understand protein synthesis, let us consider its similarity to an everyday activity such as baking a cake (**Figure 9.6**). To bake a cake, you would consult a recipe book (genome) for the specific recipe (gene) to make your cake (protein). You may copy the recipe (mRNA) out of the book so that the original recipe (gene) does not become stained or damaged. The original recipe (gene) is left in the book (genome) on a shelf (nucleus) so that you can make another copy when you need it. The original recipe (gene) can be copied again and again. The copy of the recipe (mRNA) is placed on the kitchen counter (ribosome) while you assemble the ingredients (amino acids). The ingredients (amino acids) for your cake (protein) include flour, sugar, butter, milk, and eggs. The ingredients are measured in measuring spoons and cups (tRNAs). (While in baking you might use the same cups and spoons for several ingredients, in protein synthesis we use tRNAs that are dedicated to one specific ingredient.) The measuring spoons and cups bring the ingredients to the kitchen counter. Like the ingredients in a cake that can be used in many ways to produce a variety of foods, amino acids can be combined in different orders to produce different proteins. The ingredients (amino acids) are always added according to the instructions specified by the original recipe (gene).

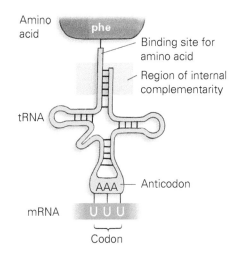

Figure 9.5 Transfer RNA (tRNA). Transfer RNAs translate the language of nucleotides into the language of amino acids. The tRNA that binds to UUU carries only one amino acid (phe, the three letter abbreviation for phenylalanine).

Visualize This: The structure of a tRNA molecule involves regions where the RNA strand forms complementary bonds with itself, causing the RNA to fold up on itself. What nitrogenous bases might be involved in bonding in such regions of internal complementarity?

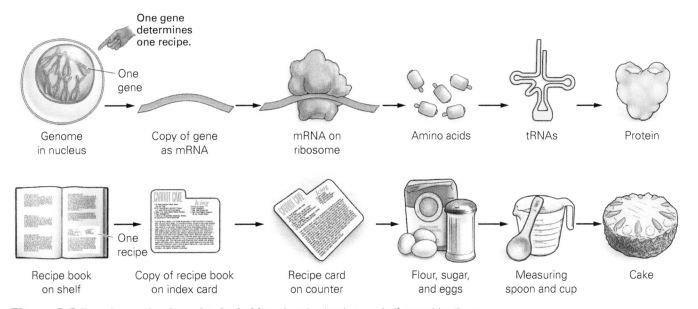

Figure 9.6 Protein synthesis and cake baking. A recipe book on a shelf resembles the genome in the nucleus of a cell. Copying one recipe onto an index card is similar to copying the gene to produce mRNA. The recipe card is placed on the kitchen counter, and the mRNA is placed on the ribosome. The ingredients for a cake are flour, sugar, butter, milk, and eggs. The ingredients to make a protein are various amino acids. The ingredients are measured in measuring spoons and cups (tRNAs) that are dedicated to one specific ingredient. The measuring spoons and cups bring the ingredients to the kitchen counter. The ingredients (amino acids) are always added according to the instructions specified by the original recipe (gene).

Figure 9.7 Translation. During translation, mRNA directs the synthesis of a protein.

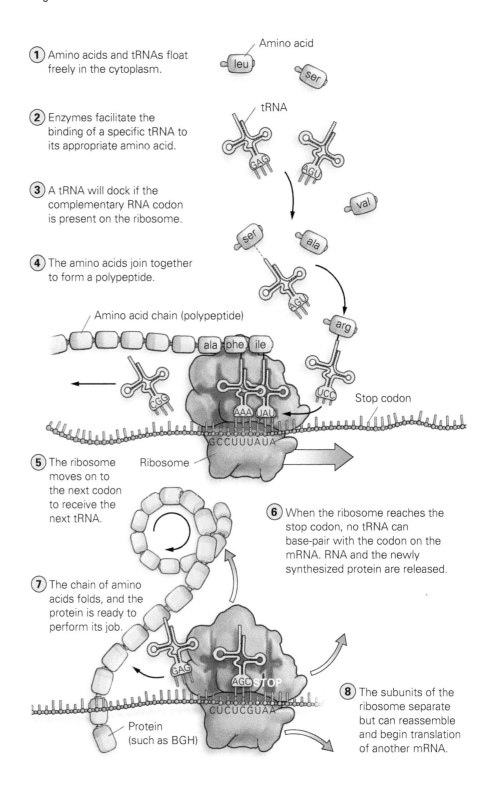

① Amino acids and tRNAs float freely in the cytoplasm.

② Enzymes facilitate the binding of a specific tRNA to its appropriate amino acid.

③ A tRNA will dock if the complementary RNA codon is present on the ribosome.

④ The amino acids join together to form a polypeptide.

⑤ The ribosome moves on to the next codon to receive the next tRNA.

⑥ When the ribosome reaches the stop codon, no tRNA can base-pair with the codon on the mRNA. RNA and the newly synthesized protein are released.

⑦ The chain of amino acids folds, and the protein is ready to perform its job.

⑧ The subunits of the ribosome separate but can reassemble and begin translation of another mRNA.

Amino acid

tRNA

Amino acid chain (polypeptide)

Stop codon

Ribosome

Protein (such as BGH)

Within cells, the sequence of bases in the DNA dictates the sequence of bases in the RNA, which in turn dictates the order of amino acids that will be joined together to produce a protein. Protein synthesis ends when a codon that does not code for an amino acid, called a **stop codon,** moves through the ribosome. When a stop codon is present in the ribosome, no new amino acid can be added, and the growing protein is released. Once released, the protein folds up on itself and moves to where it is required in the cell. A summary of the process of translation is shown in Figure 9.7.

The process of translation allows cells to join amino acids in the sequence coded by the gene. Scientists can determine the sequence of amino acids that a gene calls for by looking at the **genetic code.**

Genetic Code. The genetic code determines which mRNA codons code for which amino acids (**Table 9.1**). As Table 9.1 shows, there are 64 codons, 61 of which code for amino acids. Three of the codons are stop codons that occur near the end of an mRNA. Because stop codons do not code for an amino acid, protein synthesis ends when a stop codon enters the ribosome. In the table, you can see that the codon AUG functions both as a start codon (and thus is found near the beginning of each mRNA) and as a codon dictating that the amino acid methionine (met) be incorporated into the protein being synthesized. This initial methionine is often removed later. Notice also that the same amino acid can be coded for by more than one codon. For example, the amino acid threonine (thr) is incorporated into a protein in response to the codons ACU, ACC, ACA, and ACG. The fact that more than one codon can code for the same amino acid is referred to as *redundancy* in the genetic code. There is, however, no situation where a given codon can call for more than one amino acid. For example, AGU codes for serine (ser) and nothing else. Therefore, there is no *ambiguity* in the genetic code regarding which amino acid any codon will call for. The genetic code is also *universal* in the sense that organisms typically decode the same gene to produce the same protein. This is why genes can be moved from one organism to another, as was the case with the gene for luminescence in the mice of the chapter opener photo, which is a gene normally expressed in jellyfish.

TABLE 9.1

The genetic code.

It is possible to determine which amino acid is coded for by each mRNA codon using a chart called the genetic code. Look at the left-hand side of the chart for the first-base nucleotide in the codon; there are 4 rows, 1 for each possible RNA nucleotide—A, C, G, or U. By then looking at the intersection of the second-base columns at the top of the chart and the first-base rows, you can narrow your search for the codon to 4 different codons. Finally, the third-base nucleotide in the codon on the right-hand side of the chart determines the amino acid that a given mRNA codon codes for. Note the 3 codons UAA, UAG, and UGA that do not code for an amino acid; these are stop codons. The codon AUG is a start codon, found at the beginning of most protein-coding sequences.

First base	Second base				Third base
	U	**C**	**A**	**G**	
U	UUU } Phenylalanine (phe) UUC UUA } Leucine (leu) UUG	UCU UCC } Serine (ser) UCA UCG	UAU } Tyrosine (tyr) UAC UAA **Stop codon** UAG **Stop codon**	UGU } Cysteine (cys) UGC UGA **Stop codon** UGG Tryptophan (trp)	U C A G
C	CUU CUC } Leucine (leu) CUA CUG	CCU CCC } Proline (pro) CCA CCG	CAU } Histidine (his) CAC CAA } Glutamine (gln) CAG	CGU CGC } Arginine (arg) CGA CGG	U C A G
A	AUU AUC } Isoleucine (ile) AUA AUG Methionine (met) **Start codon**	ACU ACC } Threonine (thr) ACA ACG	AAU } Asparagine (asn) AAC AAA } Lysine (lys) AAG	AGU } Serine (ser) AGC AGA } Arginine (arg) AGG	U C A G
G	GUU GUC } Valine (val) GUA GUG	GCU GCC } Alanine (ala) GCA GCG	GAU } Aspartic acid (asp) GAC GAA } Glutamic acid (glu) GAG	GGU GGC } Glycine (gly) GGA GGG	U C A G

Figure 9.8 Substitution mutation.
A single nucleotide change from the normal DNA sequence (a) to the mutated sequence (b) can result in the incorporation of a different amino acid. If the substituted amino acid has chemical properties different from those of the original amino acid, then the protein may assume a different shape and thus lose its ability to perform its job.

(a) Normal DNA sequence

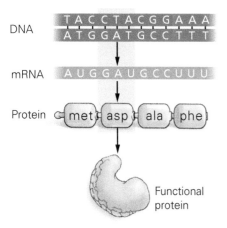

(b) Mutated DNA sequence

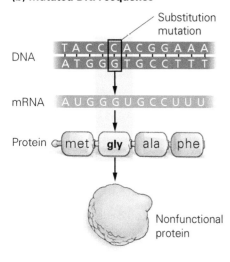

Figure 9.9 Neutral and frameshift mutations. (a) Neutral mutations result in the incorporation of the same amino acid as was originally called for. (b) The insertion (or deletion) of a nucleotide can result in a frameshift mutation.
Visualize This: Is the top or bottom strand of the DNA serving as the template for RNA synthesis in this figure?

Stop & Stretch What amino acids would be coded for by the mRNA CCU-AAU?

Mutations

Changes to the DNA sequence, called **mutations,** can affect the order or types of amino acids incorporated into a protein during translation. Mutations to a gene can result in the production of different forms, or alleles, of a gene. Different alleles result from changes in the DNA that alter the amino acid order of the encoded protein. Mutations can result in the production of either a nonfunctional protein or a protein different from the one previously required. If this protein does not have the same amino acid composition, it may not be able to perform the same job (Figure 9.8). For instance, a substitution of a single nucleotide results in the incorporation of a new amino acid in the hemoglobin protein and compromises the ability of cells to carry oxygen, producing sickle-cell disease.

There are also cases when a mutation has no effect on a protein. These cases may occur when changes to the DNA result in the production of a mRNA codon that codes for the same amino acid as was originally required. Due to the redundancy of the genetic code, a mutation that changes the mRNA codon from say ACU to ACC will have no impact because both of these codons code for the amino acid threonine. This is called a **neutral mutation** (Figure 9.9a). In addition, mutations can result in the substitution of one amino acid for another with similar chemical properties, which may have little or no effect on the protein.

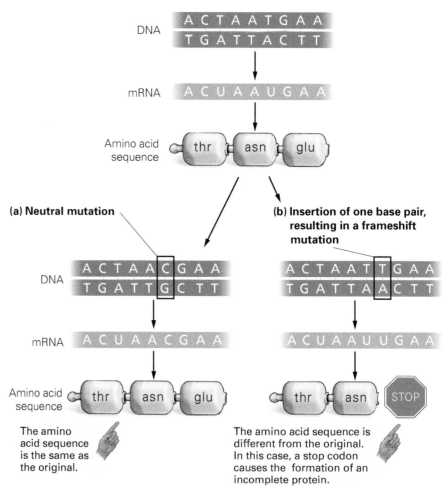

(a) Eukaryotic protein synthesis

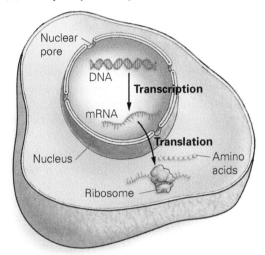

(b) Prokaryotic protein synthesis

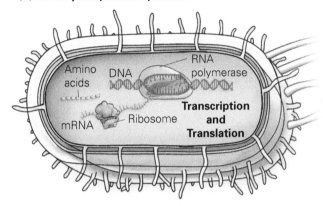

Figure 9.10 Protein synthesis in eukaryotic and prokaryotic cells. (a) In eukaryotes, transcription occurs in the nucleus and translation in the cytoplasm. (b) In prokaryotic cells, which lack nuclei, transcription and translation occur simultaneously.

Inserting or deleting a single nucleotide can have a severe impact because the addition (or deletion) of a nucleotide can change the groupings of nucleotides in every codon that follows (**Figure 9.9b**). Changing the triplet groupings is called altering the **reading frame.** All nucleotides located after an insertion or deletion will be regrouped into different codons, producing a **frameshift mutation.** For example, inserting an extra letter *H* after the fourth letter of the sentence, "The dog ate the cat," could change the reading frame to the nonsensical statement, "The dHo gat eth eca t." Inside cells, this often results in the incorporation of a stop codon and the production of a shortened, nonfunctional protein.

Stop & Stretch What effect does the formation of new alleles have on natural selection?

Cells in all organisms undergo this process of protein synthesis, with different cell types selecting different genes from which to produce proteins. **Figure 9.10a** shows the coordination of these two processes as they occur in cells with nuclei, that is, eukaryotic cells. In eukaryotic cells, transcription and translation are spatially separate, with transcription occurring in the nucleus and translation occurring in the cytoplasm. Cells lacking a membrane-bound nucleus and organelles are called prokaryotic cells. Prokaryotic cells (such as bacterial cells) also undergo protein synthesis, but transcription and translation occur at the same time and in the same location instead of occurring in separate places. As an mRNA is being transcribed, ribosomes attach and begin translating (**Figure 9.10b**).

An Overview: Gene Expression

Different cell types transcribe and translate different genes. Each cell in your body, except sperm or egg cells, has the same complement of genes you inherited from your parents but expresses only a small percentage of those genes. For example, because your muscles and nerves each perform a specialized suite of jobs, muscle cells turn on or express one suite of genes and nerve cells another (**Figure 9.11**). Turning a gene on or off, or modulating it more subtly, is called **regulating gene expression.** The expression of a given gene is regulated so that it is turned on and turned off in response to the cell's needs. *See the next section for **A Closer Look** at regulating gene expression.*

(a) Muscle cells

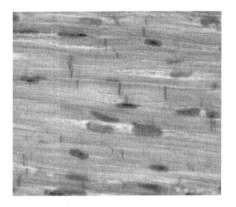

(b) Nerve cells

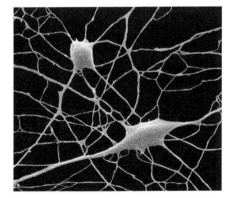

Figure 9.11 Gene expression differs from cell to cell. (a) A muscle cell performs different functions and expresses different genes than (b) a nerve cell. Both of these cells have the same suite of genes in their nucleus but express different subsets of genes.

A Closer Look:
Regulating Gene Expression

Gene expression can be regulated through the rate of transcription or translation. It is also possible to regulate how long an mRNA or protein remains functional in a cell.

Regulation of Transcription

Gene expression is most commonly regulated by controlling the rate of transcription. Regulation of transcription can occur at the promoter, the sequence of nucleotides adjacent to a gene to which the RNA polymerase binds to initiate transcription. When a cell requires a particular protein, the RNA polymerase enzyme binds to the promoter for that particular gene and transcribes the gene.

Prokaryotic and eukaryotic cells both regulate gene expression by regulating transcription but have different strategies for doing so. Prokaryotic cells typically regulate gene expression by blocking transcription via proteins called **repressors** that bind to the promoter and prevent the RNA polymerase from binding. When the gene needs to be expressed, the repressor will be released from the promoter so that the RNA polymerase can bind (**Figure 9.12a**). This is the main mechanism by which simple single-celled prokaryotes regulate gene expression.

The more complex eukaryotic cells have evolved more complex mechanisms to control gene expression. To control transcription, eukaryotic cells more commonly enhance gene expression using proteins called **activators** that help the RNA polymerase bind to the promoter, thus facilitating gene expression (**Figure 9.12b**). The rate at which the polymerase binds to the promoter is also affected by substances that are present in the cell. For example, the presence of alcohol in a liver cell might result in increased transcription of a gene involved in the breakdown of alcohol.

(a) Repression of transcription

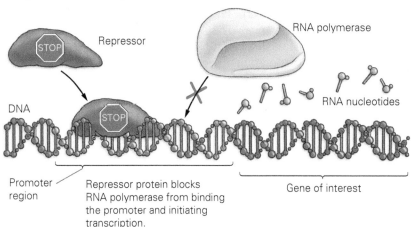

Promoter region — Repressor protein blocks RNA polymerase from binding the promoter and initiating transcription.

Gene of interest

(b) Activation of transcription

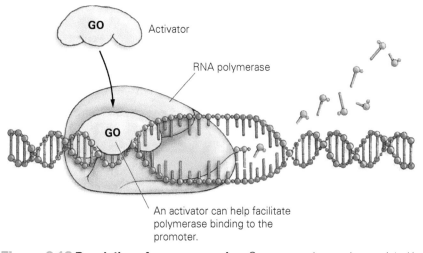

An activator can help facilitate polymerase binding to the promoter.

Figure 9.12 Regulation of gene expression. Gene expression can be regulated by (a) repression or (b) activation. Prokaryotes more typically use repression and eukaryotes activation of transcription to regulate gene expression.

Stop & Stretch How would a mutation in a repressor that makes it nonfunctional likely affect the individual with this mutation?

Regulation by Chromosome Condensation

It is also possible to regulate gene expression by condensing all or part of a chromosome. This prevents RNA polymerase from being able to access genes. The inactivation of an X chromosome turns off the expression of X-linked genes in organisms that have two X chromosomes (Chapter 8). Entire chromosomes are also inactivated when they condense during mitosis.

Regulation by mRNA Degradation

Eukaryotic cells can also regulate the expression of a gene by regulating how long a messenger RNA is present in the cytoplasm. Enzymes called *nucleases* roam the cytoplasm, cutting RNA molecules by binding to one end and breaking the bonds between nucleotides. If a particular mRNA has a long adenosine nucleotide "tail," it will survive longer in the cytoplasm and be translated more times. All mRNAs are eventually degraded in this manner; otherwise, once a gene had been transcribed one time, it would be expressed forever.

Regulation of Translation

It is also possible to regulate many of the steps of translation. For example, the binding of the mRNA to the ribosome can be slowed or hastened, as can the movement of the mRNA through the ribosome.

Regulation of Protein Degradation

Once a protein is synthesized, it will persist in the cell for a characteristic amount of time. Like the mRNA that provided the instructions for its synthesis, the life of a protein can be affected by cellular enzymes called *proteases* that degrade the protein. Speeding up or slowing down the activities of these enzymes can change the amount of time that a protein is able to be active inside a cell.

Genetic engineering permits precise control of gene expression in many different circumstances. For example, the problem of regulating gene expression is easily solved in the case of rBGH. Farmers can simply decide how much protein to inject into the bloodstream of a cow. However, they must first synthesize the protein.

9.2 Producing Recombinant Proteins

The first step in the production of the rBGH protein is to transfer the BGH gene from the nucleus of a cow cell into a bacterial cell. Bacteria are single-celled prokaryotes that copy themselves rapidly. They can thrive in the laboratory if they are allowed to grow in a liquid broth containing the nutrients necessary for survival. Bacteria with the BGH gene can serve as factories to produce millions of copies of this gene and its protein product. Making many copies of a gene is called **cloning** the gene.

Cloning a Gene Using Bacteria

The following three steps are involved in moving a BGH gene into a bacterial cell (**Figure 9.13** on the next page).

Step 1. **Remove the Gene from the Cow Chromosome.** The gene is sliced out of the cow chromosome on which it resides by exposing the cow DNA to enzymes that cut DNA. These enzymes, called **restriction enzymes,** act like highly specific molecular scissors. Most restriction enzymes cut DNA only at specific sequences, called *palindromes,* such as

Note that the bottom middle sequence is the reverse of the top middle sequence. Many restriction enzymes cut the DNA in a staggered pattern, leaving "sticky ends," such as

The unpaired bases form bonds with any complementary bases with which they come in contact. The enzyme selected by the scientist cuts on both ends of the *BGH* gene but not inside the gene.

A particular restriction enzyme cuts DNA at a specific sequence. Therefore, scientists need some information about the entire suite of genes present in a particular organism, called the **genome,** to determine which restriction enzyme cutting sites surround the gene of interest. Cutting the DNA generates many different fragments, only one of which will carry the gene of interest.

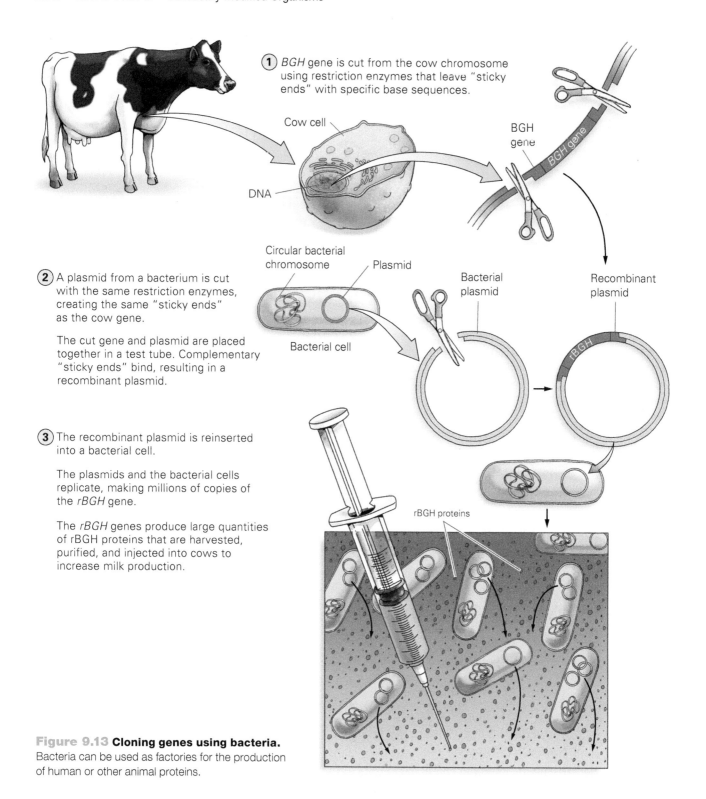

① *BGH* gene is cut from the cow chromosome using restriction enzymes that leave "sticky ends" with specific base sequences.

Cow cell

DNA

BGH gene

BGH gene

Circular bacterial chromosome

Plasmid

Bacterial plasmid

Recombinant plasmid

② A plasmid from a bacterium is cut with the same restriction enzymes, creating the same "sticky ends" as the cow gene.

The cut gene and plasmid are placed together in a test tube. Complementary "sticky ends" bind, resulting in a recombinant plasmid.

Bacterial cell

rBGH

③ The recombinant plasmid is reinserted into a bacterial cell.

The plasmids and the bacterial cells replicate, making millions of copies of the *rBGH* gene.

The *rBGH* genes produce large quantities of rBGH proteins that are harvested, purified, and injected into cows to increase milk production.

rBGH proteins

Figure 9.13 Cloning genes using bacteria. Bacteria can be used as factories for the production of human or other animal proteins.

Step 2. Insert the BGH Gene into the Bacterial Plasmid. Once the gene is removed from the cow genome, it is inserted into a bacterial structure called a plasmid. A plasmid is a circular piece of DNA that normally exists separate from the bacterial chromosome and can replicate independently of the bacterial chromosome. Think of the plasmid as a ferry that carries the gene into the bacterial cell, where it can be replicated. To incorporate the BGH gene into the plasmid, the plasmid is also cut with the same restriction enzyme used to cut the gene. Cutting both the plasmid and gene with the same enzyme allows the sticky ends that are generated to base-pair with each other

(A to T and G to C). When the cut plasmid and the cut gene are placed together in a test tube, they re-form into a circular plasmid with the extra gene incorporated.

The bacterial plasmid has now been genetically engineered to carry a cow gene. At this juncture, the BGH gene is referred to as the rBGH gene, with the r indicating that this product is genetically engineered, or recombinant, because it has been removed from its original location in the cow genome and recombined with the plasmid DNA.

Step 3. **Insert the Recombinant Plasmid into a Bacterial Cell.** The recombinant plasmid is now inserted into a bacterial cell. Bacteria can be treated so that their cell membranes become porous. When they are placed into a suspension of plasmids, the bacterial cells allow the plasmids back into the cytoplasm of the cell. Once inside the cell, the plasmids replicate themselves, as does the bacterial cell, making thousands of copies of the *rBGH* gene. Using this procedure, scientists can grow large amounts of bacteria capable of producing BGH.

Once scientists successfully clone the BGH gene into bacterial cells, the bacteria produce the protein encoded by the gene. Then the scientists are able to break open the bacterial cells, isolate the BGH protein, and inject it into cows. Bacteria can be genetically engineered to produce many proteins of importance to humans. For example, bacteria are now used to produce the clotting protein missing from people with hemophilia, as well as human insulin for people with diabetes.

Close to one-third of all dairy cows in the United States now undergo daily injections with rBGH. These injections increase the volume of milk that each cow produces by around 20%. Prior to marketing the recombinant protein to dairy farmers, scientists had to demonstrate that its product would not be harmful to cows or to humans who consume the cows' milk. This involved obtaining approval from the U.S. Food and Drug Administration (FDA).

The FDA is the governmental organization charged with ensuring the safety of all domestic and imported foods and food ingredients (except for meat and poultry, which are regulated by the U.S. Department of Agriculture). The manufacturer of any new food that is not **generally recognized as safe (GRAS)** must obtain FDA approval before marketing its product. Adding substances to foods also requires FDA approval unless the additive is GRAS.

According to the FDA, there is no detectable difference between milk from treated and untreated cows and no way to distinguish between the two. In 1993, the FDA deemed the milk from rBGH-treated cows as safe for human consumption.

Stop & Stretch Not all anti-GM organism activists are convinced that milk produced by rBGH-treated cows is identical to that from untreated cows. Given the effects of rBGH, how might the milk produced by these cows be different, even if it is safe?

In addition, because the milk from treated and untreated cows is indistinguishable, the FDA does not require that milk obtained from rBGH-treated cows be labeled in any manner. However, many distributors of milk from untreated cows label their milk as "hormone free," even though there is no evidence of the hormone in milk from treated cows (**Figure 9.14**).

But what about the welfare of the cows? There is some evidence that cows treated with rBGH are more susceptible to certain infections than untreated cows. Due in part to concerns about the health of these animals, Europe and Canada have banned rBGH use.

The rBGH story is a little different from that of genetically modified crop foods because rBGH protein is produced by bacteria and then administered to cows. When foods are genetically modified, the genome of the food itself is altered.

Figure 9.14 Hormone-free milk. Some manufacturers label their products as hormone free.

Figure 9.15 Artificial selection in corn. Selective-breeding techniques resulted in the production of modern corn (right) from ancient teosinite corn.

Figure 9.16 Golden rice. Golden rice has been genetically engineered to produce beta-carotene, which causes the rice to look more gold in color than the unmodified rice. This picture shows a mixture of modified and unmodified rice.

9.3 Genetically Modified Foods

Whether you realize it or not, you have probably been eating genetically modified foods for your entire life. Some genetic modifications involve moving genes between organisms in labs. Other modifications have occurred over the last several thousand years due to farmers' use of selective breeding techniques—breeding those cattle that produce the most milk or crossing crop plants that are easiest to harvest (Figure 9.15). While this artificial selection does not involve moving a gene from one organism to another, it does change the overall frequency of certain alleles for a gene in the population.

Unless you eat only certified organic foods, you have been eating food that has been modified. This may lead you to wonder why and how plants are genetically modified, what impact eating them has on your health, and whether growing them affects the environment.

Why Genetically Modify Crop Plants?

Crop plants are genetically modified to increase their shelf life, yield, and nutritive value. For example, tomatoes have been engineered to soften and ripen more slowly. The longer ripening time means that tomatoes stay on the vine longer, thus making them taste better. The slower ripening also increases the amount of time that tomatoes can be left on grocery store shelves without becoming overripe and mushy.

Genetic engineering techniques increase crop yield when plants are manipulated to be resistant to pesticides, herbicides, drought, and freezing. The nutritive value of crops can also be increased through genetic engineering. A gene that regulates the synthesis of beta-carotene has been inserted into rice, a staple food for many of the world's people. Scientists hope that the engineered rice will help decrease the number of people in underdeveloped nations who become blind due to a deficiency of beta-carotene, which is necessary for the synthesis of vitamin A. Eating this genetically modified rice, called *golden rice*, increases a person's ability to synthesize the vitamin (Figure 9.16). However, golden rice has not yet been approved for human consumption.

Modifying Crop Plants with the Ti Plasmid and Gene Gun

To modify crop plants, the gene must be able to gain access to the plant cell, which means it must be able to move through the plant's rigid outer cell wall. One "ferry" for moving genes into flowering plants is a naturally occurring plasmid of the bacterium *Agrobacterium tumefaciens*. In nature, this bacterium infects plants and causes tumors called **galls** (Figure 9.17a). The tumors are induced by a plasmid, called **Ti plasmid** (Ti for tumor inducing).

(a) Gall caused by *A. tumefaciens*

(b) Using the Ti plasmid

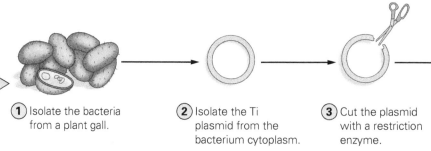

(1) Isolate the bacteria from a plant gall.

(2) Isolate the Ti plasmid from the bacterium cytoplasm.

(3) Cut the plasmid with a restriction enzyme.

Figure 9.17 Genetically modifying plants using the Ti plasmid. (a) Plants infected by *Agrobacterium tumefaciens* in nature show evidence of the infection by producing tumorous galls. (b) The Ti plasmid from *A. tumefaciens* serves as a shuttle for incorporating genes into plant cells. The recombinant plasmid is then used to infect developing plant cells, producing a genetically modified plant. When the plant cell reproduces, it may pass on the engineered gene to its offspring.

Genes from different organisms can be inserted into the Ti plasmid by using the same restriction enzyme to cut the Ti plasmid and the gene, resulting in identical sticky ends, and then connecting the gene and plasmid. The recombinant plasmid can then be inserted into a bacterium. The *A. tumefaciens* bacterium with the recombinant Ti plasmid is then used to infect plant cells. During infection, the recombinant plasmid is transferred into the host-plant cell (Figure 9.17b). For genetic engineering purposes, scientists use only the portion of a plasmid that does not cause tumor formation.

Moving genes into other agricultural crops such as corn, barley, and rice can also be accomplished by using a device called a **gene gun**. A gene gun shoots metal-coated pellets covered with foreign DNA into plant cells (Figure 9.18). A small percentage of these genes may be incorporated into the plant's genome. When a gene from one species is incorporated into the genome of another species, a **transgenic organism** is produced. A transgenic organism is more commonly referred to as a **genetically modified organism (GMO).**

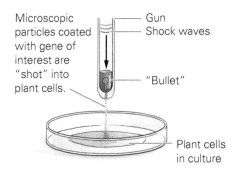

Figure 9.18 Genetically modifying plants using a gene gun. A gene gun shoots a plastic bullet loaded with tiny DNA-coated pellets into a plant cell. The bullet shells are prevented from leaving the gun, but the DNA-covered pellets penetrate the cell wall, cell membrane, and nuclear membrane of some cells.

Effect of GMOs on Health

Concerns about the potential negative health effects of consuming GM crops have led some citizens to fight for legislation requiring that genetically modified foods be labeled so that consumers can make informed decisions about the foods they eat. Manufacturers of GM crops argue that labeling foods is expensive and that the labels will be viewed by consumers as a warning, even in the absence of any proven risk. Those manufacturers believe that GM food labeling will decrease sales and curtail further innovation.

As the labeling controversy continues, most of us are already eating GM foods. Scientists estimate that over half of all foods in U.S. markets, including virtually all processed foods, contain at least small amounts of GM foods. Products that do not contain GMOs are often labeled to promote that fact. Concern about GM foods is not limited to their consumption. Many people are also concerned about the effects of GM crop plants on the environment.

GM Crops and the Environment

Many genetically modified crops have been engineered to increase their yield. For centuries, farmers have tried to increase yields by killing the pests that damage crops and by controlling the growth of weeds that compete for nutrients, rain, and sunlight. In the United States, farmers typically spray high volumes of chemical pesticides and herbicides directly onto their fields

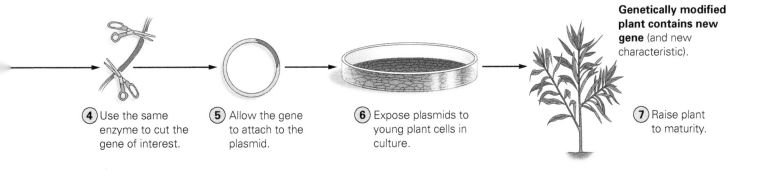

(4) Use the same enzyme to cut the gene of interest.

(5) Allow the gene to attach to the plasmid.

(6) Expose plasmids to young plant cells in culture.

Genetically modified plant contains new gene (and new characteristic).

(7) Raise plant to maturity.

Figure 9.19 Application of chemicals. It remains to be seen how genetic technologies might affect the use of herbicides and pesticides.

(Figure 9.19). This practice concerns people worried about the health effects of eating foods that have been treated by these often toxic or cancer-causing chemicals. In addition, both pesticides and herbicides may leach through the soil and contaminate drinking water.

To help decrease farmers' reliance on pesticides, agribusiness companies have engineered plants that are genetically resistant to pests. Unfortunately, this is a short-term solution because the pest populations will eventually evolve resistance. Application of a pesticide does not always kill all of the targeted organisms. The few pests that have preexisting resistance genes and are not susceptible survive and produce resistant offspring. Eventually, widespread resistance develops, and a new pesticide must be developed and applied.

The continued need for the development of new pesticides in farming is paralleled by farmers' reliance on herbicides. Roundup®, or glyophosphate, is a herbicide used to control weeds in soybean fields. Glyophosphate inhibits the synthesis of certain amino acids required by growing plants. Herbicide-resistant crop plants, such as Roundup Ready® soybeans, have been engineered to be resistant to Roundup. Roundup Ready plants are engineered to synthesize more of these amino acids than unengineered plants and thus can grow in the presence of the herbicide while nearby plants and weeds will die. Therefore, farmers can spray their fields of genetically engineered soybeans with herbicides that will kill everything but the crop plant. Some people worry that this resistance gene will allow farmers to spray more herbicide on their crops because there is no chance of killing the GM plant, thereby exposing consumers and the environment to even more herbicide.

There is also concern that GM crop plants may transfer engineered genes from modified crop plants to their wild or weedy relatives. Wind, rain, birds, and bees carry genetically modified pollen to related plants near fields containing GM crops (or even to farms where no GM crops are being grown). Many cultivated crops have retained the ability to interbreed with their wild relatives; in these cases, genes from farm crops can mix with genes from the wild crops.

While there is hope that genetic engineers will be able to help solve hunger problems by making farming more productive, there are also concerns about possible negative health and environmental effects of GM foods. It remains to be seen whether genetic engineering will constitute a lasting improvement to agriculture.

9.4 Genetically Modified Humans

Humans can also be genetically modified. Scientists may someday be able to cure or alleviate the symptoms of diseases by manipulating genes. It may also be possible for physicians to diagnose genetic defects in early embryos and fix them, thereby eliminating genetic disease in children and adults.

Stem Cells

Stem cells are unspecialized or **undifferentiated** precursor cells that have not yet been programmed to perform a specific function. Stem cells can be embryonic in origin or can be found in some adult tissues.

Imagine that you are remodeling an old home, and you have a type of material that you can mold into anything you might need for the remodeling job—brick, tile, pipe, plaster, and so forth. Having a supply of this material would help you fix many different kinds of damage. Scientists believe that stem cells may serve as this type of all-purpose repair material in the body. If cells are nudged in a particular developmental direction in the laboratory, they can be directed to become a particular tissue or organ. Using stem cells to produce healthy tissues as replacements for damaged tissues is a type of **therapeutic cloning.** Tissues and organs grown from stem cells in the laboratory may someday be used to replace organs damaged in accidents or that are gradually failing due to **degenerative diseases.** Degenerative diseases start with the slow breakdown of an organ and progress to organ failure. In addition, when one organ is not working properly, other organs are affected. Degenerative diseases include diabetes, liver and lung diseases, heart disease, multiple sclerosis, and Alzheimer's and Parkinson's diseases.

Stem cells could provide healthy tissue to replace those tissues damaged by spinal cord injury or burns. New heart muscle could be produced to replace muscle damaged during a heart attack. A diabetic could have a new pancreas, and people suffering from some types of arthritis could have replacement cartilage to cushion their joints. Thousands of people waiting for organ transplants might be saved if new organs were grown in the lab.

Stem cells can be isolated from early embryos that are left over after fertility treatments. *In vitro* (Latin, meaning "in glass") fertilization procedures often result in the production of excess embryos because many egg cells are harvested from a woman who wishes to become pregnant. These egg cells are then mixed with her partner's sperm in a petri dish, resulting in the production of many fertilized eggs that grow into embryos. A few of the embryos are then implanted into the woman's uterus. The remaining embryos are stored so that more attempts can be made if pregnancy does not result or if the couple desires more children. When the couple achieves the desired number of pregnancies, the remaining embryos are discarded or, with the couple's consent, used for stem cell research. The crux of the controversy surrounding stem cells involves the idea of some that early embryos constitute life, and that using them for research is unethical. Others argue that early embryos will not become life unless they are implanted in a woman's uterus.

Early embryonic cells are harvested because stem cells are **totipotent**, meaning they can become any other cell type immediately after fertilization (Figure 9.20). As the embryo develops, its cells become less and less able to produce other cell types. As a human embryo grows, the early cells start dividing and forming different, specialized cells such as heart cells, bone cells, and muscle cells. Once formed, specialized non-stem cells can divide only to produce replicas of themselves. They cannot backtrack and become a different type of cell.

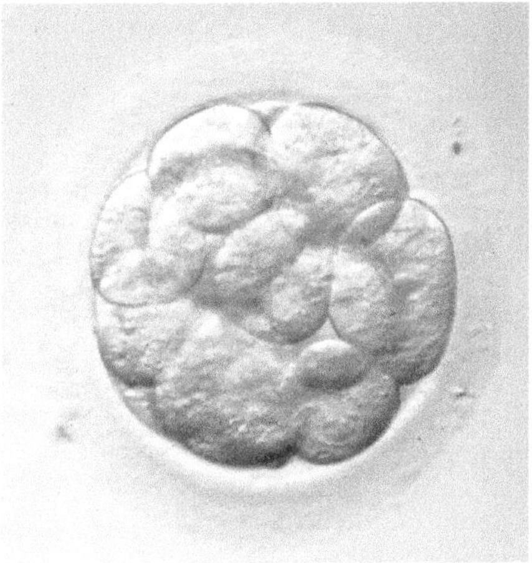

Figure 9.20 Human embryo. Stem cells can be obtained from an early embryo such as the one in this petri dish.

Stop & Stretch For parents known to carry a harmful genetic mutation, fertility clinics will take a cell from an 8-cell preembryo to test it for the presence of the defective gene. Taking a single cell causes no apparent harm to the developing embryo, and the remaining cells are allowed to continue undergoing development. Once a preembryo is found that does not carry the mutation, it can be implanted into the female parent's uterus. What does the fact that you can remove a cell from an embryo without harming subsequent development tell you about the capabilities of preembryonic cells?

There are also stem cells present in adult tissues, probably so that these tissues can repair themselves. Scientists have found adult stem cells in many different tissues, and work is under way to determine whether these cells can be used for transplants and treatment of degenerative diseases.

Genetic modifications may one day include replacing defective or non-functional alleles of a gene with a functional copy of the gene. This requires detailed information about the human genome, much of which has been obtained.

The Human Genome Project

The **Human Genome Project** involves determining the nucleotide-base sequence (A, C, G, or T) of the entire human genome and the location of each of the 20,000 to 25,000 human genes.

The scientists involved in this multinational effort also sequenced the genomes of the mouse, the fruit fly, the roundworm, baker's yeast, and a common intestinal bacterium named *Escherichia coli* (*E. coli*). Scientists thought it was important to sequence the genomes of organisms other than humans because these **model organisms** are easy to manipulate in genetic studies and because important genes are often found in many different organisms. In fact, 90% of human genes are also present in mice; 50% are in fruit flies, and 31% are in baker's yeast. Therefore, understanding how a certain gene functions in a model organism helps us understand how the same gene functions in humans.

To sequence the human genome, scientists first isolated DNA from white blood cells. They then cleaved the chromosomes into more manageable sizes using restriction enzymes, cloned them into plasmids, and determined the base sequence using automated DNA sequencers. These sequencing machines distinguish between nucleotides based on structural differences in the nitrogenous bases. Sequence information was then uploaded to the Internet. Scientists working on this, or any other project, could search for regions of sequence information that overlapped with known sequences, a process called *chromosome walking* (Figure 9.21). Using overlapping regions, scientists in laboratories all over the world worked together to patch together DNA sequence information. DNA sequence information obtained by means of the Human Genome Project may someday enable medical doctors to take blood samples from patients and determine which genetic diseases are likely to affect them.

Many people worry about having these types of tests performed because of the potential for insurance companies or employers to obtain the personal information and use it against them. Yet there is a positive side to having genetic information available. Once the genetic basis of a disease has been worked out—that is, how the gene of a healthy person differs from the gene of a person with a genetic disease—the information can be used to develop treatments or cures.

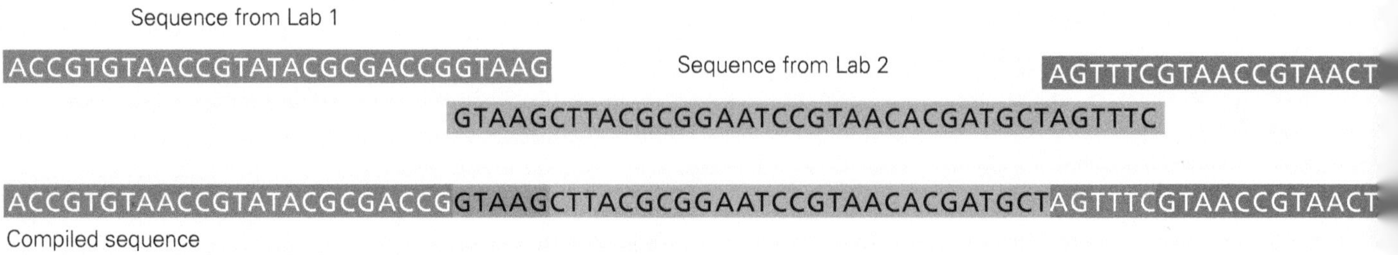

Figure 9.21 Chromosome walking. Scientists from many different labs worked together to sequence the human genome. Sequence information is uploaded to a common database, and scientists search for overlapping regions to fill in gaps in the sequence, much like assembling a jigsaw puzzle.

Sequence from Lab 1

ACCGTGTAACCGTATACGCGACCGGTAAG

Sequence from Lab 2

AGTTTCGTAACCGTAACT

GTAAGCTTACGCGGAATCCGTAACACGATGCTAGTTTC

ACCGTGTAACCGTATACGCGACCGGTAAGCTTACGCGGAATCCGTAACACGATGCTAGTTTCGTAACCGTAACT

Compiled sequence

Gene Therapy

Scientists who try to replace defective human genes (or their protein products) with functional genes are performing **gene therapy.** Gene therapy may someday enable scientists to fix genetic diseases in an embryo. To do so, the scientists would supply the embryo with a normal version of a defective gene; this so-called **germ-line gene therapy** would ensure that the embryo and any cells produced by cell division would replicate the new, functional version of the gene. Thus, most of the cells would have the corrected version of the gene, and when these genetically modified individuals had children, they would pass on the corrected version of the gene.

Another type of gene therapy, called **somatic cell gene therapy,** can be performed on body cells to fix or replace the defective protein in only the affected cells. Using this method, scientists introduce a functional version of a defective gene into an affected individual cell in the laboratory, allow the cell to reproduce, and then place the copies of the cell bearing the corrected gene into the diseased person.

This treatment may seem like science fiction, but it is likely that this method of treating genetic diseases will be considered a normal procedure in the not-too-distant future. In fact, genetic engineers already have successfully treated a genetic disorder called **severe combined immunodeficiency (SCID),** a disease caused by a genetic mutation that results in the absence of an important enzyme and severely weakens the individual's immune system. Because their immune systems are compromised, people with SCID are incapable of fighting off any infection, and they often suffer severe brain damage from the high temperatures associated with unabated infection. Any exposure to infection can kill or disable someone with SCID, so most patients must stay inside their homes and often live inside protective bubbles that separate them from everyone, even family members.

To devise a successful treatment for SCID, or any disease treated with gene therapy, scientists had to overcome a major obstacle—getting the therapeutic gene to the right place.

Proteins break down easily and are difficult to deliver to the proper cells, so it is more effective to replace a defective gene than to continually replace a defective protein. Delivering a normal copy of a defective gene only to the cell type that requires it is a difficult task. SCID, a disorder that has been treated successfully, was chosen by early gene therapists in part because defective immune system cells could be removed from the body, treated, and returned to the body.

Immune system cells that require the enzyme missing in SCID patients circulate in the bloodstream. Blood removed from a child with SCID is infected with nonpathogenic (non-disease-causing) versions of a virus. This virus is first engineered to carry a normal copy of the defective gene in SCID patients. After the immune system cells are infected with the virus, these recombinant

Sequence from Lab 3

`ACTACCGTTACGGATATGCTTACTGTAC`

Sequence from Lab 4

`TATGCTTACTGTACCTTCAAAACTGACTTTGGCTAACCGTACTCTGTACCTTCAAAACTGACTTTT`

`ACTACCGTTACGGATATGCTTACTGTACCTTCAAAACTGACTTTGGCTAACCGTACTCTGTACCTTCAAAACTGACTTTT`

(a) Gene therapy for SCID patients

(b) SCID survivor

Virus

Normal allele

Immune system cell

1 Remove immune system cells from patient.

2 Infect the cells with a virus carrying the normal allele.

3 Return cells carrying the normal allele.

Figure 9.22 Gene therapy in an SCID patient. (a) A virus carrying the normal gene is allowed to infect immune system cells that have been removed from a person with SCID. The virus inserts the normal copy of the gene into some of the cells, and these cells are then injected into the SCID patient. (b) Ashi DiSilva, the first gene therapy patient.

cells, which now bear copies of the functional gene, are returned to the SCID patient (Figure 9.22a).

In 1990, a 4-year-old girl named Ashi DiSilva (Figure 9.22b) was the first patient to receive gene therapy for SCID. Ashi's parents were willing to face the unknown risks to their daughter because they were already far too familiar with the risks of SCID—the couple's two other children also had SCID and were severely disabled. Ashi is now a healthy adult with an immune system that is able to fight off most infections.

However, Ashi must continue to receive treatments because blood cells, whether genetically engineered or not, have limited life spans. When most of Ashi's engineered blood cells have broken down, she must be treated again; thus, she undergoes this gene therapy a few times each year. Since Ashi's gene therapy turned out well, many other SCID patients have been successfully treated and can live normal lives. Unfortunately, Ashi's gene therapy does not prevent her from passing on the defective allele to her biological children because this therapy is not "fixing" the allele in cells of her ovaries.

Although things worked out well for Ashi, successful gene therapy is far from routine. Two of 11 French boys treated with gene therapy for SCID developed leukemia that is thought to be related to their treatment, and an American teenager, Jesse Gelsinger, died from complications of experimental gene therapy meant to cure his relatively mild genetic disorder.

In addition to the risks involved in conducting any experimental therapy, not many genetic diseases can be treated with gene therapy. Gene therapy to date has focused on diseases caused by single genes for which defective cells can be removed from the body, treated, and reintroduced to the body. Most genetic diseases are caused by many genes, affect cells that cannot be removed and replaced, and are influenced by the environment.

Most people support the research of genetic engineers in their attempts to find better methods for delivering gene sequences to the required locations and for regulating the genes once they are in place. A far more controversial type of genetic engineering involves making an exact copy of an entire organism by a process called reproductive **cloning.**

Cloning Humans

Human cloning occurs commonly in nature via the spontaneous production of identical twins. These clones arise when an embryo subdivides itself into 2 separate embryos early in development. This is not the type of cloning that many people find objectionable; people are more likely to be upset by cloning that involves selecting which traits an individual will possess.

Natural cloning of an early embryo to make identical twins does not allow any more selection for specific traits than does fertilization. However, in the future it may be possible to select adult humans who possess desired traits and clone them. Because cloning does not actually alter an individual's genes, it is more of a reproductive technology than a genetic engineering technology. However, it may someday be possible to alter the genes of a cloned embryo.

Cloning offspring from adults with desirable traits has been successfully performed on cattle, goats, mice, cats, pigs, rabbits, and sheep. In fact, the animal that brought cloning to the attention of the public was a ewe named Dolly (Figure 9.23).

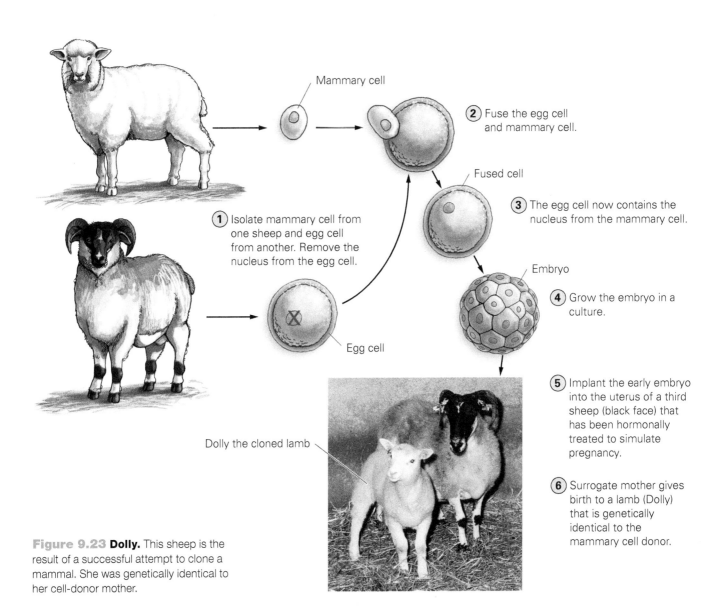

Mammary cell

2 Fuse the egg cell and mammary cell.

Fused cell

1 Isolate mammary cell from one sheep and egg cell from another. Remove the nucleus from the egg cell.

3 The egg cell now contains the nucleus from the mammary cell.

Egg cell

Embryo

4 Grow the embryo in a culture.

5 Implant the early embryo into the uterus of a third sheep (black face) that has been hormonally treated to simulate pregnancy.

6 Surrogate mother gives birth to a lamb (Dolly) that is genetically identical to the mammary cell donor.

Dolly the cloned lamb

Figure 9.23 Dolly. This sheep is the result of a successful attempt to clone a mammal. She was genetically identical to her cell-donor mother.

Dolly was cloned when Scottish scientists took the nucleus from a mammary gland cell of an adult female sheep and fused it with an egg cell that had previously had its nucleus removed. Treated egg cells were then placed in the uterus of an adult ewe that had been hormonally treated to support a pregnancy. Scientists had to try many times before this **nuclear transfer** technique worked. In all, 277 embryos were constructed before one was able to develop into a live lamb. Dolly was born in 1997.

The research that led to Dolly's birth was designed to provide a method of ensuring that cloned livestock would have the genetic traits that made them most beneficial to farmers. Sheep that produced the most high-quality wool and cattle that produced the best beef would be cloned. This technique is more efficient than allowing two prize animals to breed because each animal gives only half of its genes to the offspring. There is no guarantee that the offspring of 2 prize animals will have the desired traits. Even when a genetic clone is produced, there is no guarantee that the clone produced will be identical in the appearance and behavior to the original because of environmental differences.

No one knows if nuclear transfer will work in humans—or if cloning is safe. If Dolly is a representative example, cloning animals may not be safe. Dolly was euthanized at age six to relieve her from the discomfort of arthritis and a progressive lung disease, conditions usually found only in older sheep. The fact that Dolly developed these conditions has led scientists to hypothesize that she had aged prematurely.

Stop & Stretch Devise a hypothesis about why Dolly may have aged more rapidly than a sheep that was produced naturally. How could you test this hypothesis?

The debate about human cloning mimics the larger debate about genetic engineering (**Table 9.2**). As a society, we need to determine whether the potential for good outweighs the potential harm for each application of these technologies.

TABLE 9.2

Some pros and cons of genetic engineering.

Why the work of genetic engineers is important

- GM crops may make farms more productive.
- GM crops may be made to taste better.
- GM crops can be made to have longer shelf lives.
- GM crops can be made to contain more nutrients.
- Genetic engineers hope to cure diseases and save lives.

Why the work of genetic engineers is controversial

- GM crops encourage agribusiness, which may close down some small farms.
- GM animals and crops may cause health problems in consumers.
- GM crops might have unexpected adverse effects on the environment.
- Lack of genetic diversity of GM crops could lead to destruction of food supply worldwide by pest or environmental change.
- Present research might lead to the unethical genetic modification of humans.

SAVVY READER

Political Cartoons

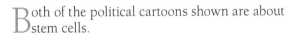

Both of the political cartoons shown are about stem cells.

1. Can we trust the scientific accuracy of cartoons?

2. Is the question of whether an embryo constitutes life one that science can answer?

3. Do you think an early embryo should have the same rights as a fetus or a baby?

Chapter Review

Learning Outcomes

LO1 Define the term *gene expression* (Section 9.1).

- Genes carry instructions for synthesizing proteins. Gene expression involves turning a gene on that includes the processes of transcription and translation (pp. 196–197).

LO2 Describe how messenger RNA (mRNA) is synthesized during the process of transcription (Section 9.1).

- Transcription occurs in the nucleus of eukaryotic cells when an RNA polymerase enzyme binds to the promoter, located at the start site of a gene, and makes an mRNA that is complementary to the DNA gene (pp. 197–198).

LO3 Describe how proteins are synthesized during the process of translation (Section 9.1).

- Translation occurs in the cytoplasm of eukaryotic cells and involves mRNA, ribosomes, and transfer RNA (tRNA). Messenger RNA carries the code from the DNA, and ribosomes are the site where amino acids are assembled to synthesize proteins. Transfer RNA carries amino acids, which bind to triplet nucleotide sequences on the mRNA called codons. A particular tRNA carries a specific amino acid. Each tRNA has its unique anticodon that binds to the codon and carries instructions for its particular amino acid (pp. 198–201).

L04 Use the genetic code to identify the amino acid for which a DNA segment or a codon codes (Section 9.1).

- The amino acid coded for by a particular codon can be determined using the genetic code (p. 201).

L05 Define the term *mutation*, and explain how mutations can affect protein structure and function (Section 9.1).

- Mutations are changes to DNA sequences that can affect protein structure and function (p. 202).
- Neutral mutations are changes to the DNA that do not result in a different amino acid being incorporated. Insertions or deletions of nucleotides can result in frameshift mutations that change the protein drastically (pp. 202–203).

L06 Explain why and how some genes are expressed in some cell types but not others (Section 9.1).

- A given cell type expresses only a small percentage of the genes that an organism possesses (p. 203).
- Turning the expression of a gene up or down is accomplished in different ways in prokaryotes and eukaryotes. Prokaryotes typically block the promoter with a repressor protein to keep gene expression turned off. Eukaryotes regulate gene expression in any of 5 ways: (1) increasing transcription through the use of proteins that stimulate RNA polymerase binding; (2) varying the time that DNA spends in the uncondensed, active form; (3) altering the mRNA life span; (4) slowing down or speeding up translation; and (5) affecting the protein life span (pp. 204–205).

L07 Outline the process of cloning a gene using bacteria (Section 9.2).

- Bacteria can be used to clone genes by placing the gene of interest into a plasmid, which then makes millions of copies of the gene as the plasmid replicates itself inside its bacterial host. Bacteria can then express the gene by transcribing an mRNA copy and translating the mRNA into a protein (pp. 205–207).

L08 Describe how foods are genetically modified, and list some of the health and environmental concerns surrounding this practice (Section 9.3).

- A Ti plasmid or gene gun can be used to insert a particular gene into plant cells (pp. 208–209).
- Although there have been no documented incidents of negative health effects from GM food consumption, there is concern that GM foods may be unhealthy. Concerns about GM crops include their impacts on surrounding organisms, the evolution of resistances, and transfer of modified genes to wild and weedy relatives (pp. 209–210).

L09 Describe the science behind and significance of stem cells, the Human Genome Project, gene therapy, and cloning (Section 9.4).

- Stem cells are undifferentiated cells that can be reprogrammed to act as a variety of cell types. Stem cells may someday allow scientists to treat degenerative diseases (pp. 210–212).
- The Human Genome Project resulted in the production of the complete DNA sequence of humans and several other organisms (p. 212).
- Gene therapy involves replacing defective genes or their products in an embryo or in the affected adult tissue. Gene therapy is considered experimental but may hold promise once scientists determine how to target genes to the right locations and express them in the proper amounts (pp. 213–215).
- Nuclear transfer has been used to clone animals with desirable agricultural traits. It may someday be possible to clone humans, but it is unclear if these humans would be healthy (pp. 215–216).

Roots to Remember (MB)™

The following roots of words come mainly from Latin and Greek and will help you decipher terms:

chromo- means color. Chapter term: chromosome

-ic is a common ending of acids. Chapter term: nucleic acid

nucleo- or nucl- refers to a nucleus. Chapter term: nucleotide

-ose is a common ending for sugars. Chapter terms: ribose and dexyribose

-some or -somal relates to a body. Chapter terms: chromosome and ribosome

Learning the Basics

1. **LO2** List the order of nucleotides on the mRNA that would be transcribed from the following DNA sequence: CGATTACTTA.

2. **LO4** Using the genetic code (Table 9.1 on p. 201), list the order of amino acids encoded by the following mRNA nucleotides: CAACGCAUUUUG.

3. **LO3** List the subcellular structures that participate in translation.

4. **LO2** Transcription _____.

   A. synthesizes new daughter DNA molecules from an existing DNA molecule; **B.** results in the synthesis of an RNA copy of a gene; **C.** pairs thymines (T) with adenines (A); **D.** occurs on ribosomes

5. LO3 Transfer RNA (tRNA) _____.

A. carries monosaccharides to the ribosome for synthesis; **B.** is made of messenger RNA; **C.** has an anticodon region that is complementary to the mRNA codon; **D.** is the site of protein synthesis

6. LO2 During the process of transcription, _____.

A. DNA serves as a template for the synthesis of more DNA; **B.** DNA serves as a template for the synthesis of RNA; **C.** DNA serves as a template for the synthesis of proteins; **D.** RNA serves as a template for the synthesis of proteins

7. LO3 Translation results in the production of _____.

A. RNA; **B.** DNA; **C.** protein; **D.** individual amino acids; **E.** transfer RNA molecules

8. LO2 The RNA polymerase enzyme binds to _____, initiating transcription.

A. amino acids; **B.** tRNA; **C.** the promoter sequence; **D.** the ribosome

9. LO3 A particular triplet of bases in the coding sequence of DNA is TGA. The anticodon on the tRNA that binds to the mRNA codon is _____.

A. TGA; **B.** UGA; **C.** UCU; **D.** ACU

10. LO2 RNA and DNA are similar because _____.

A. they are both double-stranded helices; **B.** uracil is found in both of them; **C.** both contain the sugar deoxyribose; **D.** both are made up of nucleotides consisting of a sugar, a phosphate, and a base

11. LO6 True/False A kidney cell and a heart cell from the same cat would have the same genes.

12. LO8 True/False Scientists know that genetically modified crops cannot hurt the environment.

13. LO9 True/False Gene therapy is used commonly and with 100% success.

Analyzing and Applying the Basics

1. LO5 Take another look at the genetic code (Table 9.1, p. 201). Do you see any similarities between codons that code for the same amino acid? Based on this difference, why might a mutation that affects the nucleotide in the third position of the codon be less likely to affect the structure of the protein than a mutation that affects the codon in the first position?

2. LO5 Genes encode RNA polymerase molecules. What would happen to a cell that has undergone a mutation to its RNA polymerase gene?

3. LO7 Draw a box around the 6-base-pair site at which a restriction enzyme would most likely be cut:

ATGAATTCCGTCCG
TACTTAAGGCAGGC

Connecting the Science

1. The first "test-tube baby," Louise Brown, was born over 30 years ago. Sperm from her father was combined with an egg cell from her mother. The fertilized egg cell was then placed into the mother's uterus for the period of gestation. At the time of Louise's conception, many people were very concerned about the ethics of scientists performing these in vitro fertilizations. Do you think human cloning will eventually be as commonplace as in vitro fertilizations are now? Why or why not?

2. Do you think it is acceptable to grow genetically modified foods if health risks turn out to be low but environmental effects are high?

Answers to **Stop & Stretch, Visualize This, Savvy Reader,** and **Chapter Review** questions can be found in the **Answers** section at the back of the book.

Appendix

Metric System Conversions

To Convert Metric Units:	Multiply by:	To Get English Equivalent:
Length		
Centimeters (cm)	0.3937	Inches (in)
Meters (m)	3.2808	Feet (ft)
Meters (m)	1.0936	Yards (yd)
Kilometers (km)	0.6214	Miles (mi)
Area		
Square centimeters (cm^2)	0.155	Square inches (in^2)
Square meters (m^2)	10.7639	Square feet (ft^2)
Square meters (m^2)	1.196	Square yards (yd^2)
Square kilometers (km^2)	0.3831	Square miles (mi^2)
Hectare (ha) (10,000 m^2)	2.471	Acres (a)
Volume		
Cubic centimeters (cm^3)	0.06	Cubic inches (in^3)
Cubic meters (m^3)	35.30	Cubic feet (ft^3)
Cubic meters (m^3)	1.3079	Cubic yards (yd^3)
Cubic kilometers (km^3)	0.24	Cubic miles (mi^3)
Liters (L)	1.0567	Quarts (qt), U.S.
Liters (L)	0.26	Gallons (gal), U.S.
Mass		
Grams (g)	0.03527	Ounces (oz)
Kilograms (kg)	2.2046	Pounds (lb)
Metric ton (tonne) (t)	1.10	Ton (tn), U.S.
Speed		
Meters/second (mps)	2.24	Miles/hour (mph)
Kilometers/hour (kmph)	0.62	Miles/hour (mph)

To Convert English Units:	Multiply by:	To Get Metric Equivalent:
Length		
Inches (in)	2.54	Centimeters (cm)
Feet (ft)	0.3048	Meters (m)
Yards (yd)	0.9144	Meters (m)
Miles (mi)	1.6094	Kilometers (km)
Area		
Square inches (in^2)	6.45	Square centimeters (cm^2)
Square feet (ft^2)	0.0929	Square meters (m^2)
Square yards (yd^2)	0.8361	Square meters (m^2)
Square miles (mi^2)	2.5900	Square kilometers (km^2)
Acres (a)	0.4047	Hectare (ha) (10,000 m^2)
Volume		
Cubic inches (in^3)	16.39	Cubic centimeters (cm^3)
Cubic feet (ft^3)	0.028	Cubic meters (m^3)
Cubic yards (yd^3)	0.765	Cubic meters (m^3)
Cubic miles (mi^3)	4.17	Cubic kilometers (km^3)
Quarts (qt), U.S.	0.9463	Liters (L)
Gallons (gal), U.S.	3.8	Liters (L)
Mass		
Ounces (oz)	28.3495	Grams (g)
Pounds (lb)	0.4536	Kilograms (kg)
Ton (tn), U.S.	0.91	Metric ton (tonne) (t)
Speed		
Miles/hour (mph)	0.448	Meters/second (mps)
Miles/hour (mph)	1.6094	Kilometers/hour (kmph)

Metric Prefixes		
Prefix		**Meaning**
giga-	G	$10^9 = 1,000,000,000$
mega-	M	$10^6 = 1,000,000$
kilo-	k	$10^3 = 1,000$
hecto-	h	$10^2 = 100$
deka-	da	$10^1 = 10$
		$10^0 = 1$
deci-	d	$10^{-1} = 0.1$
centi-	c	$10^{-2} = 0.01$
milli-	m	$10^{-3} = 0.001$
micro-	μ	$10^{-6} = 0.000001$

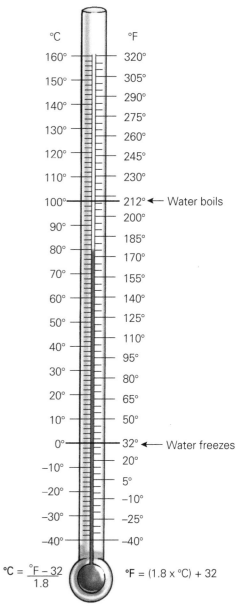

°C   °F

160° — 320°
150° — 305°
      290°
140° — 275°
130° — 260°
120° — 245°
110° — 230°
100° — 212° ← Water boils
      200°
90° — 185°
80° — 170°
70° — 155°
60° — 140°
50° — 125°
      110°
40° — 95°
30° — 80°
20° — 65°
10° — 50°
0° — 32° ← Water freezes
      20°
−10° — 5°
−20° — −10°
−30° — −25°
−40° — −40°

$$°C = \frac{°F - 32}{1.8}$$

$$°F = (1.8 \times °C) + 32$$

Periodic Table of the Elements

☐ Metals ☐ Metalloids ☐ Nonmetals

Group number — 2
Atomic number

1																	8
1 H 1.008	2											3	4	5	6	7	**2 He** 4.003
3 Li 6.941	**4 Be** 9.012											**5 B** 10.81	**6 C** 12.01	**7 N** 14.01	**8 O** 16.00	**9 F** 19.00	**10 Ne** 20.18
11 Na 22.99	**12 Mg** 24.31				Transition elements							**13 Al** 26.98	**14 Si** 28.09	**15 P** 30.97	**16 S** 32.07	**17 Cl** 35.45	**18 Ar** 39.95
19 K 39.10	**20 Ca** 40.08	**21 Sc** 44.96	**22 Ti** 47.87	**23 V** 50.94	**24 Cr** 52.00	**25 Mn** 54.94	**26 Fe** 55.85	**27 Co** 58.93	**28 Ni** 58.69	**29 Cu** 63.55	**30 Zn** 65.41	**31 Ga** 69.72	**32 Ge** 72.64	**33 As** 74.92	**34 Se** 78.96	**35 Br** 79.90	**36 Kr** 83.80
37 Rb 85.47	**38 Sr** 87.62	**39 Y** 88.91	**40 Zr** 91.22	**41 Nb** 92.91	**42 Mo** 95.94	**43 Tc** (98)	**44 Ru** 101.1	**45 Rh** 102.9	**46 Pd** 106.4	**47 Ag** 107.9	**48 Cd** 112.4	**49 In** 114.8	**50 Sn** 118.7	**51 Sb** 121.8	**52 Te** 127.6	**53 I** 126.9	**54 Xe** 131.3
55 Cs 132.9	**56 Ba** 137.3	**57 La** 138.9	**72 Hf** 178.5	**73 Ta** 180.9	**74 W** 183.8	**75 Re** 186.2	**76 Os** 190.2	**77 Ir** 192.2	**78 Pt** 195.1	**79 Au** 197.0	**80 Hg** 200.6	**81 Tl** 204.4	**82 Pb** 207.2	**83 Bi** 209.0	**84 Po** (209)	**85 At** (210)	**86 Rn** (222)
87 Fr (223)	**88 Ra** (226)	**89 Ac** (227)	**104 Rf** (261)	**105 Db** (262)	**106 Sg** (266)	**107 Bh** (264)	**108 Hs** (269)	**109 Mt** (268)	**110 Ds** (271)	111 — (272)	112 — (285)	113 — (284)	114 — (289)	115 — (288)			

Lanthanides

58 Ce 140.1	**59 Pr** 140.9	**60 Nd** 144.2	**61 Pm** (145)	**62 Sm** 150.4	**63 Eu** 152.0	**64 Gd** 157.3	**65 Tb** 158.9	**66 Dy** 162.5	**67 Ho** 164.9	**68 Er** 167.3	**69 Tm** 168.9	**70 Yb** 173.0	**71 Lu** 175.0

Actinides

90 Th 232.0	**91 Pa** 231.0	**92 U** 238.0	**93 Np** (237)	**94 Pu** (244)	**95 Am** (243)	**96 Cm** (247)	**97 Bk** (247)	**98 Cf** (251)	**99 Es** 252	**100 Fm** 257	**101 Md** 258	**102 No** 259	**103 Lr** 260

Name	Symbol	Name	Symbol	Name	Symbol	Name	Symbol
Actinium	Ac	Einsteinium	Es	Mendelevium	Md	Samarium	Sm
Aluminum	Al	Erbium	Er	Mercury	Hg	Scandium	Sc
Americium	Am	Europium	Eu	Molybdenum	Mo	Seaborgium	Sg
Antimony	Sb	Fermium	Fm	Neodymium	Nd	Selenium	Se
Argon	Ar	Fluorine	F	Neon	Ne	Silicon	Si
Arsenic	As	Francium	Fr	Neptunium	Np	Silver	Ag
Astatine	At	Gadolinium	Gd	Nickel	Ni	Sodium	Na
Barium	Ba	Gallium	Ga	Niobium	Nb	Strontium	Sr
Berkelium	Bk	Germanium	Ge	Nitrogen	N	Sulfur	S
Beryllium	Be	Gold	Au	Nobelium	No	Tantalum	Ta
Bismuth	Bi	Hafnium	Hf	Osmium	Os	Technetium	Tc
Bohrium	Bh	Hassium	Hs	Oxygen	O	Tellurium	Te
Boron	B	Helium	He	Palladium	Pd	Terbium	Tb
Bromine	Br	Holmium	Ho	Phosphorus	P	Thallium	Tl
Cadmium	Cd	Hydrogen	H	Platinum	Pt	Thorium	Th
Calcium	Ca	Indium	In	Plutonium	Pu	Thulium	Tm
Californium	Cf	Iodine	I	Polonium	Po	Tin	Sn
Carbon	C	Iridium	Ir	Potassium	K	Titanium	Ti
Cerium	Ce	Iron	Fe	Praseodymium	Pr	Tungsten	W
Cesium	Cs	Krypton	Kr	Promethium	Pm	Uranium	U
Chlorine	Cl	Lanthanum	La	Protactinium	Pa	Vanadium	V
Chromium	Cr	Lawrencium	Lr	Radium	Ra	Xenon	Xe
Cobalt	Co	Lead	Pb	Radon	Rn	Ytterbium	Yb
Copper	Cu	Lithium	Li	Rhenium	Re	Yttrium	Y
Curium	Cm	Lutetium	Lu	Rhodium	Rh	Zinc	Zn
Darmstadtium	Ds	Magnesium	Mg	Rubidium	Rb	Zirconium	Zr
Dubnium	Db	Manganese	Mn	Ruthenium	Ru		
Dysprosium	Dy	Meitnerium	Mt	Rutherfordium	Rf		

Answers

Chapter 1

STOP & STRETCH

p. 7: Given the information here, it is most likely that someone brought a box of doughnuts to share, and that Homer ate a large number of them. This hypothesis is based on inductive reasoning.

p. 7: No, because even though Homer is a likely culprit, there are probably other individuals in your workplace who may have eaten some or all of the doughnuts.

p. 10: The independent variable is the presence of *H. pylori* infection, the dependent variable is number of ulcers per animal.

p. 11: Participants could have been exposed to different cold viruses; early arrivals may have been infected by a severe virus that was spreading in the hospital, while later arrivals may have been infected by a milder virus.

p. 14: The independent variable is stress, and the dependent variable is susceptibility to developing a cold when exposed to a virus.

p. 19: According to the poll, the actual percentage of the population who favor candidate A has a 95% probability of being between 44% and 50%, while the percentage who favor B is likely between 48% and 54%. Since these ranges overlap, it is currently unclear whether most of population favor A or B.

p. 21: It is much easier to relate to individual experiences than to a more abstract piece of information. Anecdotes are the way humans have learned from each other throughout history; polling or statistical data are much more recent.

VISUALIZE THIS

Figure 1.3: Even with a well-designed experiment, it is possible that an alternative factor can explain the results, and researchers need to consider if that might be the case in their experiment.

Figure 1.4: Preventing virus attachment and invasion of cells. Preventing activation of the immune system response. Other answers may be acceptable.

Figure 1.10: There may not have been enough participants at a particular stress level to allow effective analysis of the correlation. If individuals at stress level 3 have the same cold susceptibility as those at stress level 4, that should cause us to question the correlation because we'd expect those at stress level 4 to have greater susceptibility.

Figure 1.15: Hypothesis: true. Difference between control and experimental groups: large. Sample size: large. This result is very likely to be both statistically and practically significant.

Figure 1.16: On the figure, it would the follow the path through "hypothesis true", "difference between groups large", and "sample size large," meaning that the result is highly likely to be both statistically and practically significant.

SAVVY READER

1. Red flags for questions 1, 2, 9, 10, and 11.
2. Most other questions from the checklist do not pertain to this particular excerpt.

LEARNING THE BASICS

1. A placebo allows members of the control group to be treated exactly like the members of the experimental group *except* for their exposure to the independent variable.
2. b; **3.** b; **4.** d; **5.** a; **6.** d; **7.** e; **8.** c; **9.** b; **10.** c; **11.** b; **12.** e

ANALYZING AND APPLYING THE BASICS

1. No, another factor that causes type 2 diabetes could lead to obesity as well, or it is possible that type 2 diabetes causes obesity, not vice versa.
2. Neither the participants nor the researchers were blind, meaning that both subject expectation and observer bias might influence the results. If participants felt that vitamin C was likely to be effective, they may underreport their cold symptoms. If the clinic workers thought one or the other treatment was less effective, they might let that bias show when they were talking to the participants, influencing the results.
3. See the answer to number 1; BDNF may increase as a result of exercise, or excess BDNF might cause an increased interest in exercise.

Chapter 2

STOP & STRETCH

p. 37: The carbons should each be bonded to two hydrogens to complete their valence shells.

p. 48: Yes, because ribosomes are not bounded by membranes and because prokaryotic cells need to synthesize proteins

VISUALIZE THIS

Figure 2.4: Oxygen

Figure 2.7: 100×

Figure 2.15: Red spheres are sugars; purple are phospahtes. Purines (blue) are double ring structures and thus longer in the diagram, while pyrimidines (yellow) are shorter.

Figure 2.18: Central vacuole, cell wall, and chloroplast

Figure 2.20: Yes

1. Not really; how would a negative charge produce hydroxyl ions? While it is true that hydroxyl ions are negatively charged, if you remove the hydroxyl ion from water you would be left with the positively charged hydrogen ion. Hydroxyl ions and the hydrogen ions would bind to each other to produce water again.
2. Red flags—this is an untested assertion put out by a company that stands to gain financially.
3. Amazing, once and for all, hugely improved etc.
4. No
5. Error bars

LEARNING THE BASICS
1. Carbohydrates, lipids, proteins, nucleic acids
2. See Table 2.1.
3. Phospholipid bilayer with embedded proteins
4. b; **5.** a; **6.** a; **7.** d; **8.** d; **9.** d; **10.** d **11.** B; **12.** C

ANALYZING AND APPLYING THE BASICS
1. No. A virus is not capable of reproduction without help from the host cell.
2. Oxygen partial negative charge; carbon partial positive.
3. Silicon, like carbon, has four spaces in its valence shell and would also be tetravalent.

Chapter 3

STOP & STRETCH
p. 58: These food pairings together provide all the essential amino acids.
p. 67: The concentration of dissolved particles outside of cells will increase after drinking this beverage and water can leave the cells.
p. 67: Active transport.

VISUALIZE THIS
Figure 3.4: The top tail of part b.

SAVVY READER
1. Published by a University, analyzes claims made by a manufacturer.
2. You should be more skeptical of a marketing claim. The University has nothing to lose if you don't use a supplement while a company loses money.
3. You should be more skeptical of a study published by scientists hired by a manufacturer of a product. Studies published by government and university scientists gave undergone peer review. A claim made on a web site has not necessarily undergone peer review.

LEARNING THE BASICS
1. Nutrients serve as energy stores and as building blocks for cellular structures.

2. Carbon dioxide, oxygen, and water
3. d; **4.** c; **5.** b; **6.** b; **7.** d; **8.** d; **9.** e; **10.** b

ANALYZING AND APPLYING THE BASICS
1. Eat lots of protein rich foods like beans.
2. Fatigue is one of the ways your body responds to toxic levels of alcohol. Energy drinks may override fatigue signals, allowing consumption of alcohol past a level at which one would normally pass out.
3. We would likely produce fewer toxins if we knew we could not simply wash other vitamins, minerals, and fiber away.

Chapter 4

STOP & STRETCH
p. 76: 200 calories per day.
p. 83: More oxygen is available for the conversion of lactic acid to pyruvic acid.
p. 87: The amount of the total cholesterol that is LDL.

VISUALIZE THIS
Figure 4.2: No
Figure 4.3: 278.4 Calories
Figure 4.12: *Glyco* means sugar, and *lysis* means cutting.
Figure 4.13: A double bond is broken to allow for the bonding of hydrogen.
Figure 4.17: Oxygen exists in cells as molecular oxygen (O_2). Because only one oxygen atom is required to make water (H_2O), the 1/2 is placed before O_2.

SAVVY READER
1. No evidence is presented for any claims.
2. No, pathogenic bacteria, plant toxins, snake venom, hurricanes and ice storms are all natural but can be very bad for humans.
3. No, there needs to be evidence presented from the results of controlled studies.
4. No. No.

LEARNING THE BASICS
1. All of the building up and breaking down chemical reactions that occur in a cell
2. When a substrate binds to the active site of an enzyme, the enzyme clamps down on the substrate, causing increased stress on the bonds holding the substrate together.
3. Reactants of photosynthesis are carbon dioxide and water. Reactants of respiration are oxygen and glucose. Products of photosynthesis are oxygen and glucose. Products of respiration are carbon dioxide and water.
4. c; **5.** b; **6.** c; **7.** e; **8.** b; **9.** c; **10.** c

ANALYZING AND APPLYING THE BASICS
1. No. Fats and proteins also feed into cellular respiration at various points in the cycle and fats actually store more energy than carbohydrates.

2. An enzyme will be digested in the stomach, like any other protein.

3. Being very muscular can lead to an inaccurate BMI because muscle weighs more than fat.

Chapter 5

STOP & STRETCH

p. 95: Because the moon has no or very little greenhouse gas, heat from the sun is not retained to be reradiated overnight, which is why the nighttime temperatures are cold. The warm temperatures during the day occur because water is not present to absorb some of the heat energy without changing temperature.

p. 100: The chlorophyll is reabsorbed because it is so valuable to the plant—its components can be stored and the chlorophyll can be reproduced in the spring.

p. 101: The products of respiration are used during photosynthesis. The products of photosynthesis are reactants in respiration.

p. 102: ATP is a short-lived molecule that cannot "store" energy. Therefore it cannot be used to transport energy to parts of a plant where photosynthesis cannot occur (e.g. roots) or provide energy that can be used when the sun is not shining.

p. 103: Nitrogen, sulfur, and phosphorous

VISUALIZE THIS

Figure 5.1: Increased carbon dioxide will absorb more heat and keep it in the atmosphere, allowing more heat to radiate back to Earth.

Figure 5.2: Water levels would rise and shorelines would erode.

Figure 5.4: Human activity

Figure 5.6: The increase in concentration is steeper during the period 2000-2005 when compared to 1960-1970.

Figure 5.8: The left axis is for the blue line and the right for the red line. They are on the same graph to better illustrate the relationship between carbon dioxide level and temperature over time.

Figure 5.10: The absorption of light would be small in the lower wavelengths (red) and thus the line would be closer to the x axis on the left of the graph. The absorption of light should be higher in the middle and perhaps highest wavelengths and thus the line would be far from the x axis in the middle and right side of the graph.

Figure 5.12: The reactions are similar in that energized electrons are used to create a gradient of hydrogen ions, which is then used to generate ATP. However, the energized electrons in photosynthesis come from chlorophyll that has been energized by the sun; in respiration the electrons come from NADP, which was produced by the breakdown of sugar. At the end of the chain, electrons in photosynthesis are used to produce sugar; in respiration, the electrons are picked up by oxygen, which reacts with hydrogen to produce water.

Figure 5.13: Without sunlight, ATP and NADPH inputs from the light reactions would be absent and the cycle could not function.

SAVVY READER

1. Rising seas from global warming are causing flooding on low-lying Panamanian islands.

2. Yes, it may be that the flooding is caused by coral mining.

3. No, because other alternative hypotheses have not been ruled out as causes for the flooding.

LEARNING THE BASICS

1. Reactants of photosynthesis are carbon dioxide and water. Products of photosynthesis are oxygen and glucose.

2. c; **3.** d; **4.** e; **5.** d; **6.** a; **7.** c; **8.** a; **9.** d; **10.** d

ANALYZING AND APPLYING THE BASICS

1. Because colder water has the capacity to absorb more heat before water molecules can evaporate into the atmosphere, colder temperatures mean that less water circulates in the water cycle.

2. The water cycle would remain basically unchanged. However, the carbon cycle would be quite different. The only way carbon dioxide would move from the atmosphere to Earth's surface would be by absorption into water and perhaps incorporation into rocks. It would return to the atmosphere via diffusion from water and volcanic activity.

3. High rates of photosynthesis are continually pumping oxygen into the air. So even though the molecule is being "used up" through reactions with other compounds, its continual production ensures high levels in the atmosphere.

Chapter 6

STOP & STRETCH

p. 114: If a cell has many mutations, it is better that it is not allowed to continue to replicate.

p. 120: DNA is semi-conservatively replicated; that is, each strand serves as a template for the synthesis of a new strand. Photocopying is completely conservative. One copy is conserved and the other is new.

p. 122: Descent with modification explains the existence of similar processes in distantly related organisms.

p. 128: Tissues and organs will cease functioning as cells reach their cell division limit.

p. 131: Because it damages DNA.

p. 135: Both meiosis I and mitosis produce two daughter cells. Meiosis I differs from mitosis in that meiosis I separates homologous pairs of chromosomes from each other. Mitosis does not do this.

VISUALIZE THIS

Figure 6.5: Four DNA molecules would be produced. Two of those would be all purple; two would be half purple, half red.

Figure 6.10: The cell should be prevented from passing the G2 checkpoint.

Figure 6.18: The X chromosome is the larger of the two and carries more genetic information.

Figure 6.22: No

Figure 6.25: 4

SAVVY READER

1. No
2. Check some highly regarded sources. Ideally peer reviewed literature. Otherwise check reputable websites and see if there is a consensus.
3. Subtract
4. In addition to controlling for type of cancer, age of patient, etc., it is imperative to know what percentage of people not taking the supplement were alive one year after taking the supplement.
5. more skeptical; more skeptical
6. An advanced degree is not a guarantee that a person will act in your best interest.
7. Desperation is one reason.

LEARNING THE BASICS

1. Mitosis occurs in somatic cells, meiosis in cells that give rise to gametes. Mitosis produces daughter cells that are genetically identical to parent cells, while meiosis produces daughter cells with novel combinations of chromosomes and half as many chromosomes as compared to parent cells.
2. Rapid cell division
3. d; **4.** a; **5.** c; **6.** d; **7.** b, **8.** c; **9.** d
   10. a, diploid; b, diploid; c, haploid; d, haploid
   11. Crossing over and random alignment of the homologues
   12. Oncogenes are mutated versions of proto-oncogenes.
   13. Cytokinesis in plant cells requires the building of a cell wall.

ANALYZING AND APPLYING THE BASICS

1. No. Only mutations in cells that produce gametes are passed to offspring.
2. No, but more cell divisions means more opportunities for mutations to occur.
3. Cancers that are thought to be localized (not yet spread) are treated with radiation.

Chapter 7

STOP & STRETCH

p. 151: A mutation in a promoter portion of the DNA may affect whether the gene is copied or translated. A mutation in a structural portion of a chromosome may weaken the chromosome or change its ability to be transcribed. A mutation in "nonsense" DNA is likely to have no effect.

p. 152: Eight different gametes. The relationship is 2^n, where n is the number of chromosome pairs.

p. 157: Its effects are not seen until later adulthood, after an individual is likely to have had children.

p. 159: Eight possible gamete types (e.g. T Y R; T y R; t Y R, etc.) and 64 possible cells inside the Punnett square.

p. 161: Immigration from countries where average height is shorter may be changing the genetics of height in the United States such that there are a larger number of "small" alleles today than in the past. Dietary changes may also be leading to reduced growth or early puberty, which stunts growth in height.

p. 167: Even a trait that is highly heritable can be strongly influenced by the environment. If the current environment results in some individuals having very low IQ scores relative to others, it is certainly possible that a changed environment would lead to smaller differences among individuals. See the case of the "maze-dull" and "maze-bright" rats.

VISUALIZE THIS

Figure 7.3: Many possibilities

Figure 7.6: Blood group gene from his dad and eye color gene from his dad; Blood group gene from his mom and eye color gene from his mom.

Figure 7.14: 50%

Figure 7.15: 9/16 or 56%

Figure 7.16: No, the majority of men were not 5 feet 10 inches. There is a large range of heights in this population.

Figure 7.19: Points would be more widely scattered, and the correlation line would be flatter.

SAVVY READER

1. There is some evidence that we will select friends that are similar to us in some genetically-influenced behaviors and dissimilar to us in other genetically-influenced behaviors. Whether the friendship itself is genetically determined is difficult to tease out from the fact that the personality trait is partially genetically-determined.
2. If people with a genetic predisposition to alcoholism tend to have a genetic predisposition to become friends, part of understanding how to prevent and treat alcoholism may include understanding the social dynamics of these friendships.
3. Clearly, your choice of friends is influenced by the environment. This study supports the idea that there is a genetic basis favoring particular friendships as well.

LEARNING THE BASICS

1. Genotype refers to the alleles you possess for a particular gene or set of genes. Phenotype is the physical trait itself, which may be influenced by genotype and environmental factors.
2. Multiple genes or multiple alleles influencing a trait, environmental effects that affect the expression of a trait, or a mixture of both factors
3. b; **4.** a; **5.** b; **6.** a; **7.** b; **8.** c; **9.** e; **10.** d

GENETICS PROBLEMS

1. All have genotype Tt and are tall.
2. One-quarter are TT and are tall; one-half are Tt and are tall; and one-quarter are tt and are short.

3. Both are alleles for the same trait; P is dominant, and p is recessive.

4. Both parents must be heterozygous, Aa.

5. 50%

6. a. yellow; b.YY and yy

7. Yellow, round parent: Yy Rr; green, wrinkled parent: yy rr.

8. a. 50%; b. 50%; c. 25%; d. yes

9. a. dominant; b. The phenotype does not inevitably surface when the allele is present.

ANALYZING AND APPLYING THE BASICS

1. The trait could be recessive, and both parents could be heterozygotes. If the trait is coded for by a single gene with two alleles, one would expect 25% of the children (1 of 4) to have blue eyes. That 50% (2 of 4) have blue eyes does not refute Mendel because the probability of any genotype is independent for each child and is not dependent on the genotype of his or her siblings. (Consider that some families may have 4 boys and no girls; this does not refute the idea that the Y chromosome is carried by only 50% of a man's sperm cells.)

2. No, most quantitative traits are influenced by the environment. Even if differences in genes account for most of the differences among individuals in a trait, if the environment changed, the trait would very likely change as well.

3. No, heritability doesn't explain why any two individuals differ. For instance, John and Jerry could be identical twins, but Jerry might suffered a minor accidental head injury that reduced his IQ.

Chapter 8

STOP & STRETCH

p. 177: $I^C I^C$ type C; $I^A I^C$ type AC; $I^B I^C$ type BC; $I^C I$ type C; along with previously described AB types

p. 181: Her father must have the disease, and her mother is a carrier or has the disease.

p. 183: Crossing over; they may be infertile.

p. 187: They may or may not carry different alleles of gene. The VNTRs are adjacent to the allele, they are not part of the allele.

p. 188: Because each parent only contributes half of their genetic information to a child

VISUALIZE THIS

Figure 8.4: 50% (1/2); 50% (1/2)

Figure 8.6: A cross involving a hemophilic female

Figure 8.9: This pedigree would have a square connected to a circle connected to a square in the first generation (the parents). The second generation (their kids) would have a square from one set of parents and a circle from the second set of parents.

SAVVY READER

1. 50%; No.

2. It is difficult to imagine how this could happen.

3. No. Your gametes do not choose which traits to select. Your environment does.

LEARNING THE BASICS

1. More than two alleles are present in the population. Each individual can have (at most) two different alleles.

2. Females inherit one X chromosome from their mom in the egg cell and one from their dad in the sperm cell. Males inherit an X chromosome from their mom in the egg cell and a Y chromosome from their dad in the sperm cell.

3. DNA is isolated, then the amount of DNA can be amplified using PCR. DNA is cut with restriction enzymes and the fragments are loaded on a gel, to which an electrical current is applied. The separated fragments of DNA produce a pattern unique to an individual.

4. $I^a i$ (dad), $I^B i$ (mom)

5. ¼A1, ¼A2, ⅛AB2, ⅛AB1, ⅛B1, ⅛B2

6. c; **7.** b; **8.** c; **9.** e; **10.** d; **11.** b; **12.** d

ANALYZING AND APPLYING THE BASICS

1. Alleles that are co-dominant are equally expressed. An allele that is incompletely dominant does not completely mask the presence of the recessive allele.

2. Answers will vary. Just be certain that each band found in a daughter is present in one of the parents. About 50% of the bands in a given daughter will be present in each parent. The sisters should have around 50% of their bands in common.

3. Pedigrees will vary. However, the first cousins should have one set of common grandparents. If a grandparent is a carrier of a rare recessive allele, he or she could pass the same rare recessive allele to each of the cousins. If these cousins (both carriers of a rare recessive allele) mate, their child has a 1 in 4 chance of having the recessive disease.

Chapter 9

STOP & STRETCH

p. 202: Proline-Asparagine

p. 203: New alleles, or mutations, are a source of variation for natural selection to act upon.

p. 204: Gene expression would no longer be negatively regulated. A gene may be on all the time.

p. 207: There could be more growth hormone in the milk of treated cows.

p. 211: Preembryonic cells are capable of becoming any cell type.

p. 216: Dolly's chromosomes were that of an older sheep and therefore shortened. Compare the lengths of chromosomes from Dolly's cells with those of a newborn lamb.

VISUALIZE THIS

Figure 9.1: The second carbon of the sugar, ribose, is bonded to -OH while the same carbon in deoxyribose is attached to hydrogen. There is a CH3 group in thymine that is not present in uracil.

Figure 9.3: Sample answer starting after the fork in the DNA: AGCTGGGCAGGTAC. From this particular DNA, the mRNA would be UCGACCCGUCCAUG.

Figure 9.5: Adenine paired with Uracil and Cytosine paired with Guanine

Figure 9.9: Bottom

SAVVY READER

1. No. Cartoons are not peer reviewed
2. No. This is not an issue that can be resolved by hypothesis testing.
3. Answers will vary.

LEARNING THE BASICS

1. GCUAAUGAAU
2. gln, arg, ile, leu

3. mRNA, ribosome, amino acids, tRNAs
4. b; **5.** c; **6.** b; **7.** c; **8.** c; **9.** b; **10.** d ; **11.** T; **12.** F; **13.** F

ANALYZING AND APPLYING THE BASICS

1. The same amino acid is often coded for by codons that differ in the third position. For instance, UUU and UUC both code for phenylalanine.
2. Transcription would be slowed or stopped in that cell.
3. GAATTC; CTTAAG

Glossary

ABO blood system A system for categorizing human blood based on the presence or absence of carbohydrates on the surface of red blood cells. (Chapter 8)

abscisic acid A hormone in plants associated with dormancy in seeds and buds. (Chapter 23)

abscission The dropping off of leaves, flowers, fruits, or other plant organs. (Chapter 23)

accessory organs Organs including the pancreas, liver, and gallbladder, which aid the digestive system. (Chapter 16)

acetylcholine A neurotransmitter with many functions, including facilitating muscle movements, and thought to be involved in development of Alzheimer's disease. (Chapter 21)

acetylcholinesterase An enzyme in nerve cells that breaks down acetylcholine. (Chapter 21)

acid A substance that increases the concentration of hydrogen ions in a solution. (Chapter 2)

acid rain Rain (or other precipitation) that is unusually acidic; caused by air pollution in the form of sulfur and nitrogen dioxides. (Chapter 16)

acquired immune deficiency syndrome (AIDS) Syndrome characterized by severely reduced immune system function and numerous opportunistic infections. Results from infection with HIV. (Chapter 18)

acrosome An organelle at the tip of the sperm cell containing enzymes that help the sperm penetrate the egg cell. (Chapter 20)

actin A protein found in muscle tissue that, together with myosin, facilitates contraction. (Chapters 16, 19)

action potential Wave of depolarization in a neuron propagated to the end of the axon—also called a nerve impulse. (Chapter 21)

activation energy The amount of energy that reactants in a chemical reaction must absorb before the reaction can start. (Chapter 4)

activator A protein that serves to enhance the transcription of a gene. (Chapter 8)

active immunity Immunity that results from the production of antibodies, as differentiated from passive immunity. (Chapter 19)

active site Substrate-binding region of an enzyme. (Chapter 4)

active smoker An individual who smokes tobacco. (Chapter 18)

active transport The ATP-requiring movement of substances across a membrane against their concentration gradient. (Chapter 3)

adaptation Trait that is favored by natural selection and increases an individual's fitness in a particular environment. (Chapters 10, 11)

adaptive radiation Diversification of one or a few species into large and very diverse groups of descendant species. (Chapter 13)

ADD See attention deficit disorder. (Chapter 22)

adenine Nitrogenous base in DNA, a purine. (Chapters 2, 4, 5, 9)

adenosine diphosphate (ADP) A nucleotide composed of adenine, a sugar, and two phosphate groups. Produced by the hydrolysis of the terminal phosphate bond of ATP. (Chapter 4)

adenosine triphosphate (ATP) A nucleotide composed of adenine, the sugar ribose, and three phosphate groups that can be hydrolyzed to release energy. Form of energy that cells can use. (Chapters 2, 3, 4)

adhesion The sticking together of unlike materials—often water and a particular surface. (Chapter 24)

adipose tissue Fat-storing connective tissue. (Chapter 17)

adrenal gland Either of two endocrine glands, one located atop each kidney, that secrete adrenaline in response to stress or excitement, help maintain water and salt balance, and secrete small amounts of sex hormones. (Chapter 20)

adrenaline A hormone secreted by the adrenal glands in response to stress or excitement. (Chapter 20)

aerobic An organism, environment, or cellular process that requires oxygen. (Chapter 4)

aerobic respiration Cellular respiration that uses oxygen as the electron acceptor. (Chapter 4)

agarose gel A jelly-like slab used to separate molecules on the basis of molecular weight. (Chapter 8)

agriculture Cultivation of crops, raising of livestock; farming. (Chapter 23)

algae Photosynthetic protists. (Chapter 13)

alimentary canal Part of the digestive system that forms a tube extending from the mouth to the anus. Also called the digestive tract. (Chapter 17)

allele Alternate versions of the same gene, produced by mutations. (Chapters 6, 7, 9, 11, 15)

allele frequency The percentage of the gene copies in a population that are of a particular form, or allele. (Chapters 12, 15)

allergy An abnormally high sensitivity to allergens such as pollen or microorganisms. Can cause sneezing, itching, runny nose, and watery eyes. (Chapter 19)

allopatric Geographic separation of a population of organisms from others of the same species. Usually in reference to speciation. (Chapter 12)

alternation of generations A reproductive cycle in which a haploid phase, called a gametophyte, alternates with a diploid phase, called a sporophyte. (Chapter 23)

alternative hypothesis Factor other than the tested hypothesis that may explain observations. (Chapter 1)

alveoli (singular: alveolus) Sacs inside lungs, making up the respiratory surface in land vertebrates and some fish. (Chapter 18)

Alzheimer's disease Progressive mental deterioration in which there is memory loss along with the loss of control of bodily functions, ultimately resulting in death. (Chapter 22)

amenorrhea Abnormal cessation of menstrual cycle. (Chapters 4, 20)

amino acid Monomer subunit of a protein. Contains an amino, a carboxyl, and a unique side group. (Chapters 2, 3, 9)

amnion The fluid-filled sac in which a developing embryo is suspended. (Chapter 21)

anabolic steroid A derivative, usually synthetic, of the steroid hormone testosterone. (Chapter 20)

anaerobic Said of an organism, environment, or cellular process that does not require oxygen. (Chapter 4)

anaerobic respiration A process of energy generation that uses molecules other than oxygen as electron acceptors. (Chapter 4)

anaphase Stage of mitosis during which microtubules contract and separate sister chromatids. (Chapter 6)

anchorage dependence Phenomenon that holds normal cells in place. Cancer cells can lose anchorage dependence and migrate into other tissues or metastasize. (Chapter 6)

androgen A masculinizing hormone, such as testosterone, secreted by the adrenal glands. (Chapters 20, 21)

anecdotal evidence Information based on one person's personal experience. (Chapter 1)

angiogenesis Formation of new blood vessels. (Chapter 6)

angiosperm Plant in the phyla Angiospermae, which produce seeds borne within fruit. (Chapters 13, 23, 24)

animal an organism that obtains energy and carbon by ingesting other organisms and is typically motile for part of its life cycle (Chapter 13)

Animalia Kingdom of Eukarya containing organisms that ingest others and are typically motile for at least part of their life cycle. (Chapters 10, 13)

annual plant Plant that completes its life cycle in a single growing season. (Chapters 16, 23, 24)

annual growth rate Proportional change in population size over a single year. Growth rate is a function of the birth rate minus the death rate of the population. (Chapter 14)

anorexia Self-starvation. (Chapter 4)

antagonistic muscle pair A set of muscles whose actions oppose each other. (Chapter 20)

anther The pollen-containing structure on the stamen of a flower. (Chapter 23)

antibiotic A chemical that kills or disables bacteria. (Chapter 11, 13, 19)

antibiotic resistant Characteristic of certain bacteria; a physiological characteristic that permits them to survive in the presence of particular antibiotics. (Chapter 11)

antibody Protein made by the immune system in response to the presence of foreign substances or antigens. Can serve as a receptor on a B cell or be secreted by plasma cells. (Chapter 19)

antibody-mediated immunity Immunity that occurs via secreted antibodies, versus immunity mediated by T cells. (Chapter 19)

anticodon Region of tRNA that binds to an mRNA codon. (Chapter 9)

antigen Short for antibody-generating substances, an antigen is a molecule that is foreign to the host and stimulates the immune system to react. (Chapter 19)

antigen receptor Protein in B- and T-cell membrane that bind to specific antigens. (Chapter 19)

antioxidant Certain vitamins and other substances that protect the body from the damaging effects of free radicals. (Chapter 3)

antiparallel Feature of DNA double helix in which nucleotides face "up" on one side of the helix and "down" on the other. (Chapter 2)

apical dominance The influence exerted by the terminal bud in suppressing the growth of axillary or lateral buds. Mediated by the hormone, auxin. (Chapter 24)

apical meristem The actively dividing cells at the tip of stems and roots in plants. (Chapter 23)

appendicular skeleton The part of the skeleton composed of the bones of the hip, shoulder, and limbs. (Chapter 20)

aquaporin A transport protein in the membrane of a plant or animal cell that facilitates the diffusion of water across the membrane (osmosis). (Chapter 3)

aquatic Of, or relating to, water. (Chapter 16)

Archaea Domain of prokaryotic organisms made up of species known from extreme environments. (Chapter 13)

artery Blood vessel that carries oxygenated blood from the heart to body tissues. (Chapter 18)

artificial selection Selective breeding of domesticated animals and plants to increase the frequency of desirable traits. (Chapter 11)

asexual reproduction A type of reproduction in which one parent gives rise to genetically identical offspring. (Chapters 6, 12, 21)

assortative mating Tendency for individuals to mate with other individuals who are like themselves. (Chapter 12)

asthma A respiratory disease characterized by spasmodic inflammation of the air passages in the lungs and overproduction of mucus. Often triggered by air contaminants. (Chapter 18)

asymptomatic Stage in an infection that is characterized by relatively unnoticeable or absent symptoms of illness. (Chapter 19)

atherosclerosis Accumulation of fatty deposits within blood vessels; hardening of the arteries. (Chapter 18)

atom The smallest unit of matter that retains the properties of an element. (Chapter 2)

atomic number The number of protons in the nucleus of an atom. Unique to each element, this number is designated by a subscript to the left of the symbol for the element. (Chapter 2)

ATP synthase Enzyme found in the mitochondrial membrane that helps synthesize ATP. (Chapter 4)

atrium An upper chamber of the heart that receives blood from the body or lungs and pumps it to a ventricle. (Chapter 18)

attention deficit disorder (ADD) Syndrome characterized by forgetfulness; distractibility, fidgeting; restlessness; impatience; difficulty sustaining attention in work, play, or conversation; or difficulty following instructions and completing tasks in more than one setting. (Chapter 22)

autoimmune disease Any of the diseases that result from an attack by the immune system on normal body cells. (Chapter 19)

autosome Non-sex chromosome, of which there are 22 pairs in humans. (Chapters 6, 8)

auxin A class of plant hormones that control cell elongation, among other effects. (Chapter 24)

AV valves Heart valves between the atria and the ventricles. (Chapter 18)

axial skeleton Part of the skeleton that supports the trunk of the body and consists largely of the bones making up the vertebral column or spine and much of the skull. (Chapter 20)

axillary bud A bud located at the junction between a plant stem and leaf. (Chapters 23, 24)

axon Long, wire-like portion of the neuron that ends in a terminal bouton. (Chapter 22)

B lymphocyte (B cell) The type of white blood cell responsible for antibody-mediated immunity. (Chapter 19)

background extinction rate The rate of extinction resulting from the normal process of species turnover. (Chapter 15)

bacteria Domain of prokaryotic organisms. (Chapters 2, 4, 9, 13, 15, 19)

ball and socket joint Type of joint in the hips and shoulders that enables arms and legs to move in three dimensions. (Chapter 20)

bark The outer layer of a woody stem, consisting of cork, cork cambium, and phloem. (Chapter 23)

basal metabolic rate Resting energy use of an awake, alert person. (Chapter 3)

base A substance that reduces the concentration of hydrogen ions in a solution. (Chapter 2)

base-pairing rule A rule governing the pairing of nitrogenous bases. In DNA, adenine pairs with thymine and cytosine with guanine. (Chapter 2)

behavioral isolation Prevention of mating between individuals in two different populations based on differences in behavior. (Chapter 11)

benign Tumor that stays in one place and does not affect surrounding tissues. (Chapter 6)

bias Influence of research participants' opinions on experimental results. (Chapter 1)

biennial Plant that completes its life cycle over the course of two growing seasons. (Chapter 23)

bile Mixture of substances produced in the liver that aids in digestion by emulsifying fats. (Chapter 16)

binary fission An asexual form of bacterial reproduction. (Chapters 18, 20)

bioaccumulation A phenomenon that results in the concentration of persistent (i.e., slow to degrade) pollutants in the bodies of animals at high levels of a food chain. (Chapter 16)

biodiversity Variety within and among living organisms. (Chapters 13, 15)

biogeography The study of the geographic distribution of organisms. (Chapter 10)

biological classification Field of science attempting to organize biodiversity into discrete, logical categories. (Chapters 10, 13)

biological diversity Entire variety of living organisms. (Chapter 13)

biological evolution See evolution. (Chapter 10)

biological population Individuals of the same species that live and breed in the same geographic area. (Chapters 10, 12)

biological race Populations of a single species that have diverged from each other. Biologists do not agree on a definition of "race." See also subspecies. (Chapter 12)

biological species concept Definition of a species as a group of individuals that can interbreed and produce fertile offspring but typically cannot breed with members of another species. (Chapter 12)

biology The study of living organisms. (Chapter 1)

biomagnification Concentration of toxic chemicals in the higher levels of a food web. (Chapter 23)

biomass The mass of all individuals of a species, or of all individuals on a level of a food web, within an ecosystem. (Chapter 15)

biome A broad ecological community defined by a particular vegetation type (e.g. temperate forest, prairie), which is typically determined by climate factors. (Chapter 16)

biophilia Humans' innate desire to be surrounded by natural landscapes and objects. (Chapter 15)

biopiracy Using the knowledge of the native people in developing countries to discover compounds for use in developed countries. (Chapter 13)

bioprospecting Hunting for new organisms and new uses of old organisms. (Chapter 13)

biopsy Surgical removal of some cells, tissue, or fluid to determine whether cells are cancerous. (Chapter 6)

bipedal Walking upright on two limbs. (Chapter 10)

bladder An organ of the excretory system that stores urine after it is excreted from the kidneys. (Chapter 18)

blade The broad part of a leaf. (Chapter 23)

blastocyst An embryonic stage consisting of a hollow ball of cells. (Chapter 21)

blind experiment Test in which subjects are not aware of exactly what they are predicted to experience. (Chapter 1)

blood The combination of cells and liquid that flows through blood vessels in the cardiovascular system; made up of red blood cells, plasma, white blood cells, and platelets. (Chapters 17, 18, 19)

blood clotting The process by which a blood clot, a mass of the protein fibrin and dead blood cells, forms in the region of blood vessel damage. (Chapter 18)

blood pressure The force of the blood as it travels through the arteries; partially determined by artery diameter and elasticity. (Chapter 18)

blood vessel The structure that carries blood via the circulatory system throughout the body; arteries, capillaries, and veins. (Chapter 18)

body mass index (BMI) Calculation using height and weight to determine a number that correlates to an estimate of a person's amount of body fat with health risks. (Chapter 3)

bolus In digestion, a soft ball of chewed food. (Chapter 17)

bone A type of connective tissue consisting of living cells in a matrix rich in collagen and calcium. (Chapter 17)

bone marrow Network of soft connective tissue that fills bones and is involved in the production of red blood cells. (Chapters 18, 19, 20)

boreal forest A biome type found in regions with long, cold winters and short, cool summers. Characterized by coniferous trees. (Chapter 16)

botanist Plant biologist. (Chapter 13)

brain stem Region of the brain that lies below the thalamus and hypothalamus; it governs reflexes and some involuntary functions such as breathing and swallowing. (Chapter 22)

bronchi The large air passageways from the trachea into the lungs. (Chapter 18)

bronchiole The branching air passageway inside the lungs. (Chapter 18)

bronchitis Inflammation of the bronchi and bronchioles in the lungs. (Chapter 18)

budding An asexual mechanism of propagation in which outgrowths of the parent form and pinch off, producing new individuals. (Chapter 21)

bulb Plant organ consisting of short stems and multiple leaves modified for food storage. (Chapter 23).

bulbourethral gland Either of the two glands at the base of the penis that secrete acid-neutralizing fluids into semen. (Chapter 21)

bulimia Binge eating followed by purging. (Chapter 3)

C₃ plant Plant that uses the Calvin cycle of photosynthesis to incorporate carbon dioxide into a 3-carbon compound. (Chapter 5)

C₄ plant Plant that performs reactions incorporating carbon dioxide into a 4-carbon compound that ultimately provides carbon dioxide for the Calvin cycle. (Chapter 5)

calcitonin A hormone produced by the thyroid that helps lower blood calcium levels. (Chapter 20)

calcium Nutrient required in plant cells for the production of cell walls and for bone strength and blood clotting in humans. (Chapter 3)

Calorie A kilocalorie or 1000 calories. (Chapter 3)

calorie Amount of energy required to raise the temperature of one gram of water by 1 °C. (Chapter 3)

Calvin cycle A series of reactions that occur in the stroma of plants during photosynthesis that utilize NADPH and ATP to reduce carbon dioxide and produce sugars. (Chapter 5)

CAM plant A plant that uses Crassulacean acid metabolism, a variant of photosynthesis during which carbon dioxide is stored in sugars at night (when stomata are open) and released during the day (when stomata are closed) to prevent water loss. (Chapter 5)

Cambrian explosion Relatively rapid evolution of the modern forms of multicellular life that occurred approximately 550 million years ago. (Chapter 13)

cancer A disease that occurs when cell division escapes regulatory controls. (Chapter 6)

capillary The smallest blood vessel of the cardiovascular system, connecting arteries to veins and allowing material exchange across their thin walls. (Chapter 18)

capillary bed A branching network of capillaries supplying a particular organ or region of the body. (Chapter 18)

capsid Protein coat that surrounds a virus. (Chapter 19)

capsule Gelatinous outer covering of bacterial cells that aids in attachment to host cells during an infection. (Chapter 19)

carbohydrate Energy-rich molecule that is the major source of energy for the cell. Consists of carbon, hydrogen, and oxygen in the ratio CH_2O. (Chapters 2, 3)

carcinogen Substance that causes cancer or increases the rate of its development. (Chapters 6, 18)

cardiac cycle The cycle of contraction and relaxation that a normally functioning heart undergoes over the course of a single heart beat. (Chapter 18)

cardiac muscle Muscle that forms the contractile wall of the heart. (Chapter 17)

cardiovascular disease Malfunction of the cardiovascular system, including, but not limited to, heart attack, stroke, and hypertension. (Chapter 18)

cardiovascular system The organ system made up of the heart and circulatory system, including arteries, capillaries, and veins. (Chapter 18)

carpel The female structure of a flower, containing the ovary, style, and stigma. (Chapter 23)

carrier Individual who is heterozygous for a recessive allele. (Chapter 8)

carrying capacity Maximum population that the environment can support. (Chapter 14)

cartilage Connective tissue found in the skeletal system that is rich in collagen fibers. (Chapter 17)

catalyst A substance that lowers the activation energy of a chemical reaction, thereby speeding up the reaction. (Chapter 4)

catalyze To speed up the rate of a chemical reaction. Enzymes are biological catalysts. (Chapter 4)

caudate nucleus Structure within each cerebral hemisphere that functions as part of the pathway that coordinates movement patterns, learning, and memory. (Chapter 22)

cell Basic unit of life, an organism's fundamental building-block units. (Chapters 2, 3)

cell body Portion of the neuron that houses the nucleus and organelles. (Chapter 22)

cell cycle An ordered sequence of events in the life cycle of a eukaryotic cell from its origin until its division to produce daughter cells. Consists of M, G_1, S, and G_2 phases. (Chapter 6)

cell division Process a cell undergoes when it makes copies of itself. Production of daughter cells from an original parent cell. (Chapter 6)

cell plate A double layer of new cell membrane that appears in the middle of a dividing plant cell and divides the cytoplasm of the dividing cell. (Chapter 6)

cell wall Tough but elastic structure surrounding plant and bacterial cell membranes. (Chapters 2, 6, 19, 23, 24)

cell-mediated immunity Type of specific immune response carried out by T cells. (Chapter 19)

cellular respiration Metabolic reactions occurring in cells that result in the oxidation of macromolecules to produce ATP. (Chapter 4)

cellulose A structural polysaccharide found in cell walls and composed of glucose molecules. (Chapters 2, 6)

central nervous system (CNS) Includes brain and spinal cord and is responsible for integrating, processing, and coordinating information taken in by the senses. It is the seat of functions such as intelligence, learning, memory, and emotion. (Chapter 22)

central vacuole A membrane-enclosed sac in a plant cell that functions to store many different substances. (Chapters 2, 23, 24)

centriole A structure in animal cells that helps anchor for microtubules during cell division. (Chapters 2, 6)

centromere Region of a chromosome where sister chromatids are attached and to which microtubules bind. (Chapter 6)

cereal grains Staples of the human diet that are from the grass family, namely wheat, corn, oats, rye, and barley. (Chapter 23)

cerebellum Region of the brain that controls balance, muscle movement, and coordination. (Chapter 22)

cerebral cortex Deeply wrinkled outer surface of the cerebrum where conscious activity and higher thought originate. (Chapter 22)

cerebrospinal fluid Protective liquid bath that surrounds the brain within the skull. (Chapter 22)

cerebrum Portion of the brain in which language, memory, sensations, and decision making are controlled. The cerebrum has two hemispheres, each of which has four lobes. (Chapter 22)

cervix The lower narrow portion of the uterus at the top end of the vagina. (Chapter 21)

chaparral A biome characteristic of climates with hot, dry summers and mild, wet winters and a dominant vegetation of aromatic shrubs. (Chapter 16)

checkpoint Stoppage during cell division that occurs to verify that division is proceeding correctly. (Chapter 6)

chemical reaction A process by which one or more chemical substances is transformed into one or more different chemical substances. (Chapter 2)

chemotherapy Using chemicals to try to kill rapidly dividing (cancerous) cells. (Chapter 6)

chlorophyll Green pigment found in the chloroplast of plant cells. (Chapter 5)

chloroplast An organelle found in plant cells that absorbs sunlight and uses the energy derived to produce sugars. (Chapters 2, 5, 13, 24)

cholesterol A steroid found in animal cell membranes that affects membrane fluidity. Serves as the precursor to estrogen and testosterone. (Chapters 2, 3, 20)

chondrocyte A type of cartilage cell that produces collagen and proteoglycans. (Chapter 17)

chromosome Subcellular structure composed of a long single molecule of DNA and associated proteins, housed inside the nucleus. (Chapters 6, 7, 8, 9, 12)

chyme Partially digested food and enzymes that are passed from the stomach to the intestine. (Chapter 17)

circulatory system The vessels that transport blood, nutrients, and waste around the body. (Chapters 6, 17, 18, 19)

citric acid cycle A chemical cycle occurring in the matrix of the mitochondria that breaks the remains of sugars down to produce carbon dioxide. (Chapter 4)

cladistic analysis A technique for determining the evolutionary relationships among organisms that relies on identification and comparison of newly evolved traits. (Chapter 13)

classification system Method for organizing biological diversity. (Chapters 10, 13)

cleavage In embryology, the period of rapid cell division that occurs during animal development. (Chapter 21)

climate The average temperature and precipitation as well as seasonality. (Chapter 16)

climax community The group of species that is stable over time in a particular set of environmental conditions. (Chapter 16)

clinical trial Controlled scientific experiment to determine the effectiveness of novel treatments. (Chapter 6)

clitoris Sensitive erectile tissue found in the external genitalia of females that functions in sexual arousal. (Chapter 21)

clonal population Population of identical cells copied from the immune cell that first encounters an antigen. The entire clonal population has the same DNA arrangement, and all cells in a clonal population carry the same receptor on their membrane. (Chapter 19)

cloning Producing copies of a gene or an organism that are genetically identical. (Chapter 9)

closed circulatory system A circulatory system in which vessels are continuous and retain the circulating fluid. (Chapter 18)

clumped distribution A spatial arrangement of individuals in a population where large numbers are concentrated in patches with intervening, sparsely populated areas separating them. (Chapter 14)

codominant Two different alleles of a gene that are equally expressed in the heterozygote. (Chapters 7, 8)

codon A triplet of mRNA nucleotides. Transfer RNA molecules bind to codons during protein synthesis. (Chapter 9)

coenzyme (or cofactor) Substances such as vitamins that help enzymes catalyze chemical reactions. (Chapter 4)

coevolution Occurs when change in one biological species triggers a change in an associated species. Coevolution commonly occurs between predator and prey or parasite and host. (Chapter 13)

cohesion The tendency for molecules of the same material to stick together. (Chapters 2, 24)

collecting duct A structure in the kidney that accepts filtrate from multiple nephrons and transmits it to the renal pelvis. Each kidney contains hundreds of collecting ducts. The amount of water retained by the kidneys is largely controlled at the collecting ducts. (Chapter 18)

colon Tube between the small intestine and anus of the digestive system. (Chapter 17)

combination drug therapy The use of more than one drug simultaneously to treat a disease. Often used for disease organisms that mutate quickly or are difficult to control to combat the problem of drug resistance. (Chapter 11)

commensalism In ecology, a relationship between two species in which one is benefitted and the other is neither harmed nor benefitted. (Chapter 15)

common descent The theory that all living organisms on Earth descended from a single common ancestor that appeared in the distant past. (Chapter 10)

community A group of interacting species in the same geographic area. (Chapter 15)

compact bone The hard, outer shell of bones. (Chapter 20)

competition Interaction that occurs when two species of organisms both require the same resources within a habitat; competition tends to limit the size of populations. (Chapter 15)

competitive exclusion Reduction or elimination of one species in an environment resulting from the presence of another species that requires the same or similar resources. (Chapter 15)

complementary Complementary bases pair with each other by hydrogen bonding across the DNA helix. Adenine is complementary to thymine and cytosine to guanine. (Chapter 2)

complement protein Type of blood protein with which an antibody-antigen complex can combine in order to kill bacterial cells. Enhances the immune response on many levels. (Chapter 19)

complementary base pair Nitrogenous bases that hydrogen bond to each other. In DNA, adenine is complementary to thymine, and cytosine is complementary to guanine. In RNA, adenine is complementary to uracil and guanine to cytosine. (Chapters 2, 8, 9)

complete digestive tract A digestive tube running from the mouth to anus. (Chapter 16)

complete protein Dietary protein that contains all the essential amino acids. (Chapter 3)

complex carbohydrate Carbohydrate consisting of two or more monosaccharides. (Chapter 3)

compound A substance consisting of two or more elements in a fixed ratio. (Chapter 2)

confidence interval In statistics, a range of values calculated to have a given probability (usually 95%) of containing the true population mean. (Chapter 1)

connective tissue Animal tissue that functions to bind and support other tissues. Composed of a small number of cells embedded in a matrix. (Chapter 17)

consilience The unity of knowledge. Used to describe a scientific theory that has multiple lines of evidence to support it. (Chapter 10)

contact inhibition Property of cells that prevents them from invading surrounding tissues. Cancer cells may lose this property. (Chapter 6)

contagious Spreading from one organism to another. (Chapter 18)

continuous variation A range of slightly different values for a trait in a population. (Chapter 7)

control Subject for an experiment who is similar to experimental subject except is not exposed to the experimental treatment. Used as baseline values for comparison. (Chapter 1)

convergent evolution Evolution of the same trait or set of traits in different populations as a result of shared environmental conditions rather than shared ancestry. (Chapters 10, 12)

copulation Sexual intercourse. (Chapter 21)

coral reef Highly diverse biome found in warm, shallow salt water, dominated by the limestone structures created by coral animals. (Chapters 13, 16)

cork A tissue produced by the cork cambium that provides protection to a woody stem by forming outer bark. (Chapter 23)

cork cambium A meristematic tissue in bark that arises from phloem cells and creates cork cells before being shed. (Chapter 23)

corpus callosum Bundle of nerve fibers at the base of the cerebral fissure that provides a communication link between the cerebral hemispheres. (Chapter 22)

corpus luteum Hormone-producing tissue (the ovarian follicle after ovulation) that makes progesterone and estrogen and degenerates about 12 days after ovulation if fertilization does not occur. (Chapter 21)

correlation Describes a relationship between two factors. (Chapters 1, 7, 12)

cotyledon The first leaves produced by an embryonic seed plant. (Chapter 23)

countercurrent exchange The process by which streams of fluid move in parallel but opposite directions, maximizing material, gas, or heat exchange between them. (Chapter 18)

covalent bond A type of strong chemical bond in which two atoms share electrons. (Chapter 2)

critical night length The night length that a plant must experience in order to trigger a photoperiodic response. (Chapter 24)

crop rotation The practice of growing different crops at different times on the same field in order to maintain soil fertility and decrease pest damage. (Chapter 23)

cross In genetics, the mating of two organisms. (Chapter 7)

crossing over Gene for gene exchange of genetic information between members of a homologous pair of chromosomes. (Chapter 6)

cryptorchidism A developmental defect in which the testes fail to descend to the scrotum. (Chapter 21)

cultural control Control of agricultural pests through environmental techniques, such as crop rotation and maintenance of native vegetation. (Chapter 23)

cuticle The waxy layer on the outer surface of plant epidermal cells. (Chapter 23)

cyst Noncancerous, fluid-filled growth. (Chapter 6)

cytokinesis Part of the cell cycle during which two daughter cells are formed by the cytoplasm splitting. (Chapter 6)

cytoplasm The entire contents of the cell (except the nucleus) surrounded by the plasma membrane. (Chapters 2, 9)

cytosine Nitrogenous base, a pyrimidine. (Chapters 2, 6, 8)

cytoskeleton A network of tubules and fibers that branch throughout the cytoplasm. (Chapter 2)

cytosol The semifluid portion of the cytoplasm. (Chapters 2, 4)

cytotoxic T cell Immune-system cell that attacks and kills body cells that have become infected with a virus before the virus has had time to replicate. (Chapter 19)

data Information collected by scientists during hypothesis testing. (Chapter 1)

daughter cells The offspring cells that are produced by the process of cell division. (Chapter 6)

death rate Number of deaths averaged over the population as a whole. (Chapter 14)

deciduous Pertaining to woody plants that drop their leaves at the end of a growing season. (Chapters 16, 24)

decomposer An organism, typically bacteria and fungi in the soil, whose action breaks down complex molecules into simpler ones. (Chapter 15)

decomposition The breakdown of organic material into smaller molecules. (Chapter 16)

deductive reasoning Making a prediction about the outcome of a test; "if/then" statements. (Chapter 1)

defensive protein Nonspecific protein of the immune system including interferons and complement proteins. (Chapter 19)

deforestation The removal of forest lands, often to enable the development of agriculture. (Chapters 5, 15)

degenerative disease Disease characterized by progressive deterioration. (Chapter 9)

dehydration Loss of water. (Chapters 3, 24)

deleterious In genetics, said of a mutation that reduces an individual's fitness. (Chapter 15)

demographic momentum Lag between the time that humans reduce birth rates and the time that population numbers respond. (Chapter 14)

demographic transition The period of time between when death rates in a human population fall (as a result of improved technology) and when birth rates fall (as a result of voluntary limitation of pregnancy). (Chapter 14)

denature (1) In proteins, the process where proteins unravel and change their native shape, thus losing their biological activity. (2) For DNA, the breaking of hydrogen bonds between the two strands of the double-stranded DNA helix, resulting in single-stranded DNA. (Chapter 8)

dendrite Short extension of the neuron that receives signals from other cells. (Chapter 22)

density-dependent factor Any of the factors related to a population's size that influence the current growth rate of a population—for example, communicable disease or starvation. (Chapter 14)

density-independent factor Any of the factors unrelated to a population's size that influence the current growth rate of a population—for example, natural disasters or poor weather conditions. (Chapter 14)

deoxyribonucleic acid (DNA) Molecule of heredity that stores the information required for making all of the proteins required by the cell. (Chapters 2, 6, 7, 8, 9, 11, 13)

deoxyribose The five-carbon sugar in DNA. (Chapters 2, 6, 7, 9)

dependent variable The variable in a study that is expected to change in response to changes in the independent variable. (Chapter 1)

depolarization Reduction in the charge difference across the neuronal membrane. (Chapter 22)

depression Disease that involves feelings of helplessness and despair, and sometimes thoughts of suicide. (Chapter 22)

desert Biome found in areas of minimal rainfall. Characterized by sparse vegetation. (Chapter 16)

desertification The process by which formerly productive land is converted to unproductive land, typically by overgrazing of cattle or unsustainable agricultural practices, but also increasingly by the effects of climate change. (Chapter 16)

determinate growth Growth of limited duration. (Chapter 23)

developed country A term coined by the United Nations to describe a country with high per-capita income and significant industrial development. (Chapter 14)

development All of the progressive changes that produce an organism's body. (Chapters 13, 21)

diabetes Disorder of carbohydrate metabolism characterized by impaired ability to produce or respond to the hormone insulin. (Chapter 3)

diaphragm (1) Dome-shaped muscle at the base of the chest cavity. Contraction of this muscle helps draw air into the lungs. (Chapter 18) (2) A birth control device consisting of a flexible contraceptive disk that covers the cervix to prevent the entry of sperm. (Chapter 21)

diastole The stage of the cardiac cycle when the heart relaxes and fills with blood. (Chapter 18)

diastolic blood pressure The lowest blood pressure in the arteries, occurring during diastole of the cardiac cycle. (Chapters 4, 18)

dicot The class of angiosperms characterized by having two cotyledons. (Chapter 23)

differentiation Structural and functional divergence of cells as they become specialized. (Chapter 21)

diffusion The spontaneous movement of substances from a region of their own high concentration to a region of their own low concentration. (Chapters 3, 18)

digestive system or tract See alimentary canal. (Chapter 17)

dihybrid cross A genetic cross involving the alleles of two different genes. For example AaBb x AaBb. (Chapter 8)

diploid cell A cell containing homologous pairs of chromosomes (2n). (Chapters 6, 21)

directional selection Natural selection for individuals at one end of a range of phenotypes. (Chapter 11)

disaccharide A double sugar consisiting of two monosaccharides joined together by a glycosidic linkage. (Chapter 2)

diverge See divergence. (Chapter 12)

divergence Occurs when gene flow is eliminated between two populations. Over time, traits found in one population begin to differ from traits found in the other population. (Chapters 10, 12, 13)

diversifying selection Natural selection for individuals at both ends of a range of phenotypes but against the "average" phenotype. (Chapter 11)

dizygotic twins Fraternal twins (non-identical) that develop when two different sperm fertilize two different egg cells. (Chapter 7)

DNA See deoxyribonucleic acid. (Chapters 2, 6, 7)

DNA fingerprinting Powerful genetic identification technique that takes advantage of differences in DNA sequences between all people other than identical twins. (Chapter 8)

DNA polymerase Enzyme that facilitates base pairing during DNA synthesis. (Chapter 6)

DNA replication The synthesis of two daughter DNA molecules from one original parent molecule. Takes place during the S phase of interphase. (Chapters 6,7)

domain Most inclusive biological category. Biologists group life into three major domains. (Chapters 10, 13)

domesticated Referring to animals or plants that are now largely dependent on humans for reproduction and survival as a result of artificial selection. (Chapter 23)

dominant Applies to an allele with an effect that is visible in a heterozygote. (Chapter 7)

dopamine Neurotransmitter active in pathways that control emotions and complex movements. (Chapter 22)

dormancy A condition of arrested growth and development in plants. (Chapter 24)

dormant In a state of dormancy; i.e. alive, but not growing or developing. (Chapter 23)

double blind Experimental design protocol when both research subjects and scientists performing the measurements are unaware of either the experimental hypothesis or who is in the control or experimental group. (Chapter 1)

double circulation The flow of blood through two circuits in an animal body, from heart to lungs, back to heart, and to the body. (Chapter 18)

double fertilization The fusion of egg and sperm and the simultaneous fusion of a sperm with nuclei in the ovule of angiosperms. (Chapter 23)

ecological footprint A measure of the natural resources used by a human population or society. (Chapter 16)

ecological niche The functional role of a species within a community or ecosystem, including its resource use and interactions with other species. (Chapter 15)

ecology Field of biology that focuses on the interactions between organisms and their environment. (Chapters 13, 14)

ecosystem All of the organisms and natural features in a given area. (Chapter 15)

ecotourism The visitation of specific geographical sites by tourists interested in natural attractions, especially animals and plants. (Chapter 15)

ectoderm The outermost of the three germ layers that arise during animal development. (Chapter 21)

ectotherm An animal that must use energy from the sun and other environmental sources to regulate its body temperature. (Chapter 17)

effector Muscle, gland, or organ stimulated by a nerve. (Chapter 22)

egg cell Gamete produced by a female organism. (Chapters 6, 7, 8, 12, 13, 20, 21)

electron A negatively charged subatomic particle. (Chapters 2, 4)

electron shell An energy level representing the distance of an electron from the nucleus of an atom. (Chapter 2)

electron transport chain A series of proteins in the mitochondrial and chloroplast membranes that move electrons during the redox reactions that release energy to produce ATP. (Chapter 4)

electronegative The tendency to attract electrons to form a chemical bond. (Chapter 2)

element A substance that cannot be broken down into any other substance. (Chapter 2)

embryo The developmental stage commencing after the first mitotic divisions of the zygote and ending when body structures begin to appear; from about the second week after fertilization to about the ninth week. (Chapters 6, 7, 9, 10, 13, 21)

emphysema A lung disease caused by the breakdown of alveoli walls; characterized by shortness of breath and an expanded chest cavity. (Chapter 18)

encephalitis Pathology, or disease, of the brain. (Chapter 19)

Endangered Species Act (ESA) U.S. law intended to protect and encourage the population growth of threatened and endangered species enacted in 1973. (Chapter 15)

endocrine disrupter Chemical that disrupts the functions of the hormone-producing endocrine system. (Chapter 21)

endocrine gland Any of the glands that secrete hormones into the bloodstream. (Chapter 20)

endocrine system The internal system of chemical signals involving hormones, the organs and glands that produce and secrete them, and the target cells that respond to them. (Chapter 20)

endocytosis The uptake of substances into cells by a pinching inward of the plasma membrane. (Chapter 3)

endoderm The innermost of the three germ layers that arise during animal development. (Chapter 21)

endometriosis The abnormal occurrence of functional endometrial tissue outside the uterus, resulting in painful menstrual cycles. (Chapter 21)

endometrium Lining of the uterus, shed during menstruation. (Chapter 21)

endoplasmic reticulum (ER) A network of membranes in eukaryotic cells. When rough, or studded with ribosomes, functions as a workbench for protein synthesis. When devoid of ribosomes, or smooth, it functions in phospholipid and steroid synthesis and detoxification. (Chapter 2)

endoskeleton A hard skeleton buried within the soft tissues of an animal. (Chapter 20)

endosperm The triploid structure formed by the fusion of a sperm and two nuclei in the ovule of an angiosperm plant. (Chapter 23)

endosymbiotic theory Theory that organelles such as mitochondria and chloroplasts in eukaryotic cells evolved from prokaryotic cells that took up residence inside ancestral eukaryotes. (Chapter 13)

endotherm An animal that uses its own metabolic energy to maintain a constant body temperature. (Chapter 17)

enveloped viruses A virus that is surrounded by a membrane. (Chapter 19)

environmental tobacco smoke (ETS) The tobacco smoke in the air that results from smoldering tobacco on the lit ends of cigarettes and pipes as well as the smoke exhaled by active smokers. (Chapter 18)

enzyme Protein that catalyzes and regulates the rate of metabolic reactions. (Chapters 2, 3, 4, 11)

epidemic Contagious disease that spreads rapidly and extensively among many individuals. (Chapter 19)

epidermal tissue The outermost layer of cells of the leaf and of young stems and roots. (Chapter 23)

epididymis A coiled tube located adjacent to the testes where sperm are stored. (Chapter 21)

epiglottis Flap that blocks the windpipe so food goes down the pharynx, not into the lungs. (Chapter 17)

epithelial tissue (epithelia) Tightly packed sheets of cells that line organs and body cavities. (Chapter 17)

equator The circle around Earth that is equidistant to both poles. (Chapter 16)

esophagus Tube that conducts food from the pharynx to the stomach. (Chapter 17)

essential amino acid Any of the amino acids that humans cannot synthesize and thus must be obtained from the diet. (Chapter 3)

essential fatty acid Any of the fatty acids that animals cannot synthesize and must be obtained from the diet. (Chapter 3)

estrogen Any of the feminizing hormones secreted by the ovary in females and adrenal glands in both sexes. (Chapters 20, 21)

estuary An aquatic biome that forms at the outlet of a river into a larger body of water such as a lake or ocean. (Chapter 16)

ethylene A plant hormone involved in fruit ripening and abscission. (Chapter 24)

eukaryote Cell that has a nucleus and membrane-bounded organelles. (Chapters 2, 9, 13)

eutrophication Process resulting in periods of dangerously low oxygen levels in water, sometimes caused by high levels of nitrogen and phosphorus from fertilizer runoff, that result in increased growth of algae in waterways. (Chapters 15, 23)

evolution Changes in the features (traits) of individuals in a biological population that occur over the course of generations. See also theory of evolution. (Chapters 1,2, 10, 11)

evolutionary classification System of organizing biodiversity according to the evolutionary relationships among living organisms. (Chapter 13)

excretion The release of urine from the kidneys into the storage organ of the bladder. (Chapter 18)

exocytosis The secretion of molecules from a cell via fusion of membrane-bounded vesicles with the plasma membrane. (Chapter 3)

exoskeleton A hard outer encasement on the surface of an animal. (Chapter 20)

experiment Contrived situation designed to test specific hypotheses. (Chapter 1)

exponential growth Growth that occurs in proportion to the current total. (Chapter 14)

external fertilization Fertilization that does not require that the parents ever have physical contact with each other. Seen in many aquatic invertebrates and most fish and amphibians. (Chapter 21)

extinction Complete loss of a species. (Chapter 15)

extinction vortex A process by which an endangered population is driven toward extinction via loss of genetic diversity, increased vulnerability to the effects of density-independent factors, and inbreeding depression. (Chapter 15)

facilitated diffusion The spontaneous passage of molecules, through membrane proteins, down their concentration gradient. (Chapter 3)

falsifiable Able to be proved false. (Chapter 1)

fat Energy-rich, hydrophobic lipid molecule composed of a three-carbon glycerol skeleton bonded to three fatty acids. (Chapters 2, 3, 4)

fatty acid A long acidic chain of hydrocarbons bonded to glycerol. Fatty acids vary on the basis of their length and on the number and placement of double bonds. (Chapters 2, 3)

femur Bone extending from the pelvis to the knee; also called thighbone. (Chapter 20)

fermentation A process that makes a small amount of ATP from glucose without using an electron transport chain. Ethyl alcohol and lactic acid are produced by this process. (Chapter 4)

fertility Ability to produce viable gametes. (Chapter 21)

fertilization The fusion of haploid gametes (in humans, egg and sperm) to produce a diploid zygote. (Chapters 6, 7, 8, 21)

fertilizer Any of a variety of growth medium amendments that increase the level of plant nutrients. (Chapter 23)

fetus The term used to describe a developing human from the ninth week of development until birth. (Chapter 21)

fever Abnormally high body temperature. (Chapter 19)

fibrin A protein produced during the process of blood clotting that makes up a net to trap and block blood flow from a damaged blood vessel. (Chapter 18)

fibroblast A protein-secreting cell found in loose connective tissue. (Chapter 17)

fibrous connective tissue A dense tissue found in tendons and ligaments composed largely of collagen fibers. (Chapter 17)

filament The stalk of a stamen. (Chapter 23)

filtration In the kidneys, the removal of plasma from the bloodstream through capillaries surrounding a nephron. (Chapter 18)

fitness Relative survival and reproduction of one variant compared to others in the same population. (Chapters 11, 15)

flagellum (*plural*: flagella) A long cellular projection that aids in motility. (Chapter 19)

flower Reproductive structure of a flowering plant. (Chapters 23, 24)

flowering plant Member of the kingdom Plantae, which produce flowers and fruit. See also angiosperm. (Chapter 13)

fluid mosaic model The accepted model for how membranes are structured with proteins bobbing in a sea of phospholipids. (Chapter 2)

follicle Structure in the ovary that contains the developing ovum and secretes estrogen. (Chapter 21)

follicle cell A flattened cell within the single layer that surrounds each primary oocyte. (Chapter 21)

follicle-stimulating hormone (FSH) Hormone secreted by the pituitary gland involved in sperm production, regulation of ovulation, and regulation of menstruation. (Chapters 20, 21)

food chain The linear relationship between trophic levels from producers to primary consumers, and so on. (Chapter 15)

food web The feeding connections between and among organisms in an environment. (Chapter 15)

foramen magnum Hole in the skull that allows passage of the spinal cord. (Chapter 10)

forest Terrestrial community characterized by the presence of trees. (Chapter 16)

fossil Remains of plants or animals that once existed, left in soil or rock. (Chapters 10, 13, 15)

fossil fuel Nonrenewable resource consisting of the buried remains of ancient plants that have been transformed by heat and pressure into coal and oil. (Chapters 5, 15)

fossil record Physical evidence left by organisms that existed in the past. (Chapter 10)

founder effect Type of sampling error that occurs when a small subset of individuals emigrates from the main population to found a new population, leading to differences in the gene pools of both. (Chapter 12)

founder hypothesis The hypothesis that the diversity of unique species in isolated habitats results from divergence from a single founding population. (Chapter 12)

frameshift mutation A mutation that occurs when the number of nucleotides inserted or deleted from a DNA sequence is not a multiple of three. (Chapter 9)

free radical A substance containing an unpaired electron that is therefore unstable and highly reactive, causing damage to cells. (Chapter 3)

frontal bone Upper front portion of the cranium, or the forehead. (Chapter 20)

frontal lobe The largest and most anterior portion of each cerebral hemisphere. (Chapter 22)

fruit A mature ovary of an angiosperm, containing the seeds. (Chapters 23, 24)

Fungi Kingdom of eukaryotes made up of members that are immobile, rely on other organisms as their food source, and are made up of hyphae that secrete digestive enzymes into the environment and that absorb the digested materials. (Chapters 10, 13, 15)

gall bladder Organ that stores bile and empties into the small intestine. (Chapter 17)

galls Tumor growths on a plant. (Chapter 9)

gamete Specialized sex cell (sperm and egg in humans) that contain half as many chromosomes as other body cells and is therefore haploid. (Chapters 6, 7, 21)

gamete incompatibility An isolating mechanism between species in which sperm from one cannot fertilize eggs from another. (Chapter 12)

gametogenesis The production of gametes. (Chapter 21)

gametophyte In terrestrial plants, the haploid, gamete-producing generation or phase. (Chapter 23)

gas exchange The passage of gases, as a result of diffusion, from one compartment to another. (Chapter 18)

gastrula The 2-layered, cup-shaped stage of embryonic development. (Chapter 21)

gel electrophoresis The separation of biological molecules on the basis of their size and charge by measuring their rate of movement through an electric field. (Chapter 8)

gene Discrete unit of heritable information about genetic traits. Consists of a sequence of DNA that codes for a specific polypeptide—a protein or part of a protein. (Chapters 6, 7)

gene expression Turning a gene on or off. A gene is expressed when the protein it encodes is synthesized. (Chapter 9)

gene flow Spread of an allele throughout a species' gene pool. (Chapter 12)

gene gun Device used to shoot DNA-coated pellets into plant cells. (Chapter 9)

gene pool All of the alleles found in all of the individuals of a species. (Chapter 12)

gene therapy Replacing defective genes (or their protein products) with functional ones. (Chapter 9)

genealogical species concept A scheme that identifies as separate species all populations with a unique lineage. (Chapter 12)

general sense Also called proprioception. Any of the senses including temperature, pain, touch, pressure, and body position, with sensory receptors scattered throughout the body. (Chapter 22)

genetic code Table showing which mRNA codons code for which amino acids. (Chapters 9, 10)

genetic drift Change in allele frequency that occurs as a result of chance. (Chapters 12, 15)

genetic variability All of the forms of genes, and the distribution of these forms, found within a species. (Chapter 15)

genetic variation Differences in alleles that exist among individuals in a population. (Chapter 7)

genetically modified organisms (GMOs) Organisms whose genome incorporates genes from another organism: also called transgenic or genetically engineered organisms. (Chapter 9)

genome Entire suite of genes present in an organism. (Chapters 9, 19)

genotype Genetic composition of an individual. (Chapters 6, 7, 8)

genus Broader biological category to which several similar species may belong. (Chapters 10, 12)

geologic period A unit of time defined according to the rocks and fossils characteristic of that period. (Chapter 13)

germ line therapy Gene therapy that changes genes in a zygote or early embryo, so that the embryo will pass on the engineered genes to their offspring. (Chapter 9)

germ theory The scientific theory that all infectious diseases are caused by microorganisms. (Chapter 1)

germination The beginning of growth by a seed. (Chapter 23)

gibberellin A plant hormone that causes the elongation of stem cells. (Chapter 24)

gill The respiratory exchange surface in many aquatic animals, consisting of thin sheets of tissue containing many capillaries. (Chapter 18)

glans penis The head of the penis. (Chapter 21)

glial cell A type of cells within the brain that does not carry messages but rather supports neurons by supplying nutrients, repairing the brain after injury, and attacking invading bacteria. (Chapter 22)

global warming Increase in average temperatures as a result of the release of increased amounts of carbon dioxide and other greenhouse gases into the atmosphere. (Chapters 5, 15)

glycolysis The splitting of glucose into pyruvate, which helps drive the synthesis of a small amount of ATP. (Chapter 4)

Golgi apparatus An organelle in eukaryotic cells consisting of flattened membranous sacs that modify and sort proteins and other substances. (Chapter 2)

gonads The male and female sex organs; testicles in human males or ovaries in human females. (Chapters 5, 20, 21)

gonadotropin-releasing hormone (GnRH) Hormone produced by the hypothalamus that stimulates the pituitary gland to release FSH and LH, thereby stimulating the activities of the gonads. (Chapters 20, 21)

gradualism The hypothesis that evolutionary change occurs in tiny increments over long periods of time. (Chapter 12)

grana Stacks of thylakoids in the chloroplast. (Chapters 2, 5)

grassland Biome characterized by the dominance of grasses, usually found in regions of lower precipitation. (Chapter 16)

gravitropism Directional plant growth in response to gravity. (Chapter 24)

gray matter Unmyelinated axons, combined with dendrites and cell bodies of other neurons that appear gray in cross section. (Chapter 22)

greenhouse effect The retention of heat in the atmosphere by carbon dioxide and other greenhouse gases. (Chapter 5)

ground tissue The plant tissues other than the vascular tissue, epidermis, and meristem. (Chapter 23)

growing season The length of time from last freeze in spring to first freeze in autumn. (Chapter 24)

growth factor Protein that stimulates cell division. (Chapter 6)

growth rate Annual death rate in a population subtracted from the annual birth rate. (Chapter 14)

guanine Nitrogenous base in DNA, a purine. (Chapters 2, 6, 8, 9)

guard cell Either of the paired cells encircling stomata that serve to regulate the size of the stomatal pore, helping to minimize water loss under dry conditions and maximize carbon dioxide uptake under wet conditions. (Chapters 5, 23)

gymnosperm Member of a group of plants made up of several phyla that produce seeds, but not true flowers. (Chapter 23, 24)

habitat Place where an organism lives. (Chapter 15)

habitat destruction Modification and degradation of natural forests, grasslands, wetlands, and waterways by people; primary cause of species loss. (Chapter 15)

habitat fragmentation Threat to biodiversity caused by humans that occurs when large areas of intact natural habitat are subdivided by human activities. (Chapter 15)

half-life Amount of time required for one-half the amount of a radioactive element that is originally present to decay into the daughter product. (Chapter 10)

haploid Describes cells containing only one member of each homologous pair of chromosomes ("); in humans, these cells are eggs and sperm. (Chapters 6, 21)

hardened Relating to plants that have undergone the appropriate physiological changes that allow them to survive freezing temperatures. (Chapter 24)

hardiness In plants, the ability to survive freezing temperatures. (Chapter 24)

Hardy-Weinberg theorem Theorem that holds that allele frequencies remain stable in populations that are large in size, randomly mating, and experiencing no migration or natural selection. Used as a baseline to predict how allele frequencies would change if any of its assumptions were violated. (Chapter 12)

heart The muscular organ that pumps blood via the circulatory system to the lungs and body. (Chapters 17, 18)

heart attack An acute condition, during which blood flow is blocked to a portion of the heart muscle, causing part of the muscle to be damaged or die. (Chapters 4, 18)

heat The total amount of energy associated with the movement of atoms and molecules in a substance. (Chapter 4)

helper T cell A type of immune-system cell that enhances cell-mediated immunity and humoral immunity by secreting a substance that increases the strength of the immune response. See also T4 cell. (Chapter 19)

hemoglobin An iron-containing protein that carries oxygen in red blood cells. (Chapter 18)

hemophilia Rare genetic disorder caused by a sex-linked recessive allele that prevents normal blood clotting. (Chapter 8)

hepatocyte Liver cell. (Chapter 17)

herbicide A chemical that kills plants. (Chapter 23)

heritability The amount of variation for a trait in a population that can be explained by differences in genes among individuals. (Chapter 7)

hermaphrodite An individual with both male and female reproductive organs. (Chapter 21)

heterozygote Individual carrying two different alleles of a particular gene. (Chapters 7, 8, 12)

heterozygous Said of a genotype containing two different alleles of a gene. (Chapter 7)

high-density lipoprotein (HDL) A cholesterol-carrying particle in the blood that is high in protein and low in cholesterol. (Chapters 4, 18)

hinge joint A joint that allows back and forth movement. (Chapter 20)

histamine Chemcial released from mast cells during allergic reactions. Causes blood vessels to dilate and become more permeable and lowers blood pressure. (Chapter 19)

HIV See human immunodeficiency virus. (Chapter 19)

homeostasis The steady state condition an organism works to maintain. (Chapters 2, 17)

hominin Referring to humans and human ancestors. (Chapters 10, 12)

homologous pair Set of two chromosomes of the same size and shape with centromeres in the same position. Homologous pairs of chromosomes carry the same genes in the same locations but may carry different alleles. (Chapters 6, 7)

homology Similarity in characteristics as a result of common ancestry. (Chapter 10)

homozygous Having two copies of the same allele of a gene. (Chapters 7, 15)

hormones A protein or steroid produced in one tissue that travels through the circulatory system to act on another tissue to produce some physiological effect. (Chapter 20)

Human Genome Project Effort to determine the nucleotide base sequences and chromosomal locations of all human genes. (Chapter 9)

human immunodeficiency virus (HIV) Agent identified as causing the transmission and symptoms of AIDS. (Chapter 19)

humoral immunity B-cell-mediated immunity that occurs when a B-cell receptor binds to an antigen. The B cell divides to produce a clonal population of memory cells; the cell also produces plasma cells. (Chapter 19)

hybrid Offspring of two different strains of an agricultural crop. See also interspecies hybrid. (Chapters 12, 23)

hydrocarbon A compound consisting of carbons and hydrogens. (Chapter 2)

hydrogen atom One negatively charged electron and one positively charged proton. (Chapters 2, 5)

hydrogen bond A type of weak chemical bond in which a hydrogen atom of one molecule is attracted to an electronegative atom of another molecule. (Chapters 2, 5)

hydrogen ion The positively charged ion of hydrogen (H+) formed by removal of the electron from a hydrogen atom. (Chapters 2, 5)

hydrogenation Adding hydrogen gas under pressure to make liquid oils more solid. (Chapter 3)

hydrophilic Readily dissolving in water. (Chapter 2)

hydrophobic Not able to dissolve in water. (Chapter 2)

hydrostatic skeleton A skeletal system composed of fluid in a closed body compartment. (Chapter 20)

hypertension High blood pressure. (Chapters 4, 18)

hyphae Thin, stringy fungal material that grows over and within a food source. (Chapter 13)

hypothalamus Gland that helps regulate body temperature; influences behaviors such as hunger, thirst, and reproduction; and secretes a hormone (GnRH) that stimulates the activities of the gonads. (Chapters 20, 21, 22)

hypothesis Tentative explanation for an observation that requires testing to validate. (Chapters 1, 10, 12)

immortal Property of cancer cells that allows them to divide more times than normal cells. (Chapter 6)

immune response Ability of the immune system to respond to an infection, resulting from increased production of B cells and T cells. (Chapter 19)

immune system The organ system that produces cells and cell products, such as antibodies, that help remove pathogenic organisms. (Chapter 19)

in vitro fertilization Fertilization that takes place when sperm and egg are combined in glass or a test tube. (Chapter 9)

inbreeding Mating between related individuals. (Chapter 15)

inbreeding depression Negative effect of homozygosity on the fitness of members of a population. (Chapter 15)

incomplete digestive system A digestive system in which there is only one opening that serves as mouth and anus. (Chapter 17)

incomplete dominance A type of inheritance where the heterozygote has a phenotype intermediate to both homozygotes. (Chapter 8)

independent assortment The separation of homologous pairs of chromosomes into gametes independently of one another during meiosis. (Chapters 7, 8)

independent variable A factor whose value influences the value of the dependent variable, but is not influenced by it. In experiments, the variable that is manipulated. (Chapter 1)

indeterminate growth Unrestricted or unlimited growth. (Chapter 23)

indirect effect In ecology, a condition where one species affects another indirectly through it intervening species. (Chapter 15)

induced fit A change in shape of the active site of an enzyme so that it binds tightly to a substrate. (Chapter 4)

inductive reasoning A logical process that argues from specific instances to a general conclusion. (Chapter 1)

infant mortality Death rate of infants and children under the age of 5. (Chapter 14)

infectious Applies to a pathogen that finds a tissue inside the body that will support its growth. (Chapter 19)

infertility The inability to conceive after one year of unprotected intercourse. (Chapter 21)

inflammatory response A line of defense triggered by a pathogen penetrating the skin or mucus membranes. (Chapter 19)

inner cell mass A cluster of cells in the blastocyst that eventually develops into the embryo. (Chapter 21)

inorganic Said of chemical compounds that do not contain carbon. (Chapter 23)

insulin A hormone secreted by the pancreas that lowers blood glucose levels by promoting the uptake of glucose by cells and the storage of glucose as glycogen in the liver. (Chapter 4)

integrated pest management (IPM) A strategy for pest control that relies on a mix of pesticides, cultural control, and better knowledge of pest populations. (Chapter 23)

interferon A chemical messenger produced by virus-infected cells that helps other cells resist infection. (Chapter 19)

intermembrane space The space between two membranes, e.g., the space between the inner and outer mitochondrial membrane. (Chapter 4)

internal fertilization The union of gametes inside the body of the female. (Chapter 21)

interneuron A neuron located between sensory and motor neurons that functions to integrate sensory input and motor output. (Chapter 22)

internode The region of a plant stem between nodes. (Chapter 23)

interphase Part of the cell cycle when a cell is preparing for division and the DNA is duplicated. Consists of G_1, S, and G_2. (Chapter 6)

interspecies hybrid Organism with parents from two different species. (Chapter 12)

intertidal zone The biome that forms on ocean shorelines between the high tide elevation and the low tide elevation. (Chapter 16)

introduced species A nonnative species that was intentionally or unintentionally brought to a new environment by humans. (Chapter 15)

invertebrate Animal without backbones. (Chapter 13)

involuntary muscle Muscle tissue whose action requires no conscious thought. (Chapter 17)

ion electrically charged atom. (Chapter 2)

ionic bond A chemical bond resulting from the attraction of oppositely charged ions. (Chapter 2)

irrigation The technique of supplying additional water to crop plants, typically via flooding or spraying. (Chapter 23)

karyotype The chromosomes of a cell, displayed with chromosomes arranged in homologous pairs and according to size. (Chapter 8)

keystone species A species that has an unusually strong effect on the structure of the community it inhabits. (Chapter 15)

kidney Major organ of the excretory system, responsible for filtering liquid waste from the blood. (Chapters 17, 18)

kingdom In some classifications, the most inclusive group of organisms; usually five or six. In other classification systems, the level below domain on the hierarchy. (Chapters 10, 13)

labia majora Paired thick folds of skin that enclose and protect the labia minor of the vulva. (Chapter 21)

labia minora Paired thin folds of the vulva; enclose the urinary and vaginal openings and clitoris. (Chapter 21)

labor Strong rhythmic contractions that force a developing baby from the uterus through the vagina during childbirth. (Chapter 21)

lactation Production of milk to nurse offspring. (Chapter 21)

lake An aquatic biome that is completely landlocked. (Chapter 16)

laparoscope A thin tubular instrument inserted through an abdominal incision and used to view organs in the pelvic cavity and abdomen. (Chapter 6)

large intestine Colon, portion of the digestive system located between the small intestine and the anus that absorbs water and forms feces. (Chapter 17)

larynx A portion of the upper respiratory tract made up primarily of stiff cartilage. Also known as the voice box. (Chapter 18)

latent Said of the inactive or dormant phase of a disease. (Chapter 19)

latent virus An inactive virus that is integrated into the host genome. (Chapter 19)

leaf The primary photosynthetic organ in plants. (Chapters 23, 24)

legumes A family of plants that produces seeds in pods, like peas and beans. (Chapter 23)

leptin A hormone by fat cells that may be involved in the regulation of appetite. (Chapter 4)

less developed country A term coined by the United Nations to describe a country with low per-capita income and often poor population health and economic prospects. (Chapter 14)

Leydig cell Cells scattered between the seminiferous tubules of the testicles that produce testosterone and other androgens. (Chapter 21)

life cycle Description of the growth and reproduction of an individual. (Chapter 7)

light reaction A series of reactions that occur on thylakoid membranes during photosynthesis and serve to convert energy from the sun into the energy stored in the bonds of ATP and evolve oxygen. (Chapter 5)

linked gene Genes located on the same chromosome. (Chapter 6)

lipid Hydrophobic molecule, including fats, phospholipids, and steroids. (Chapters 2, 3)

liver Organ with many functions, including the production of bile to aid in the absorption of fats. (Chapter 17)

lobule Subdivision of the lobes of the liver. (Chapter 17)

logistic growth Pattern of growth seen in populations that are limited by resources available in the environment. A graph of logistic growth over time typically takes the form of an S-shaped curve. (Chapter 14)

loose connective tissue Connective tissue that serves to bind epithelia to underlying tissues and to hold organs in place. (Chapter 17)

low-density lipoprotein (LDL) Cholesterol-carrying substance in the blood that is high in cholesterol and low in protein. (Chapters 4, 18)

lung The primary organ of the respiratory system; the site where gas exchange occurs. (Chapter 18)

luteinizing hormone (LH) Hormone involved in sperm production, regulation of ovulation, and regulation of menstruation. (Chapters 20, 21)

lymph node Organ located along lymph vessels that filter lymph and help defend against bacteria and viruses. (Chapters 6, 19)

lymphatic system A system of vessels and nodes that return fluid and protein to the blood. (Chapters 6, 19)

lymphocyte White blood cells that make up part of the immune system. (Chapter 19)

lysosome A membrane-bounded sac of hydrolytic enzymes found in the cytoplasm of many cells. (Chapter 2)

macroevolution Large-scale evolutionary change, usually referring to the origin of new species. (Chapter 10)

macromolecule Any of the large molecules including polysaccharides, proteins, and nucleic acids, composed of subunits joined by dehydration synthesis. (Chapters 2, 4)

macronutrient Nutrient required in large quantities. (Chapters 3, 23)

macrophage Phagocytic white blood cell that swells and releases toxins to kill bacteria. (Chapter 18)

malignant Describes a tumor that is cancerous, whether it is invasive or metastatic. (Chapter 6)

mandible Bone of the lower jaw. (Chapter 20)

marine Of, or pertaining to, salt water. (Chapter 16)

mark-recapture method A technique for estimating population size, consisting of capturing and marking a number of individuals, releasing them, and recapturing more individuals to determine what proportion are marked. (Chapter 14)

mass extinction Loss of species that is rapid, global in scale, and affects a wide variety of organisms. (Chapter 15)

matrix (1) In a mitochondrion, the semifluid substance inside the inner mitochondrial membrane, which houses the enzymes of the citric acid cycle. (Chapters 2, 4). (2) In connective tissue, a nonliving substance between cells, ranging from fluid blood plasma to fibrous matrix in tendons to solid bone matrix. (Chapter 17)

mean Average value of a group of measurements. (Chapter 1)

mechanical isolation A form of reproductive isolation between species that depends on the incompatibility of the genitalia of individuals of different species. (Chapter 12)

medulla The center of an organ or gland, such as the kidney or adrenal gland. (Chapter 17)

medulla oblongata Region of the brain stem that is a continuation of the spinal cord and conveys information between the spinal cord and other parts of the brain. (Chapter 22)

meiosis Process that diploid sex cells undergo in order to produce haploid daughter cells. Occurs during gametogenesis. (Chapters 6, 7, 8, 11)

memory cell Cell that is part of a clonal population, programmed to respond to a specific antigen, that helps the body respond quickly if the infectious agent is encountered again. (Chapter 19)

menopause Cessation of menstruation. (Chapters 20, 21)

menstrual cycle Changes that occur in the uterus and depend on intricate interrelationships among the brain, ovaries, and lining of the uterus. (Chapter 21)

menstruation The shedding of the lining of the uterus during the menstrual cycle. (Chapter 21)

meristematic tissue Undifferentiated plant tissue from which new plant cells arise. (Chapter 23)

mesoderm The middle of three germ layers that arise during animal development. (Chapter 21)

messenger RNA (mRNA) Complementary RNA copy of a DNA gene, produced during transcription. The mRNA undergoes translation to synthesize a protein. (Chapters 9, 11)

metabolic rate Measure of an individual's energy use. (Chapter 4)

metabolism All of the physical and chemical reactions that produce and use energy. (Chapters 2, 4, 5)

metaphase Stage of mitosis during which duplicated chromosomes align across the middle of the cell. (Chapter 6)

metastasis When cells from a tumor break away and start new cancers at distant locations. (Chapter 6)

microbe Microscopic organism, especially Bacteria and Archaea. (Chapters 13, 19)

microbiologists Scientists who study microscopic organisms, especially referring to those who study prokaryotes. (Chapter 13)

microevolution Changes that occur in the characteristics of a population. (Chapter 10)

micronutrient Nutrient needed in small quantities. (Chapters 3, 23)

microorganism See microbe. (Chapter 13)

microtubule Protein structure that moves chromosomes around during mitosis and meiosis. (Chapters 2, 6)

microvillus Fine fingerlike projection composed of epithelial cells that function in absorption. (Chapter 17)

micturation Release of urine from the bladder. Also known as urination. (Chapter 18)

midbrain Uppermost region of the brain stem, which adjusts the sensitivity of the eyes to light and of the ears to sound. (Chapter 22)

mineral Inorganic nutrient essential to many cell functions. (Chapter 3)

mitochondria Organelles in which products of the digestive system are converted to ATP. (Chapters 2, 4)

mitosis The division of the nucleus that produces daughter cells that are genetically identical to the parent cell. Also, portion of the cell cycle in which DNA is apportioned into two daughter cells. (Chapter 6)

model organism Any nonhuman organism used in genetic studies to help scientists understand human genes because they share genes with humans. (Chapters 1, 9)

mold A fungal form characterized by rapid, asexual reproduction. (Chapters 10, 13)

molecular clock Principle that DNA mutations accumulate in the genome of a species at a constant rate, permitting estimates of when the common ancestor of two species existed. (Chapter 10)

molecule Two or more atoms held together by covalent bonds. (Chapter 2)

monocot One of the two classes of flowering plants, also called narrow-leaved plants. Monocot seeds contain one leaf (cotyledon). (Chapter 23)

monoculture Practice of planting a single crop over a wide acreage. (Chapter 23)

monosaccharide Simple sugar. (Chapter 2)

monosomy A chromosomal condition in which only one member of a homologous pair is present. (Chapter 6)

monozygotic twins Identical twins that developed from one zygote. (Chapter 7)

morphological species concept Definition of species that relies on differences in physical characteristics among them. (Chapter 12)

morphology Appearance or outward physical characteristics. (Chapter 12)

motor neuron Neuron that carries information away from the brain or spinal cord to muscles or glands. (Chapter 22)

mulching Covering the ground around favored plants with light-blocking material to prevent weed growth. (Chapter 23)

multicellular The condition of being composed of many coordinated cells. (Chapter 13)

multiple allelism A gene for which there are more than 2 alleles segregating in the population. (Chapter 8)

multiple hit model The notion that many different genetic mutations are required for a cancer to develop. (Chapter 6)

multiple sclerosis Chronic, degenerative nervous system disease caused by breakdown of myelin. (Chapter 19)

muscle fiber Single cell that aligns with others in parallel bundles to form muscles. (Chapter 20)

muscle tissue Specialized contractile tissue. (Chapter 17)

mutagen Substance that increases the likelihood of mutation occurring; increases the likelihood of cancer. (Chapter 6)

mutation Change to a DNA sequence that may result in the production of altered proteins. (Chapters 6, 7, 9, 11)

mutualism Interaction between two species that provides benefits to both species. (Chapter 15)

mycologist Scientist who specializes in the study of fungi. (Chapter 13)

myelin sheath Protective layer that coats many axons, formed by supporting cells such as Schwann cells. The myelin sheath increases the speed at which the electrochemical impulse travels down the axon. (Chapter 22)

myofibril Structure found in muscle cells, composed of thin filaments of actin and thick filaments of myosin. (Chapter 20)

myosin A type of protein filament that, along with actin, causes muscle cells to contract. (Chapters 17, 20)

natural experiment Situation where unique circumstances allow a hypothesis test without prior intervention by researchers. (Chapter 7)

natural killer cell A cell that attacks virus-infected cells or tumor cells without being activated by an immune system cell or antibody. (Chapter 19)

natural selection Process by which individuals with certain traits have greater survival and reproduction than individuals who lack these traits, resulting in an increase in the frequency of successful alleles and a decrease in the frequency of unsuccessful ones. (Chapters 10, 11, 12, 21)

negative feedback A mechanism of maintaining homeostasis in which the product of the process inhibits the process. (Chapter 17)

nephron The functional structure within a kidney where waste filtration and urine concentration occurs. (Chapter 18)

nerve Bundle of neurons; nerves branch out from the brain and spinal cord to eyes, ears, internal organs, skin, and bones. (Chapter 22)

nerve impulse Electrochemical signal that controls the activities of muscles, glands, organs, and organ systems. (Chapter 22)

nervous system Brain, spinal cord, sense organs, and nerves that connect organs and link this system with other organ systems. (Chapter 22)

nervous tissue Tissue composed of neurons and associated cells. (Chapters 17, 22)

net primary production (NPP) Amount of solar energy converted to chemical energy by plants, minus the amount of this chemical energy plants need to support themselves. A measure of plant growth, typically over the course of a single year. (Chapter 14)

neurobiologist A scientist that studies the nervous system. (Chapter 22)

neuron Specialized message-carrying cell of the nervous system. (Chapters 17, 22)

neurotransmitter One of many chemicals released by the presynaptic neuron into the synapse, which then diffuse across the synapse and bind to receptors on the membrane of the postsynaptic neuron. (Chapter 22)

neutral mutation A genetic mutation that confers no selective advantage or disadvantage. (Chapters 9, 22)

neutron An electrically neutral particle found in the nucleus of an atom. (Chapter 2)

nutrients Atoms other than carbon, hydrogen, and oxygen that must be obtained from a plant's environment for photosynthesis to occur. (Chapters 3, 23, 24)

nicotinamide adenine dinucleotide Intracellular electron carrier. Oxidized form is NAD+; reduced form is NADH. (Chapter 4)

nicotine The active drug in tobacco that stimulates dopamine receptors in the brain. (Chapter 18)

nitrogen-fixing bacteria Organisms that convert nitrogen gas from the atmosphere into a form that can be taken up by plant roots; some species live in the root nodules of legumes. (Chapters 15, 23)

nitrogenous base Nitrogen-containing base found in DNA: A, C, G, and T and in RNA : U. (Chapters 2, 4, 6)

node Point on a plant stem where a leaf and axillary bud arise. (Chapter 23)

node of Ranvier Small indentation separating segments of the myelin sheath. Nerve impulses "jump" successively from one node of Ranvier to the next. (Chapter 22)

nodule Compartment housing nitrogen-fixing bacteria, produced on the roots of legume plants such as beans and alfalfa. (Chapter 23)

nondisjunction The failure of members of a homologous pair of chromosomes to separate from each other during meiosis. (Chapter 6)

nonpolar Won't dissolve in water. Hydrophobic. (Chapter 2)

nonrenewable resource Resource that is a one-time supply and cannot be easily replaced. (Chapter 14)

nonspecific defenses Defense system against infection that does not distinguish one pathogen from another. Includes the skin, secretions, and mucous membranes. (Chapter 19)

normal distribution Bell-shaped curve, as for the distribution of quantitative traits in a population. (Chapter 7)

notochord A long, flexible rod that runs through the axis of the vertebral body in the future position of the spinal cord. (Chapter 22)

nuclear envelope The double membrane enclosing the nucleus in eukaryotes. (Chapters 2, 6)

nuclear transfer Transfer of a nucleus from one cell to another cell that has had its nucleus removed. (Chapter)

nucleic acids Polymers of nucleotides that comprise DNA and RNA. (Chapters 2, 6)

nucleoid region The region of a prokaryotic cell where the DNA is located. (Chapter 19)

nucleotides Building blocks of nucleic acids that include a sugar, a phosphate, and a nitrogenous base. (Chapters 2, 4, 6, 8, 9)

nucleus Cell structure that houses DNA; found in eukaryotes. (Chapters 2, 6, 9, 11, 23, 24)

nutrient cycling Process by which nutrients become available to plants. Nutrient cycling in a natural environment relies on a healthy community of decomposers within the soil. (Chapter 15)

nutrients Substances that provide nourishment. (Chapter 3)

obesity Condition of having a BMI of 30 or greater. (Chapter 4)

objective Without bias. (Chapter 1)

observation Measurement of nature. (Chapters 1, 11)

occipital lobe The posterior lobe of each cerebral hemisphere, containing the visual center of the brain. (Chapter 22)

ocean A biome consisting of open stretches of salt water. (Chapter 16)

oncogene Mutant version of a cell cycle controlling proto-oncogene. (Chapter 6)

oogenesis Formation and development of female gametes, which occurs in the ovaries and results in the production of egg cells. (Chapter 21)

open circulatory system A circulatory system in which the circulating liquid is released into the body cavity and returned to the system via pores in the vessels. (Chapter 18)

opportunistic infections Disease caused by a pathogen that does not normally cause illness in a healthy host, only when the host is compromised. (Chapter 19)

organ A specialized structure composed of several different types of tissues. (Chapter 17)

organ systems Suites of organs working together to perform a function or functions. (Chapter 17)

organelle Subcellular structure found in the cytoplasm of eukaryotic cells that performs a specific job. (Chapters 2, 4)

organic When pertaining to agriculture, refers to products grown without the use of manufactured pesticides and inorganic fertilizer. (Chapter 23)

organic chemistry The chemistry of carbon-containing substances. (Chapter 2)

organic farming Agricultural technique that requires no manufactured chemical inputs (such as inorganic fertilizer or chemical pesticides), and relies on developing and maintaining the health of soil. (Chapter 22)

osmosis The diffusion of water across a selectively permeable membrane. (Chapter 3)

ossa coxae The paired bones that form the bony pelvis. (Chapter 20)

osteoblast Bone-forming cell responsible for the deposition of collagen. (Chapter 20)

osteoclast Bone-reabsorbing cell that liberates calcium. (Chapter 20)

osteocyte Highly branched cell found in bone. (Chapter 17)

osteoporosis A condition resulting in an elevated risk of bone breakage from weakened bones. (Chapter 3)

ovary (1) In animals, the paired abdominal structures that produce egg cells and secrete female hormones. (Chapters 6, 20, 21) (2) In plants, a chamber of the carpel containing the ovules. (Chapter 23)

overexploitation Threat to biodiversity caused by humans that encompasses overhunting and overharvesting. (Chapter 15)

oviduct Egg-carrying duct that brings egg cells from ovaries to uterus. (Chapter 21)

ovulation Release of an egg cell from the ovary. (Chapter 21)

ovule Structure in flowering plants consisting of eggs and accessory tissues. After fertilization, will develop into a seed. (Chapter 23)

ovum An egg; the female gamete. Cell produced during oogenesis that receives the majority of the cytoplasmic nutrients and organelles. (Chapter 21)

oxytocin Pituitary hormone that stimulates the contraction of smooth muscle of the uterus during labor and facilitates secretion of milk from the breast during nursing. (Chapter 21)

pacemaker A patch of heart tissue or an implanted device that produces a regular electrical signal, setting the rhythm for the cardiac cycle. (Chapter 18)

paleontologist Scientist who searches for, describes, and studies ancient organisms. (Chapter 9)

pancreas Gland that secretes digestive enzymes and insulin. (Chapters 17, 20)

parasite An organism that benefits from an association with another organism that is harmed by the association. (Chapters 15, 19)

parathyroid gland One of four endocrine glands located on the thyroid that secrete parathyroid hormone in order to regulate blood calcium levels. (Chapter 20)

parietal lobe Part of the brain that processes information about touch and is involved in self-awareness.(Chapter 22)

Parkinson's disease Disease that results in tremors, rigidity, and slowed movements. May be due to faulty dopamine production. (Chapter 22)

particulate Tiny airborne particle found in smoke and other pollutants. (Chapter 18)

passive immunity Immunity acquired when antibodies are passed from one individual to another, as from mother to child during breast-feeding. (Chapter 19)

passive smoker An individual who inhales environmental tobacco smoke but who does not actively consume tobacco. (Chapter 18)

passive transport The diffusion of substances across a membrane with their concentration gradient and not requiring an input of ATP. (Chapter 3)

pathogen Disease-causing organism. (Chapter 19)

pedigree Family tree that follows the inheritance of a genetic trait for many generations. (Chapter 8)

peer review The process by which reports of scientific research are examined and critiqued by other researchers before they are published in scholarly journals. (Chapter 1)

penis The copulatory structure in males. (Chapter 21)

peptide bond Covalent bond that joins the amino group and carboxyl group of adjacent amino acids. (Chapter 2)

perennial plant Plant that lives for many years. (Chapters 16, 23, 24)

peripheral nervous system (PNS) Network of nerves outside the brain and spinal cord that links the CNS with sense organs. (Chapter 22)

peristalsis Rhythmic muscle contractions that move food through the digestive system. (Chapter 17)

permafrost Permanently frozen soil. (Chapter 16)

pest Any organism that competes with humans for agricultural production or other resources. (Chapter 23)

pesticide Chemical that kills or disables agricultural pests. (Chapter 23)

pesticide treadmill The tendency for pesticide toxicity and total applications to increase once a farmer begins to employ pesticides. (Chapter 23)

petal A flower part that typically functions to attract pollinators. (Chapter 23)

petiole The stalk of a leaf. (Chapter 23)

pH A logarithmic measure of the hydrogen ion concentration ranging from 0–14. Lower numbers indicate higher hydrogen ion concentrations. (Chapter 2)

phagocytosis Ingestion of food or pathogens by cells. (Chapter 19)

pharynx Tube and muscles connecting the mouth to the esophagus; throat. (Chapter 17)

phenotype Physical and physiological traits of an individual. (Chapters 6, 7, 8)

phenotypic ratio Proportion of individuals produced by a genetic cross who possess each of the various phenotypes that cross can generate. (Chapter 7)

phloem The portion of a plant's vascular tissue modified for multidirectional transport of nutrients throughout the body of the plant. (Chapter 23)

phloem sap The liquid, often containing dissolved carbohydrates, carried in the phloem tubes in a plant. (Chapter 24)

phospholipid One of three types of lipids, phospholipids are components of cell membranes. (Chapter 2)

phospholipid bilayer The membrane that surrounds cells and organelles and is composed of two layers of phospholipids. (Chapter 2)

phosphorylation To introduce a phosphoryl group into an organic compound. (Chapter 4)

photoperiodism Response to the duration and timing of day and night length in plants. (Chapter 24)

photorespiration A series of reactions triggered by the closing of stomatal openings to prevent water loss. (Chapter 5)

photosynthesis Process by which plants, along with algae and some bacteria, transform light energy to chemical energy. (Chapter 5)

phototropism Growth in response to directional light. (Chapter 24)

phyla (singular: phylum) The taxonomic category below kingdom and above class. (Chapters 10, 13)

phylogeny Evolutionary history of a group of organisms. (Chapter 13)

phytochrome The light-sensitive chemical in plants responsible for photoperiodic responses. (Chapter 24)

pituitary gland Small gland attached by a stalk to the base of the brain that secretes growth hormone, reproductive hormones, and other hormones. (Chapters 9, 20)

pivot joint A type of joint that allows freedom of movement. (Chapter 20)

placebo Sham treatments in experiments. (Chapter 1)

placenta Membrane produced by a developing fetus that releases a hormone to extend the life of the corpus luteum. (Chapter 21)

plant hormone Substance in plants that helps tune the organisms' response to the environment. (Chapter 24)

Plantae Multicellular photosynthetic eukaryotes, excluding algae. (Chapter 13)

plasma The liquid portion of blood. (Chapter 18)

plasma cell Cell produced by a clonal population that secretes antibodies specific to an antigen. (Chapter 19)

plasma membrane Structure that encloses a cell, defining the cell's outer boundary. (Chapters 2, 3, 19)

plasmid Circular piece of bacterial DNA that normally exists separate from the bacterial chromosome and can make copies of itself. (Chapter 9)

platelet Cell in the blood that carries constituents required for the clotting response. (Chapter 18)

pleiotropy The ability of one gene to affect many different functions. (Chapter 8)

polar Describes a molecule with regions having different charges; capable of ionizing. (Chapter 2)

polar body Smaller of the two cells produced during oogenesis; it therefore does not have enough nutrients to undergo further development. (Chapter 21)

polarization The difference in charge between the inside and outside of the resting neuron cell. (Chapter 22)

poles Opposite ends of a sphere, such as of a cell (Chapter 6) or of a planet such as Earth (Chapter 16).

pollen The male gametophyte of seed plants. (Chapters 12, 13, 15, 23)

pollination The process by which pollen is transferred to the female structures of a plant. (Chapter 23)

pollinator An organism that transfers sperm (pollen grains) from one flower to the female reproductive structures of another flower. (Chapter 15)

pollution Human-caused threat to biodiversity involving the release of poisons, excess nutrients, and other wastes into the environment. (Chapter 15)

polyculture Practice of planting many different crop plants over a single farm's acreage. (Chapter 23)

polygenic trait A trait influenced by many genes.(Chapters 7, 8)

polymer General term for a macromolecule composed of many chemically bonded monomers. (Chapter 2)

polymerase An enzyme that catalyzes phosphodiester bond formation between nucleotides. (Chapter 8)

polymerase chain reaction (PCR) A laboratory technique that allows the production of many identical DNA molecules. (Chapter 8)

polyploidy A chromosomal condition involving more than two sets of chromosomes. (Chapter 12)

polysaccharide A carbohydrate composed of three or more monosaccharides. (Chapter 2)

polysaturated Containing several double bonds between carbon atoms. (Chapter 3)

polyunsaturated Relating to fats consisting of carbon chains with many double bonds unsaturated by hydrogen atoms. (Chapter 3)

pond An aquatic biome that is completely landlocked. (Chapter 16)

pons A structure located on the brain stem, between the brain and spinal cord. (Chapter 22)

population Subgroup of a species that is somewhat independent from other groups. (Chapters 12, 14)

population bottleneck Dramatic but short-lived reduction in population size followed by an increase in population. (Chapter 12)

population crash Steep decline in number that may occur when a population grows larger than the carrying capacity of its environment. (Chapter 14)

population cycle In some populations, the tendency to increase in number above the environment's carrying capacity, resulting in a crash, following by an overshoot of the carrying capacity and another crash, continuing indefinitely. (Chapter 14)

population genetics Study of the factors in a population that determine allele frequencies and their change over time. (Chapter 12)

population pyramid A visual representation of the number of individuals in different age categories in a population. (Chapter 14)

positive feedback A relatively uncommon homeostatic mechanism in which the product of a process intensifies the process, thereby promoting change. (Chapter 17)

postsynaptic neuron The neuron that responds to neurotransmitter released from the presynaptic neuron. (Chapter 22)

prairie A grassland biome. (Chapter 16)

precipitation When water vapor in the atmosphere turns to liquid or solid form and falls to Earth's surface. (Chapter 16)

predation Act of capturing and consuming an individual of another species. (Chapter 15)

predator Organism that eats other organisms. (Chapter 15)

prediction Result expected from a particular test of a hypothesis if the hypothesis were true. (Chapter 1)

pressure flow mechanism The process by which phloem sap moves from a sugar source to a sugar sink within a plant. (Chapter 24)

presynaptic The neuron that secretes neurotransmitter into a synapse, transmitting a signal. (Chapter 22)

primary consumer Organism that eats plants. (Chapter 15)

primary growth Growth occurring at the tips of a plant, originating in the apical meristems. (Chapter 23)

primary source Article reporting research results, written by researchers, and reviewed by the scientific community. (Chapter 1)

primary spermatocyte Diploid cell that begins meiosis in the production of sperm. (Chapter 21)

primers Short nucleotide sequences used to help initiate replication of nucleic acids. (Chapter 8)

probability Likelihood that something is the case or will happen. (Chapter 1)

probe Single-stranded nucleic acid that has been radioactively labeled. (Chapter 8)

processed food Food that has been modified from its original form to increase shelf life, transportability, or the like. (Chapter 3)

producer Organism that produces carbohydrates from inorganic carbon, typically via photosynthesis. (Chapter 15)

product The modified chemical that results from a chemical or enzymatic reaction. (Chapter 2)

progesterone Ovarian hormone. High levels have a negative feedback effect on the hypothalamus, causing GnRH secretion to decrease. (Chapter 21)

prokaryote Type of cell that does not have a nucleus or membrane-bounded organelles. (Chapters 2, 9, 13)

prolactin A hormone produced by the pituitary gland that stimulates the development of mammary glands.(Chapter 21)

promoter Sequence of nucleotides to which the polymerase binds to start transcription. (Chapter 9)

prophase Stage of mitosis during which duplicated chromosomes condense. (Chapter 6)

prostate A gland in human males that secretes an acid-neutralizing fluid into semen. (Chapter 21)

protein Cellular constituent made of amino acids coded for by genes. Proteins can have structural, transport, or enzymatic roles. (Chapters 2, 3, 4, 9, 11, 15)

protein synthesis Joining amino acids together, in an order dictated by a gene, to produce a protein. (Chapter 9)

Protista Kingdom in the domain Eukarya containing a diversity of eukaryotic organisms, most of which are unicellular. (Chapter 13)

proton A positively charged subatomic particle. (Chapters 2, 5)

proto-oncogenes Genes that encode proteins that regulate the cell cycle. Mutated proto-oncogenes (oncogenes) can lead to cancer. (Chapter 6)

prune To trim back the growing tips of a plant. (Chapter 24)

pseudopodia A cellular extension of amoebas used for feeding and in motility. (Chapter 19)

puberty The point in human development when male and female hormones are triggered in the body. Males produce sperm at this time, and females begin the menstrual cycle. (Chapter 20)

pulmonary circuit The path of blood through vessels from the heart, through the lungs, and back to the heart; one half of the double circulation. (Chapter 18)

pulse The volume of blood that passes into the arteries as a result of the heart's contraction. (Chapter 18)

punctuated equilibrium The hypothesis that evolutionary changes occur rapidly and in short bursts, followed by long periods of little change. (Chapter 12)

Punnett square Table that lists the different kinds of sperm or eggs parents can produce relative to the gene or genes in question and predicts the possible outcomes of a cross between these parents. (Chapter 7)

purine Nitrogenous base (A or G) with a two-ring structure. (Chapter 2)

pyrimidine Nitrogenous base (C, T, or U) with a single-ring structure. (Chapter 2)

pyruvic acid The 3-carbon molecule produced by glycolysis. (Chapter 4)

Q angle Angle of the femur in relation to a horizontal line drawn through the kneecap. (Chapter 20)

qualitative trait Trait that produces phenotypes in distinct categories. (Chapter 7)

quantitative trait Trait that has many possible values. (Chapter 7)

race See biological race. (Chapter 12)

racism Idea that some groups of people are naturally superior to others. (Chapter 12)

radiation therapy Focusing beams of reactive particles at a tumor to kill the dividing cells. (Chapter 6)

radioactive decay Natural, spontaneous breakdown of radioactive elements into different elements, or "daughter products." (Chapter 10)

radioimmunotherapy Experimental cancer treatment with the goal of delivering radioactive substances directly to tumors without affecting other tissues. (Chapter 6)

radiometric dating Technique that relies on radioactive decay to estimate a fossil's age. (Chapters 10, 15)

random alignment When members of a homologous pair line up randomly with respect to maternal or paternal origin during metaphase I of meiosis, thus increasing the genetic diversity of offspring. (Chapters 6, 7)

random assignment Placing individuals into experimental and control groups randomly to eliminate systematic differences between the groups. (Chapter 1)

random distribution The dispersion of individuals in a population without pattern. (Chapter 14)

random fertilization The unpredictability of exactly which gametes will fuse during the process of sexual reproduction. (Chapter 7)

reabsorption Reuptake of water and other essential substances by a nephron from the filtrate initially squeezed into the kidney. (Chapter 18)

reactant Any starting material in a chemical reaction. (Chapter 2)

reading frame The grouping of mRNAs into 3 base codons for translation. (Chapter 9)

receptor (1) Protein on the surface of a cell that recognizes and binds to a specific chemical signal. (Chapters 6, 10) (2) See sensory receptor. (Chapter 22)

recessive Applies to an allele with an effect that is not visible in a heterozygote. (Chapter 7)

recombinant Produced by manipulating a DNA sequence. (Chapter 9)

recombinant bovine growth hormone (rBGH) Growth hormone produced in a laboratory and injected into cows to increase their size and ability to produce milk. (Chapter 9)

red blood cell Primary cellular component of blood, responsible for ferrying oxygen throughout the body. (Chapters 6, 18)

reflex Automatic response to a stimulus. (Chapter 22)

reflex arc Nerve pathway followed during a reflex consisting of a sensory receptor, a sensory neuron, an interneuron, a motor neuron, and an effector. (Chapter 22)

remission The period during which the symptoms of a disease subside. (Chapter 6)

renal artery The vessel that carries blood to the kidney for filtering. (Chapter 18)

renal vein The vessel that carries filtered blood from the kidney. (Chapter 18)

repolarization The restoration of a charge difference across a membrane. (Chapter 22)

repressor A protein that suppresses the expression of a gene. (Chapter 9)

reproductive cloning Transferring the nucleus from a donor adult cell to an egg cell without a nucleus in order to clone the adult. (Chapter 9)

reproductive isolation Prevention of gene flow between different biological species due to failure to produce fertile offspring; can include premating barriers and postmating barriers. (Chapter 12)

reproductive organ Internal and external genitalia involved in production and delivery of gametes. (Chapters 21, 23)

respiratory surface Body surface across which gas exchange occurs. (Chapter 18)

respiratory system The organ system involved in gas exchange between an animal and its environment. In humans, the lungs and air passages. (Chapter 18)

restriction enzyme An enzyme that cleaves DNA at specific nucleotide sequences. (Chapters 7, 9)

reticular formation Extensive network of neurons that runs through the medulla of the brain and projects toward the cerebral cortex. It functions as a filter for sensory input and activates the cerebral cortex. (Chapter 22)

reuptake In neurons, the process by which neurotransmitters are reabsorbed by the neuron that secreted them. (Chapter 22)

reverse transcriptase Enzyme in RNA viruses that produces DNA by transcription of viral RNA. (Chapter 19)

Rh factor Surface molecule found on some red blood cells. (Chapter 8)

rhizomes Underground stems in plants. (Chapter 23)

ribose The five-carbon sugar in RNA. (Chapter 8)

ribosomal RNA (rRNA) RNA that makes up part of the structure of ribosomes. (Chapter 2)

ribosome Subcellular structure that helps translate genetic material into proteins by anchoring and exposing small sequences of mRNA. (Chapters 2, 9, 13)

risk factor Any exposure or behavior that increases the likelihood of disease. (Chapter 6)

Ritalin Stimulant used to treat ADD. (Chapter 22)

river Aquatic biome characterized by flowing water. (Chapter 16)

RNA (ribonucleic acid) Information-carrying molecule composed of nucleotides. (Chapters 2, 9)

RNA polymerase Enzyme that synthesizes mRNA from a DNA template during transcription. (Chapter 9)

root cap Loosely organized cells that cover the growing tip of a root and are continually shed as the root pushes through the soil. (Chapter 23)

root hair Extension of an epidermal cell on young roots, which maximize the surface area for water and nutrient uptake. (Chapter 23)

root system The plant organ system responsible for water and nutrient uptake and anchorage. (Chapter 23)

rough endoplasmic reticulum Ribosome-studded subcellular membranes found in the cytoplasm and responsible for some protein synthesis. (Chapter 2)

RuBisCo Abbreviation for ribulose bisphosphate carboxylase oxygenase, the enzyme that catalyzes the first step in the Calvin cycle of photosynthesis. (Chapter 5)

runoff Water moving across a land surface, picking up contaminants and eventually flowing into lakes, streams, or groundwater. (Chapter 23)

salinization Degradation of soil by mineral salts deposited as a result of irrigation. (Chapter 23)

salivary amylase An enzyme secreted by salivary glands of the mouth to break down starch. (Chapter 17)

salt A charged substance that ionizes in solution. (Chapter 2)

sample Small subgroup of a population used in an experimental test. (Chapter 1)

sample size Number of individuals in both the experimental and control groups. (Chapter 1)

sampling error Effect of chance on experimental results. (Chapter 1)

sarcomere Any of the repeating units in a myofibril of striated muscle, bounded by Z discs. (Chapter 20)

saturated fat Type of lipid rich in single bonds. Found in butter and other fats that are solids at room temperature. This type of fat is associated with higher blood cholesterol levels. (Chapter 3)

savanna Grassland biome containing scattered trees. (Chapter 16)

Schwann cell Cells that form the myelin sheath along the axons of nerve cells in the peripheral nervous system. (Chapter 22)

scientific method A systematic method of research consisting of putting a hypothesis to a test designed to disprove it, if it is in fact false. (Chapter 1)

scientific theory Body of scientifically accepted general principles that explain natural phenomena. (Chapters 1, 10)

scrotum The pouch of skin that houses the testes. (Chapter 21)

secondary compounds Chemicals produced by plants and some other organisms as side reactions to normal metabolic pathways and that typically have an antipredator or antibiotic function. (Chapter 13)

secondary consumers Animals that eat primary consumers; predators. (Chapter 15)

secondary growth Growth in girth of plants, as a result of cell division in lateral meristems. (Chapter 23)

secondary sources Books, news media, and advertisements as sources of scientific information. (Chapter 1)

secondary spermatocyte Haploid cell produced by the first meiotic division in the production of sperm. (Chapter 21)

secretion A step in waste removal in the kidney, in which contaminants in low concentration in the blood are actively absorbed into the urine. (Chapter 18)

seed A plant embryo packaged with a food source and surrounded by a seed coat. (Chapters 13, 23)

seed coat Layer of tissue surrounding a plant embryo and endosperm, arising from accessory tissue in the ovule. (Chapter 23)

seedling A young plant that develops from a germinating seed. (Chapter 23)

segregation Separation of pairs of alleles during the production of gametes. Results in a 50% probability that a given gamete contains one allele rather than the other. (Chapter 7)

semen Sperm and energy-rich associated fluids. (Chapter 21)

semilunar valve Heart valve controlling blood flow from the ventricles into blood vessels leading away from the heart. (Chapter 18)

seminal vesicle Either of two pouchlike glands located on both sides of the bladder that add a fructose-rich fluid to semen prior to ejaculation. (Chapter 21)

seminiferous tubule Highly coiled tube in the testicles where sperm are formed. (Chapter 21)

semipermeable In biological membranes, a membrane that allows some substances to pass but prohibits the passage of others. (Chapter 2)

sensory neuron A neuron that conducts impulses from a sense organ to the central nervous system. (Chapter 22)

sensory receptor Any of the cellular systems that collect information about the environment inside or outside the body and transmit that information to the brain. (Chapter 22)

sepal The outermost floral structure, usually enclosing the other flower parts in a bud. (Chapter 23)

Sertoli cell Testicular cell that secretes substances that aid in the development of mature sperm. (Chapter 21)

severe combined immunodeficiency (SCID) Illness caused by a genetic mutation that results in the absence of an enzyme and a severely weakened immune system. (Chapter 9)

sex chromosome Any of the sex-determining chromosomes (X and Y in humans). (Chapters 6, 7)

sex determination Determining the biological sex of an offspring. Humans have a chromosomal mechanism of sex determination in which two X chromosomes produce a female and an X and a Y chromosome produce a male. (Chapter 8)

sex hormone Any of the steroid hormones that affect development and functions of reproductive structures and secondary sex characteristics. (Chapter 20)

sex-linked gene Any of the genes found on the X or Y sex chromosomes. (Chapter 8)

sexual reproduction Reproduction involving two parents that gives rise to offspring that have unique combinations of genes. (Chapters 6, 21)

sexual selection Form of natural selection that occurs when a trait influences the likelihood of mating. (Chapter 12)

shoot system The organs of a plant that are modified for photosynthesis: stem and leaf. (Chapter 23)

signal transduction When a change in a cell or its environment is relayed through various molecules and results in a cellular response. (Chapter 20)

single nucleotide polymorphism A DNA sequence variation that occurs when members of species differ from each other at a single nucleotide (A, T, C, or G) locus. (Chapter 12)

sink A plant organ that is using carbohydrate, either for growth and metabolism or for conversion into starch. (Chapter 24)

sinoatrial (SA) node Region of the heart muscle that generates an electrical signal that controls heart rate. See pacemaker. (Chapter 18)

sister chromatid Either of the two duplicated, identical copies of a chromosome formed after DNA synthesis. (Chapter 6)

skeletal muscle Striated muscle involved with voluntary movements. (Chapter 17)

sliding filament model The theory that muscles contract when actin filaments slide across myosin filaments, shortening the sarcomere. (Chapter 20)

small intestine The narrow, twisting, upper part of the intestine where nutrients are absorbed into the blood. (Chapter 17)

smog Products of fossil fuel combustion in combination with sunlight, producing a brownish haze in still air. (Chapter 16)

smooth endoplasmic reticulum The subcellular, cytoplasmic membrane system responsible for lipid and steroid biosynthesis. (Chapter 2)

smooth muscle Nonstriated, spindle-shaped muscle cells that line organs and blood vessels. (Chapter 17)

sodium potassium pump A protein pump in a cell membrane that moves sodium out of the cell and potassium into the cell, both against their concentration gradients. (Chapter 22)

soil erosion Loss of topsoil. (Chapter 23)

soil Medium for plant growth made up of mineral particles, partially decayed organic matter, and living organisms. (Chapter 23)

solar irradiance The amount of solar energy hitting Earth's surface at any given point. (Chapter 16)

solid waste Garbage. (Chapter 16)

solstice When the sun reaches its maximum and minimum elevation in the sky. (Chapter 16)

solute The substance that is dissolved in a solution. (Chapter 2)

solution A mixture of two or more substances. (Chapter 2)

solvent A substance, such as water, that a solute is dissolved in to make a solution. (Chapter 2)

somatic cell Any of the body cells in an organism. Any cell that is not a gamete. (Chapter 6)

somatic cell gene therapy Changes to malfunctioning genes in somatic or body cells. These changes will not be passed to offspring. (Chapter 9)

somatomammotropin A hormone produced by the placenta that plays a role in stimulating milk production. (Chapter 21)

source Plant organs that are actively generating sugar. (Chapter 24)

spatial isolation A mechanism for reproductive isolation that depends on the geographic separation of populations. (Chapter 12)

special creation The hypothesis that all organisms on Earth arose as a result of the actions of a supernatural creator. (Chapter 10)

special senses Senses that have specialized organs. These include sight, hearing, equilibrium, taste, touch, and smell. (Chapter 22)

speciation Evolution of one or more species from an ancestral form: macroevolution. (Chapter 12)

species A group of individuals that regularly breed together and are generally distinct from other species in appearance or behavior. (Chapters 2, 10, 12, 13, 15)

species-area curve Graph describing the relationship between the size of a natural landscape and the relative number of species it contains. (Chapter 15)

specific defense Defense against pathogens that utilizes white blood cells of the immune system. (Chapter 19)

specificity Phenomenon of enzyme shape determining the reaction the enzyme catalyzes. (Chapters 4, 19)

sperm Gametes produced by males. (Chapters 6, 7, 8, 13, 20, 21)

spermatid Cell produced when secondary spermatocytes undergo meiosis II to produce haploid cells that no longer have duplicated chromosomes. (Chapter 21)

spermatogenesis Production of sperm. (Chapter 21)

spermatozoa Mature sperm composed of a small head containing DNA, a midpiece that contains mitochondria, and a tail (flagellum). (Chapter 21)

spinal cord Thick cord of nervous tissue that extends from the base of the brain through the spinal column. (Chapters 17, 22)

spongy bone The porous, honeycomblike material of the inner bone. (Chapter 20)

spore Reproductive cell in plants and fungi that is capable of developing into an adult without fusing with another cell. (Chapters 13, 23)

sporophyte Diploid stage in the alternation of generations in plants, which produces spores via meiosis. (Chapter 23)

stabilizing selection Natural selection that favors the average phenotype and selects against the extremes in the population. (Chapter 11)

stamen Male flower part, producing the pollen. (Chapter 23)

standard error A measure of the variance of a sample; essentially the average distance a single data point is from the mean value for the sample. (Chapter 1)

staple crop One of a number of agricultural crops that make up the majority of calories in human societies. (Chapter 23)

statistical test Mathematical formulation that helps scientists evaluate whether the results of a single experiment demonstrate the effect of treatment. (Chapter 1)

statistically significant Said of results for which there is a low probability that experimental groups differ simply by chance. (Chapter 1)

statistics Specialized branch of mathematics used in the evaluation of experimental data. (Chapter 1)

stem The part of vascular plants that is above ground, as well as similar structures found underground. (Chapter 23)

stem cell Cells that can divide indefinitely and can differentiate into other cell types. (Chapters 9, 19)

steppe Biome characterized by short grasses, found in regions with relatively little annual precipitation. (Chapter 16)

steroid naturally occurring or synthetic organic fat-soluble substance that produces physiologic effects. (Chapters 2, 20)

stigma Sticky pad on the tip of a flower's carpel where pollen is deposited. (Chapter 23)

stomach a sac-like enlargement of the alimentary canal involved in digestion. (Chapter 17)

stomata Pores on the photosynthetic surfaces of plants that allow air into the internal structure of leaves and green stems. Stomata also provide portals through which water can escape. (Chapters 5, 23, 24)

stop codon An mRNA codon that does not code for an amino acid and causes the amino acid chain to be released into the cytoplasm. (Chapter 9)

stream Biome characterized by flowing water, sometimes seasonal. Typically smaller than rivers. (Chapter 16)

striated muscle A voluntary muscle made up of elongated, multinucleated fibers. Typically skeletal and cardiac muscle that is distinguished from smooth muscle by the in-register banding patterns of actin and myosin filaments. (Chapter 17)

stroke Acute condition caused by a blood clot that blocks blood flow to an organ or other region of the body. (Chapters 4, 18)

stroma The semi-fluid matrix inside a chloroplast where the Calvin cycle of photosynthesis occurs. (Chapters 2, 5)

style Stalk connecting stigma to ovary in a flower's carpel. (Chapter 23)

subspecies Subdivision of a species that is not reproductively isolated but represents a population or set of populations with a unique evolutionary history. See also biological race, variety. (Chapter 12)

substrate The substance upon which an enzyme reacts. (Chapter 4)

succession Replacement of ecological communities over time since a disturbance, until finally reaching a stable state. (Chapter 16)

sugar-phosphate backbone Series of alternating sugars and phosphates along the length of the DNA helix. (Chapters 2, 5, 6)

supernatural Not constrained by the laws of nature. (Chapter 1)

sustainable Referring to human activities that are able to continue indefinitely with no degradation of the resources on which they depend. (Chapter 23)

symbiosis A relationship between two species. (Chapter 13)

sympatric In the same geographic region. (Chapter 12)

synapse Gap between neurons consisting of the terminal boutons of the presynaptic neuron, the space between the two adjacent neurons, and the membrane of the postsynaptic neuron. (Chapter 22)

systematist Biologist who specializes in describing and categorizing a particular group of organisms. (Chapter 13)

systemic circuit Flow of blood from the heart to body capillaries and back to the heart. (Chapter 18)

systole Portion of the cardiac cycle when the heart is contracting, forcing blood into arteries. (Chapter 18)

systolic blood pressure Force of blood on artery walls when heart is contracting. (Chapters 4, 18)

T lymphocyte (T cell) Immune-system cell that develops in the thymus gland it facilitates cell-mediated immunity. (Chapter 19)

Taq polymerase An enzyme that can withstand high temperatures, which is used during polymerase chain reactions. (Chapter 8)

target cell A cell that responds to regulatory signals such as hormones. (Chapter 20)

telomerase An enzyme that helps prevent the degradation of the tips of chromosomes, active during development and sometimes reactivated in cancer cells. (Chapter 6)

telophase Stage of mitosis during which the nuclear envelope forms around the newly produced daughter nucleus, and chromosomes decondense. (Chapter 6)

temperate forest Biome dominated by deciduous trees. (Chapter 16)

temperature A measure of the intensity of heat or kinetic energy. (Chapters 4, 16)

template Something that serves as a pattern or mold. (Chapter 8)

temporal bone Either of the paired bones forming the sides and base of the cranium. (Chapter 20)

temporal isolation Reproductive isolation between populations maintained by differences in the timing of mating or emergence. (Chapter 12)

temporal lobe Part of the cerebral hemisphere that processes auditory and visual information, memory, and emotion. Located in front of the occipital lobe. (Chapter 22)

tension In plants, negative water pressure. (Chapter 24)

terminal bouton Knoblike structure at the end of an axon. (Chapter 22)

terminal bud Apical meristem on the tip of a stem or branch. (Chapter 23)

testable Possible to evaluate through observations of the measurable universe. (Chapter 1)

testicle See testis. (Chapters 20, 21)

testis (*plural*: testes) Either of the paired male gonads, involved in gametogenesis and secretion of reproductive hormones. (Chapters 20, 21)

testosterone Masculinizing hormone secreted by the testes. (Chapters 20, 21)

thalamus Main relay center between the spinal cord and the cerebrum. (Chapter 22)

theory See scientific theory. (Chapters 1, 10)

theory of evolution Theory that all organisms on Earth today are descendants of a single ancestor that arose in the distant past. See also evolution. (Chapters 1, 2, 10)

therapeutic cloning Using early embryos as donors of stem cells for the replacement of damaged tissues and organs in another individual. (Chapter 9)

thermoregulation The ability to maintain a body temperatures within a narrow range. (Chapter 17)

thickening meristem Dividing tissue near the apical meristem in monocots that increases stem girth. (Chapter 23)

thigmotropism Growth in response to contact with a solid object. (Chapter 24)

thylakoid Flattened membranous sac located in the chloroplast stroma. Function in photosynthesis. (Chapters 2, 5)

thymine Nitrogenous base in DNA, a pyrimidine. (Chapters 2, 8)

thymus An endocrine gland located in the neck region that helps establish the immune system. (Chapters 19, 20)

thyroid gland An endocrine gland in the neck region that stimulates metabolism and regulates blood calcium levels. (Chapter 20)

Ti plasmid Tumor-inducing plasmid used to genetically modify crop plants. (Chapter 9)

tilling Turning over the soil to kill weed seedlings, with the goal of removing competitors before a crop is planted. (Chapter 23)

tissue A group of cells with a common function. (Chapter 17)

tongue A muscular structure in the mouth that has taste buds that help you taste food. It aids in breaking down food for the digestive process. (Chapter 17)

totipotent Describes a cell able to specialize into any cell type of its species, including embryonic membrane. Compare with multipotent, pluripotent. (Chapter 9)

trachea Air passage from upper respiratory system into lower respiratory system. Also called windpipe. (Chapters 17, 18)

tracheid Narrow xylem cell with pitted walls. (Chapter 24)

trans fat Contains unsaturated fatty acids that have been hydrogenated, which changes the fat from a liquid to a solid at room temperature. (Chapter 3)

transcription Production of an RNA copy of the protein coding DNA gene sequence. (Chapters 9, 11, 19)

transfer RNA (tRNA) Amino-acid-carrying RNA structure with an anti-codon that binds to an mRNA codon. (Chapter 9)

transgenic organism Organism whose genome incorporates genes from another organism; also called genetically modified organism (GMO). (Chapter 9)

translation Process by which an mRNA sequence is used to produce a protein. (Chapters 9, 11, 19)

translocation Movement of phloem sap around the body of a plant. (Chapter 24)

transpiration Movement of water from the roots to the leaves of a plant, powered by evaporation of water at the leaves and the cohesive and adhesive properties of water. (Chapters 5, 24)

trisomy A chromosomal condition in which three copies of a chromosome exist instead of the two copies of a chromosome normally present in a diploid organism. (Chapter 6)

trophic level Feeding level or position on a food chain; e.g. producers, primary consumers, etc. (Chapter 15)

trophic pyramid Relationship among the mass of populations at each level of a food web. (Chapter 15)

trophoblast The outer layer of a developing blastocyst that supplies nutrition to the embryo. (Chapter 21)

tropical forest Biome dominated by broad-leaved, evergreen trees; found in areas where temperatures never drop below the freezing point of water. (Chapter 16)

tropism In plants, directional growth. (Chapter 24)

tuber Underground storage structure in plants, made up of modified stem. (Chapter 23)

tuberculosis (TB) Degenerative lung disease caused by infection with the bacterium *Mycobacterium tuberculosis*. (Chapter 11)

tumor Mass of tissue that has no apparent function in the body. (Chapter 6)

tumor suppressor Cellular protein that stops tumor formation by suppressing cell division. When mutated leads to increased likelihood of cancer. (Chapter 6)

tundra Biome that forms under very low temperature conditions. Characterized by low-growing plants. (Chapter 16)

turgid In plants, cells that are filled with enough water so that the cell walls are deformed. (Chapter 23)

type I diabetes Diabetes that results from inability to produce insulin. (Chapter 4)

understory The level of a forest below the canopy trees and shrub cover, typically consisting of herbaceous perennial plants and tree and shrub seedlings. (Chapter 16)

undifferentiated A cell that is not specialized. (Chapter 9)

unicellular Made up of a single cell. (Chapter 13)

uniform distribution Occurs when individuals in a population are disbursed in a uniform manner across a habitat. (Chapter 14)

unsaturated fat Type of lipid containing many carbon-to-carbon double bonds; liquid at room temperature. (Chapter 3)

unsustainable Relating to practices that may compromise the ability of future generations to support a large human population with an adequate quality of life. (Chapter 23)

uracil Nitrogenous base in RNA, a pyrimidine. (Chapters 2, 9)

urban sprawl The tendency for the boundaries of urban areas to grow over time as people build housing and commercial districts farther and farther from an urban core. (Chapter 16)

ureter Tube that delivers urine from a kidney to the bladder. (Chapter 18)

urethra Urine-carrying duct that also carries sperm in males. (Chapters 18, 21)

urinary system The organ system responsible for the filtering, collection, and excretion of liquid waste; consisting of the kidneys, ureters, bladder, and urethra. (Chapter 18)

urine Liquid expressed by the kidneys and expelled from the bladder in mammals, containing the soluble waste products of metabolism. (Chapter 18)

uterus Pear-shaped muscular organ in females that can support pregnancy and that undergoes menstruation when its lining is shed. (Chapter 21)

vaccination A preparation of a weakened or killed pathogen, or portion of a pathogen, that will stimulate the immune system of a recipient to prepare a long-term defense (memory cells) against that pathogen. (Chapter 19)

vagina Muscular canal in females leading from the cervix to the vulva. (Chapter 21)

valence shell The outermost energy shell of an atom containing the valence electrons which are most involved in the chemical reactions of the atom. (Chapter 2)

variable A factor that varies in a population or over time. (Chapter 1)

variable number tandem repeat (VNTR) A DNA sequence that varies in number between individuals. Used during the process of DNA fingerprinting. (Chapter 8)

variance Mathematical term for the amount of variation in a population. (Chapter 7)

variant An individual in a population that differs genetically from other individuals in the population. (Chapter 11)

vas deferens Either of the two ducts in males that carry sperm from the epididymis to the urethra. (Chapter 21)

vascular cambium Meristematic tissue that forms in vascular bundles of dicot roots and stems and permits secondary growth. (Chapter 23)

vascular system Plant tissue system responsible for the delivery of water and dissolved solutes throughout the plant body. (Chapter 18)

vascular tissue Cells that transport water and other materials within a plant. (Chapters 13, 23)

vasodilation Widening of blood vessels, caused by relaxation of smooth muscle lining vessel walls. (Chapter 19)

vector An organism that carries a pathogen from one host to another, such as a mosquito carrying West Nile virus. (Chapter 19)

vegetative organ Plant organ that is not involved in sexual reproduction. (Chapter 23)

vegetative reproduction Asexual cloning of plants. (Chapter 21)

vein Vessel that carries blood from the body tissues back to the heart. (Chapter 18)

ventricle Chamber of the heart that pumps blood from the heart to the lungs or systemic circulation. (Chapter 18)

vertebra (*plural:* vertebrae) Bone of the spinal column through which the spinal cord passes. (Chapter 22)

vertebral column Also called the spine, the series of vertebrae and cartilaginous disks extending from the brain to the pelvis. (Chapter 22)

vertebrate Animal with a backbone. (Chapter 13)

vesicle Membrane-bounded sac-like structure. In neurons, these structures are found in the terminal bouton and store neurotransmitters. (Chapter 22)

vessel element Wide xylem cell with perforated ends, found only in angiosperms. (Chapter 24)

vestigial trait Modified with no or relatively minor function compared to the function in other descendants of the same ancestor. (Chapter 10)

villus (*plural:* villi) Small fingerlike projections on the inside of the small intestine that function in nutrient absorption.

viral envelope Layer formed around some virus protein coats (capsids) that is derived from the cell membrane of the host cell and may also contain some proteins encoded by the viral genome. (Chapter 19)

virus Infectious intracellular parasite composed of a strand of genetic material and a protein or fatty coating that can only reproduce by forcing its host to make copies of it. (Chapters 1, 9, 19)

vitamin Organic nutrient needed in small amounts. Most vitamins function as coenzymes. (Chapter 3)

voluntary Muscle normally under conscious control. Mainly skeletal muscle. (Chapter 17)

vulva The outer portion of the female external genitalia including the labia majora, labia minora, clitoris, and vaginal and urethral openings. (Chapter 21)

wastewater Liquid waste produced by residential, commercial, or industrial activities. (Chapter 16)

water One molecule of water consists of one oxygen and two hydrogen atoms. (Chapters 2, 3, 4, 16, 23, 24)

weather Current temperature and precipitation conditions. (Chapter 16)

weed Common term for nonpreferred plant. (Chapter 23)

wetland Biome characterized by standing water, shallow enough to permit plant rooting. (Chapter 16)

white blood cell General term for cell of the immune system. (Chapter 18)

white matter Nervous system tissue, especially in the brain and spinal cord, made of myelinated cells. (Chapter 22)

whole food Any food that has not undergone processing. Includes grains, beans, nuts, seeds, fruits, and vegetables. (Chapter 3)

wood Xylem cells produced by secondary growth of stems and roots. (Chapter 23)

woody plant Plant that produces stiffened stems via secondary production of xylem. (Chapter 24)

X inactivation The inactivation of one of two chromosomes in the XX female. (Chapter 8)

x-axis The horizontal axis of a graph. Typically describes the independent variable. (Chapter 1)

X-linked gene Any of the genes located on the X chromosome. (Chapter 8)

xylem Plant vascular tissue, dead at maturity, that carries water and dissolved minerals in a one-way flow from the roots to the shoots of a plant. (Chapters 23, 24)

xylem sap Water and dissolved minerals flowing in the xylem vessels. (Chapter 24)

y-axis The vertical axis of a graph. Typically describes the dependent variable. (Chapter 1)

yeast Single-celled eukaryotic organisms found in bread dough. Often used as model organisms and in genetic engineering. (Chapters 4, 10, 13)

Y-linked gene Any of the genes located on the Y chromosome. (Chapter 8)

Z disc The border of a sarcomere in muscle. (Chapter 20)

zona pellucida The substance surrounding an egg cell that a sperm must penetrate for fertilization to occur. (Chapter 21)

zoologist Scientist who specializes in the study of animals. (Chapter 13)

zygote Single cell resulting from the fusion of gametes (egg and sperm). (Chapters 6, 8, 21)

Credits

Text and Art Credits

CHAPTER 1 **Savvy Reader.** Gupta, Terry, "Get back your health with vitamin C, tea" *The Telegraph* Nashua, NH, Oct. 26, 2007. Used with permission.

Figure 1.7. Data from: Lindenmuth, G. Frank, Ph.D. and Elise B. Lindenmuth, Ph.D., "The efficacy of echinacea compound herbal tea preparation on the severity and duration of upper respiratory and flu symptoms: A randomized, double-blind placebo-controlled study," *The Journal of Alternative and Complementary Medicine,* Vol. 6: 4, pp. 327–334, 2000, Mary Ann Liebert, Inc.

Figure 1.10. Cohen, Sheldon, Ph.D., et al., "Psychological Stress and Susceptibility to the Common Cold," *New England Journal of Medicine,* Vol. 325: 9, pp. 606–612, fig. 1, Aug. 29, 1991. Copyright © 1991 Massachusetts Medical Society. Used with permission.

Figure 1.13. Data from: Mossad, Sherif B., MD, et al., "Zinc Gluconate Lozenges for Treating the Common Cold, A Randomized, Double-Blind, Placebo-Controlled Study," *Annals of Internal Medicine,* Vol. 125: 2, July 15, 1996.

CHAPTER 2 **Opener (L)** joaquin croxatto/iStockphoto

CHAPTER 3 **Savvy Reader.** "Wellness Guide to Dietary Supplements: Açaí Juice," *UC Berkeley Wellness Letter,* June 20, Copyright Clearance Center, http://www.wellnessletter.com/html/ds/dsAcai.php.Used with permission.

CHAPTER 5 **Savvy Reader.** Sean Mattson, "Rising sea drives Panama islanders to mainland" from reuters.com, July 11, 2010. Used by permission. All rights reserved. Republication or redistribution of Thomson Reuters content, including by framing or similar means, is expressly prohibited without the prior written consent of Thomson Reuters. Thomson Reuters and its logo are registered trademarks or trademarks of the Thomson Reuters group of companies around the world. © Thomson Reuters 2010. Thomson Reuters journalists are subject to an Editorial Handbook which requires fair presentation and disclosure of relevant interests.

Figure 5.6. Data from: "Trends in Atmospheric Carbon Dioxide—Mauna Loa," NOAA/ESRL Global Monitoring Division, www.esrl.noaa.gov/gmd/ccgg/trends/.

Figure 5.8. Data from: Petit, J.R., et al., 2001, Vostok Ice Core Data for 420,000 Years, IGBP PAGES/World Data Center for Paleoclimatology Data Contribution Series #2001-076. NOAA/NGDC Paleoclimatology Program, Boulder CO, USA.

CHAPTER 7 **Savvy Reader.** Jennifer Warner, "Genes May Link Friends, not Just Family" from WebMD Health News, Jan. 18, 2011. Reviewed by Laura J. Martin, MD. © 2011 WebMD, LLC. All rights reserved. Reproduced by permission.

Figure 7.19: Data from: Raberg, L., "Immune responsiveness in adult blue tits: Heritability and effects of nutritional status during ontogeny," *Oecologia,* 136: 360–4, Fig. 1, 2003.

Photo Credits

CHAPTER 1   Opener (L) Paul Bradbury/OJO Images/Getty, Opener (R) Top Photolibrary, Opener (R) Middle Edyta Pawlowska/Shutterstock, Opener (R) Bottom iStockphoto, 1.2a Eye of Science/Photo Researchers, Inc., 1.2b Anders Wiklund/Reuters/Landov, 1.4a Custom Medical Stock Photo, Inc., 1.5 The Natural History Museum, London/Alamy Images, 1.6 Zina Seletskaya/Shutterstock, 1.9a Leonard Lessin/Photolibrary, 1.9b Joel Sartore/National Geographic/Getty, 1.9c Photo Researchers, Inc., 1.17 Graca Victoria/Shutterstock.

CHAPTER 2   Opener (R) Top NASA, Opener (R) Middle NASA, Opener (R) Bottom NASA, 2.1 Photos.com, 2.2a NASA, 2.2b DLR/FU Berlin (G. Neukum)/European Space Agency, 2.16 National Cancer Institute, 2.17a Juergen Berger/Photo Researchers, Inc., 2.19a Dr. Gary Gaugler/Photo Researchers, Inc., 2.19b Eye of Science/Photo Researchers, Inc., 2.19c Brittany Carter Courville/istockphoto, 2.19d Jan Krejci/Shutterstock, 2.19e webphotographeer/istockphoto, 2.19f M. I. Walker/Photo Researchers, Inc., 2.19g Dusan Zidar/istockphoto, 2.19h Vicki Beaver/istockphoto, 2.20 King County Deartment of Natural Resources and Parks, Water and Land Resources Division.

CHAPTER 3   Opener (L) Ina Fassbender/Reuters, Opener (R) Top Thinkstock, Opener (R) Middle Tony Gentile/CORBIS, Opener (R) Bottom Fuse/Getty Images, 3.3a fonats/Shutterstock, 3.3b Craig Veltri/istockphoto.

CHAPTER 4   Opener (L) John Schults/Reuters, Opener (R) Top Newscom, Opener (R) Middle Chistopher LaMarca/Redux Pictures, Opener (R) Bottom Guang Niu/Reuters, 4.10a Professors P. Motta & T. Naguro/SPL/Photo Researchers, Inc., 4.18a Suza Scalora/Photodisc/Getty Images, 4.18b SciMAT/Photo Researchers, Inc., 4.19a Mike Blake/Reuters, 4.19b Renn Sminkey, 4.19c Michael Ochs Archives/Getty Images Inc., 4.19d Everett Collection Inc/Alamy, SR4.1a Theo Wargo/WireImage/Getty Images, SR4.1b Steve Grantiz/WireImage/Getty Images.

CHAPTER 5   Opener (L) STR News/Reuters, Opener (R) Top Fotolia, Opener (R) Middle Milos Peric/istockphoto, Opener (R) Bottom Shutterstock, 5.5 STR News/Reuters, 5.7 British Antarctic Survey/Science Photo Library/Photo Researchers, Inc., 5.9a David Furness, Keele University/Photo Researchers, Inc., 5.14 Winton Patnode/Photo Researchers, Inc, T5.1a TongRo Image Stock/Jupiter Images, T5.1b iStockphoto, T5.1c James Forte/National Geographic Society.

CHAPTER 6   Opener (L) SPL/Photo Researchers, Inc., Opener (R) Top Photos by Jim and Mary Whitmer, Opener (R) Middle Photos by Jim and Mary Whitmer, Opener (R) Bottom Photos by Jim and Mary Whitmer, 6.2a Peter Arnold, Inc./Alamy, 6.2b rossco/Shutterstock, 6.3a Biophoto Associates/Photo Researchers, Inc., 6.3b Biophoto Associates/Photo Researchers, Inc., 6.8a Photo Researchers, Inc., 6.8b Credit: Kent Wood/Photo Researchers, Inc., 6.10a Fotolia, 6.10b From: Science, September 17, 1999 in the article entitled "Antiangiogenic Activity of the Cleaved Conformation of Serpin Antithrombin" written by Judah Folkman, Michael S. O'Reilly, Steven Pirie-Shepherd and William S. Lane, 6.15 Photos by Jim and Mary Whitmer, 6.16a CNRI/SPL/Photo Researchers, Inc., 6.16b CNRI/SPL/Photo Researchers, Inc., 6.16c CNRI/SPL/Photo Researchers, Inc.

CHAPTER 7   Opener (L) Tony Brain/Science Photo Library/Photo Researchers, Inc., Opener (R) Top Melissa Schalke/Shutterstock, Opener (R) Middle Digital Vision/Getty Images, Opener (R) Bottom Thinkstock/Superstock, 7.2a Photo courtesy of Gerald McCormack, 7.2b Ninan, C.A., 1958. Studies on the cytology and phylogeny of the pteridophytes VI. Observations on the Ophioglossaceae. Cytologia, 23: 291–316., 7.8 Alamy Images, 7.10 Photo by Linda Gordon, 7.11 Simon Fraser/Royal Victoria Infrimary/Photo Researchers, Inc., 7.12 Dr. Kathryn Lovell, Ph.D., 7.17a Photos courtesy of New York plastic surgeon, Dr. Darrick E. Antell. www.Antell-MD.com, 7.17b Photos courtesy of New York plastic surgeon, Dr. Darrick E. Antell. www.Antell-MD.com, 7.18a Getty Images, 7.18b Photonica/Getty Images, 7.21 Angel Franco/The New York Times/Redux Pictures, 7.23 Juice Images/Alamy.

CHAPTER 8   Opener (L) World History Archive/Alamy, Opener (R) Top V. Velengurin/R.P.G./Corbis, Opener (R) Middle Photo courtesy of Dr. Sergey Nikitin, Opener (R) Bottom Adam Gault/Alamy, 8.2a iStockphoto, 8.2b luri/Shutterstocki, 8.2c AGStockUSA/Alamy, 8.5 Andrew Syred/Photo Researchers, Inc., T8.3a Johan Pienaar/Shutterstock, T8.3b Shutterstock, T8.3c Vasilkin/Shutterstock, T8.3d Arnon Ayal/Shutterstock, T8.3e James King-Holmes/Photo Researchers, Inc., 8.8a Joellen L Armstrong/Shutterstock, 8.8b Oleg Kozlov & Sophy Kozlova/Shutterstock, 8.8c Justyna Furmanczyk/Shutterstock, 8.10a Biophoto Associates/Photo Researchers, Inc., 8.10b Fotolia, 8.10c Ellen B. Senisi/Photo Researchers, Inc., 8.15 David Parker/Photo Researchers, Inc., 8.17b Pictorial Press Ltd/Alamy.

CHAPTER 9   Opener (L) Sion Touhig/Getty Images, Inc., Opener (R) Top brue/istockphoto, Opener (R) Middle Newscom, Opener (R) Bottom H. Kuboto, R. Brinster and J.E. Hayden, RBP, School of Veterinary Medicine, University of Pennsylvania. Proc. Natl. Acad. Sci. USA 101:16489–16494, 2004, 9.11a Biology Media/Photo Researchers, Inc., 9.11b David McCarthy/Science Photo Library/Photo Researchers, Inc., 9.14 Pearson Education, 9.15 From Doebley, J. Plant Cell, 2005 Nov; 17 (11): 2859–72. Courtesy of John Doebley/University of Wisconsin, 9.16 Jon Christensen, 9.17a Leonard Lessin/Photo Researchers, Inc., 9.19 Corbis/Superstock, 9.20 Pascal Goetgheluck/SPL/Photo Researchers, Inc., 9.22b Ted Thai/Time Life Pictures/Getty Images, 9.23 Photograph courtesy of The Roslin Institute, SR9-1a (Best & Wittiest) Jimmy Margulies © North America Syndicate, SR9-1b Rob Rogers: © The Pittsburgh Post-Gazette/Dist. by United Features Syndicate, Inc.

Index

Page references that refer to figures are in **bold**. Page references that refer to tables are in *italic*.

PCBs. *See* Polychlorinated biphenyls
PCR. *See* Polymerase chain reaction
Peacock, 300, **300**
Peas
　genes and, **154**
　traits studied by Mendel, *155*
Pediculus pubis, 522
Pedigrees, 183–184, 183–185, *185*
　analysis of, **183**
　with different modes of inheritance, **184**
　Romanov family, **190**
Peer review, 21
Pelvic inflammatory disease (PID), *521*
Penis, 518, **518**
Peptide bond formation, **39**
Perennials, 569, **607**, 608
Peripheral nervous system (PNS), 547
Peristalsis, 431
Permafrost, 404
Pest control, for maximizing plant growth, **578**, 578–585, **579**, **580**, **581**, **582**, *582*, **583**
Pesticides, 208, 210, **210**, 587–588
　cosmetic use of, **589**
　health effects of, 588
　impact of modern agriculture on, *591–592*
Petals, 327, 571, **571**, **572**
Petroleum, 586–587
pH, 35
　scale, **35**
Phagocytes, **477**, 477–478, **478**
Phagocytosis, 477, **477**
Pharynx, 431, 443, **443**
Phenotype, 154–155
　effect of the environment on, **161**
　x inactivation and, **182**
Phenotypic ratio, 159
Phenylalanine, **39**
Phenylthiocarbamide (PTC), 292
Phloem sap, 606
Phospholipid bilayer, 44
Phospholipids, **40**, 40–41
Phosphorus, *63, 582*
Phosphorylation, 77
Photoperiodism, 609–612, **610, 611**
Photorespiration, 104
Photosynthesis, 92–109
　adaptations, 599–600, **600**
　definition of, 99
　global warming and, 103–104
　light reactions of, **102**
　overview, 100–101, **101**
　steps of, 101–103
Phototropism, 612, **612**
Phyla, 318
Phytochrome, 610
Phytophthora infestans, **383**
Phytosynthetic plankton, 408, **408**
PID. *See* Pelvic inflammatory disease
Pigeons, **255**

Pigs, **195**
Pima Indians, 85–86
Pineal gland, *495*
Pituitary gland, **493**, *494*, 495–496, **496**
Placebo, 11–12, 19
Placenta, 528
Plantae kingdom, *312*
Plant physiology, 596–617
　adaptations that affect transpiration, 599–601
　freezing inside plant cells, **604**
　hormones, **613**, *613*, 613–614, **614**
　managing translocation, **607**, 607–609, **609**
　overview, 597
　overwatering, 601–602
　photoperiodism, 609–612, **610, 611**
　plants on land, 603–604
　translocation of sugars and nutrients, 606, **607**
　transpiration, 598–599, **599**
　tropisms, **612**, 612–613
　water inside plant cells, **604**, 604–606
Plants, 49, 255
　absorption spectra of leaf pigments, **100**
　alternation of generations, **574**, 574–575
　cells, 48
　comparison of cytokineses in animal and plant cells, **124**
　diversity, *326,* 566–568, **567**
　ferns, **232**
　as food, 566–576
　founder effect, **299**
　genetically modified, **208**, 208–209, **209**
　genetics and, **149, 154**
　grasses, 264
　growth requirements, **576**, 576–578, **577, 578**
　hardiness of, 604–605, **605**
　osmosis in cells, 66
　plant tissue systems, 568, **568**
　prickly pear cactus, **226**
　reproduction, **571**, 571–576, **572, 573, 574, 575**
　stomata status of, *105*
　structure of, **568**, 568–570, **569, 570**
　vestigial structures in, **232**
Plant structure and growth, **564**, 564–595
　damage reduction, **589**, 589–591, **590**, *591–592*
　food plant diversity, 566–568, **567**
　future of agriculture, **586**, 586–589, **587, 588**
　hybrids and genetic engineering, 585–586
　organic food, 593
　overview, 565–566
　plant growth requirements, **576**, 576–578, **577, 578**
　plant reproduction, **571**, 571–576, **572, 573, 574, 575**
　plants as food, 566–576

plant structure, **568**, 568–570, **569, 570**
　requirements for plant growth, **579**
　water, nutrients, and pest control for maximizing plant growth, **578**, 578–585, **579, 580, 581, 582**, *582*, **583, 584**
Plasma membrane, 44, *45*
Plasmodium, 475
Platyhelminthes, *321*
Pleiotropy, 178
Pneumocystis jiroveci, 484
PNS. *See* Peripheral nervous system
Poetry, 275
Poison dart frog, **320**
Polar bodies, 525
Polarity, 33
　in water, **33**
Polarization, 553
Poles, 121, 391
Political cartoon, **217**
Pollination, 571, **571, 572**
Pollution, **361**, 365
　fertilizer, 587
Polychlorinated biphenyls (PCBs), 537
Polyculture, 584, **584**, 585
Polygenic traits, 161, 176
Polymerase chain reaction (PCR), **186**, 186–187, 307
Polyploidy, 282
Polysaccharides, 38
Polyunsaturated fat, 59
Ponds, *405,* 406–407, **407**
Pons, 551, **551**
Population, 340, 482
　structure of, 340–342
The Population Bomb (Ehrlich), 339
Population ecology, 338–355
　of birds, 352
　capacity of, **348**, 348–349
　crash of, **349**, 349–350
　demographic transition, **344**, 344–345
　exponential population growth, **342**, 342–345, **343**
　future of human population, **349**, 349–352, **351**
　growing human population, 340–345
　growth rate and doubling time, **343, 351**
　limits to population growth, **345**, 345–349, **346, 347, 348**
　overshooting and crashing, **349**, 350
　overview, 339–340
　patterns of population dispersion, **341**
　reducing population growth through opportunity, 352, **352**
　structure of, 340–342, **341**
Population pyramid, 350, **351**
Populations, 280
Porifera, *321*
Positive feedback, 434
Potassium, *63, 582*
Potatoes, 382–383, **383**